Thermal Processing of Food Products by Steam and Hot Water

Thermal Processing of Food Products by Steam and Hot Water

Unit Operations and Processing Equipment in the Food Industry

Edited by

Seid Mahdi Jafari

Department of Food Materials and Process Design Engineering, Gorgan University of Agricultural Sciences and Natural Resources, Gorgan, Iran

Nutrition and Bromatology Group, Department of Analytical Chemistry and Food Science, Faculty of Science, Universidade de Vigo, Ourense, Spain

College of Food Science and Technology, Hebei Agricultural University, Baoding, China

ELSEVIER

WP
WOODHEAD PUBLISHING
An imprint of Elsevier

Woodhead Publishing is an imprint of Elsevier
50 Hampshire Street, 5th Floor, Cambridge, MA 02139, United States
The Boulevard, Langford Lane, Kidlington, OX5 1GB, United Kingdom

ISBN: 978-0-12-818616-9 (print)
ISBN: 978-0-12-819706-6 (online)

For information on all Woodhead publications
visit our website at https://www.elsevier.com/books-and-journals

Publisher: Nikki Levy P. (ELS-SDG)
Acquisitions Editor: Nina Bandeira (ELS-BRA)
Editorial Project Manager: Andrea R. Dulberger (ELS-CMA)
Production Project Manager: Manju Paramasivam
Cover Designer: Matthew Limbert

Typeset by STRAIVE, India

Contents

Section C Different thermal processing operations in the food industry

Contributors

Shabbir Ahmad Department of Food Science and Technology, MNS-University of Agriculture, Multan, Pakistan

Ghani Albaali King Talal Faculty of Business and Technology (KTFBT), Princess Sumaya University for Technology, Amman, Jordan

Elham Ansarifar Social Determinants of Health Research Center, Department of Public Health, School of Health, Birjand University of Medical Science, Birjand, Iran

Maryam Azizi-Lalabadi Research Center for Environmental Determinants of Health (RCEDH), Kermanshah University of Medical Sciences, Kermanshah, Iran

Deniz Cekmecelioglu Department of Food Engineering, Middle East Technical University, Ankara, Turkey

Sara Hedayati Nutrition Research Center; Department of Food Hygiene and Quality Control, School of Nutrition and Food Sciences, Shiraz University of Medical Sciences, Shiraz, Iran

Seid Mahdi Jafari Department of Food Materials and Process Design Engineering, Gorgan University of Agricultural Sciences and Natural Resources, Gorgan, Iran; Nutrition and Bromatology Group, Department of Analytical Chemistry and Food Science, Faculty of Science, Universidade de Vigo, Ourense, Spain

Aiman Karim State Key Laboratory of Food Science and Technology, Jiangnan University, Wuxi, Jiangsu, China

Gokcem Tonyali Karsli Department of Food Engineering, Middle East Technical University, Ankara, Turkey

Navneet Kumar Department of Processing and Food Engineering, Anand College of Agricultural Engineering and Technology, Agricultural University, Godhra, Gujarat, India

Chi-Ching Lee Istanbul Sabahattin Zaim University, Faculty of Engineering and Natural Sciences, Department of Food Engineering, Istanbul, Turkey

Zhang Lianfu State Key Laboratory of Food Science and Technology, Jiangnan University, Wuxi, Jiangsu, China

Dharani Loganathan Department of Chemical Engineering, KPR Institute of Engineering and Technology, Coimbatore, Tamil Nadu, India

Periyasamy Manikandan Department of Chemical Engineering, Kongu Engineering College, Perundurai, Tamil Nadu, India

Najme Rezaee Moghaddam Laboratory Instrumental Expert, IMEN KIA SANAT Engineering Company, Shiraz, Iran

Pooja Nandakumar Department of Plastics Technology, Central Institute of Petrochemicals Engineering & Technology, Chennai, Tamil Nadu, India

Deepa Priya Natarajan Department of Chemical Engineering, Sri Venkateswara College of Engineering, Sriperumbudur, Tamil Nadu, India

Azadeh Ranjbar Nedamani Department of Farm Machinery, Faculty of Agricultural Engineering, Sari Agricultural Sciences and Natural Resources University, Sari, Mazandaran, Iran

Asia Noreen Enzyme Biotechnology Lab, Department of biochemistry, University of Agriculture, Faisalabad, Pakistan

Tong Qunyi State Key Laboratory of Food Science and Technology, Jiangnan University, Wuxi, Jiangsu, China

Hosahalli S. Ramaswamy Department of Food Science and Agricultural Chemistry, Macdonald Campus of McGill University, Ste Anne de Bellevue, QC, Canada

Abdur Rehman State Key Laboratory of Food Science and Technology, Jiangnan University, Wuxi, Jiangsu, China

Felipe Richter Reis Instituto Federal de Educação, Ciência e Tecnologia do Paraná, Campus Colombo, Colombo, PR, Brazil

Navin C. Shahi Department of Post-Harvest Process and Food Engineering, College of Technology, Govind Ballabh Pant University of Agriculture and Technology, Pantnagar, Uttarakhand, India

Lovepreet Singh Sidhu Department of Food Science and Agricultural Chemistry, Macdonald Campus of McGill University, Ste Anne de Bellevue, QC, Canada

İbrahim Agah Taştemir Istanbul Sabahattin Zaim University, Faculty of Engineering and Natural Sciences, Department of Interior Architecture and Environmental Design, Istanbul, Turkey

Muhammad Usman Beijing Advance Innovation Center for Food Nutrition and Human Health, School of Food and Chemical Technology, Beijing Technology and Business University, Beijing, China

Hamed Vatankhah Department of Food Science and Agricultural Chemistry, Macdonald Campus of McGill University, Ste Anne de Bellevue, QC, Canada

Aman Mohammad Ziaiifar Department Food Process, Gorgan University of Agricultural Science and Natural Resources, Gorgan, Iran

Section A

An overview of thermal food processing

Introduction to thermal food processes by steam and hot water

1

Aiman Karim[a], Abdur Rehman[a], Zhang Lianfu[a], Asia Noreen[b], Shabbir Ahmad[c], Muhammad Usman[d], and Seid Mahdi Jafari[e]
[a]State Key Laboratory of Food Science and Technology, Jiangnan University, Wuxi, Jiangsu, China, [b]Enzyme Biotechnology Lab, Department of biochemistry, University of Agriculture, Faisalabad, Pakistan, [c]Department of Food Science and Technology MNS-University of Agriculture, Multan, Pakistan, [d]Beijing Advance Innovation center for Food Nutrition and Human Health, School of Food and Chemical Technology, Beijing Technology and Business University, Beijing, China, [e]Department of Food Materials and Process Design Engineering, Gorgan University of Agricultural Sciences and Natural Resources, Gorgan, Iran

1.1 Introduction

Processed food sources now attract people's attention beyond the basic criteria of shelf-stability and safety. More emphasis is being placed on clearly labeled, high-quality, and value-added food sources with practical applications. Administrative requirements on manufacturers, as well as public or global assemblies that advise and also apply execution instructions or approaches for achieving security and quality assurance, have improved the quality and safety of processed foods. Similarly, the buyer's need for affordable, high-value-added items has been substantial (Awuah, Ramaswamy, & Economides, 2007). In addition, autoxidation in food and biological systems causes a variety of negative impacts and consequences in food stability and preservation, as well as human health. Human health is currently thought to be influenced by oxidative food damage, which leads to the development of major illnesses in humans such as heart disease, obesity, hypertensive, metabolic syndrome, tumors, and so on. As a result, it is critical to preserve foods in order to reduce waste and maintain human health by employing a variety of commercial and creative processing processes, or a combination of them (Tavman, Otles, Glaue, & Gogus, 2019a).

To date, heat is now the most used method for processing food, because of its capability to inactivate bacteria and deterioration enzymes. Thermal processing of foods, often known as traditional preservation, represent a large portion of the food market. Thermal processing is based on the utilization of thermal energy (heat), in which the food is heated to a specified temperature by a hot fluid, held for a set amount of time to improve product features (processing time), and then cooled by a cold fluid to stop the thermal activities. The use of some thermal processing procedures has been promoted by significant technical advancements in recent years, as well as simpler monitoring of the equipment involved (Augusto, Soares, & Castanha, 2018; Karel & Lund, 2003).

Thermal Processing of Food Products by Steam and Hot Water. https://doi.org/10.1016/B978-0-12-818616-9.00001-8

One of the most essential unit operations in the food company is thermal processing of food and food products. It has a lot of processing and preservation uses, so it is really interesting. Pasteurization, sterilization, blanching, boiling, steaming microwaving, and roasting, are just a few of the techniques that may be used on liquid or solid foods. On the other hand, thermal processing alters the physical, chemical, and organoleptic qualities of foods (Liu et al., 2020; Rattan & Ramaswamy, 2014). The most common processing technologies in the food industry are in-line pasteurization and sterilization of liquids using heat exchangers and hold tubes prior to filling into containers, and in-container pasteurization or sterilization (canning) using pressurized retorts to heat filled and sealed food containers. Blanching is a moderate kind of heat treatment used to liberate trapped air and inactivate enzymes in fresh foods prior to preservation methods (Teixeira, 2014). This chapter will give an overview of the thermal processing methods used in the food industry to achieve these various types of heat treatments using steam or hot water, as well as the concepts of heat transmission and how they relate to the design and implementation of food processing equipment. More details about each thermal processing technique and relevant heat exchangers have been presented in the following chapters of this book.

1.2 An overview of thermal processing technologies by steam/hot water

For more than a century, the use of heat for thermal processing to inactivate microbes has been one of the most extensively used technologies for food preservation, leading to an overall health of a huge portion of the global population. A batch or continuous heating procedure might be used. The goal should be to heat and cool the product as rapidly as feasible in all thermal procedures. This has financial implications as well as the potential to increase quality. The rate of heat transfer is related to the temperature difference when heat or energy is transferred from a high to a low temperature. The most common heating medium is saturated steam or hot water. Steam and hot water will be above air pressure at temperatures >100°C. The degree of thermal treatments varies greatly, ranging from moderate operations like thermization and pasteurization to more extreme processes like in-container sterilization (Lewis & Jun, 2011).

There are two types of heating processes: direct and indirect. Indirect heating is the most common method, in which the heat transfer fluid and the liquid food are divided by a barrier; for in-container sterilization, this will be the bottle wall, while for continuous processing, it will be the heat exchanger plate or tube wall. Steam is the heating medium in direct operations, and the steam comes into direct touch with the product (Lewis & Heppell, 2000). The conventional and successful way for microbial decontamination is thermal processing, which is still a frequently used method for destroying pathogenic and spoilage-causing bacteria from foods. Nicolas Appert patented research on the first heat-based stabilizing technique in the 19th century. The science-based heat technique was established in food preservation later in the early twentieth century (Aaliya et al., 2021; Khan, Tango, Miskeen, Lee, & Oh, 2017). Thermal processing commonly employs direct hot steam/air, hot water bath, shell, tube, and plate heat exchangers. The energy (heat) is transferred into a product by conduction or convection in traditional thermal processing (Aaliya et al., 2021; Bahrami, Baboli, Schimmel, Jafari, & Williams, 2020). The convectional thermal processes are pasteurization, commercial sterilization, blanching and hot filling.

1.2.1 Pasteurization of packaged foods

Pasteurization is a moderate heating procedure that eliminates bacteria from food and inactivates enzymes at temperatures between 60°C and 80°C ($<$ boiling temperature). Pasteurization has two basic goals: to eliminate pathogenic microorganisms from foods, so limiting disease, and to remove spoilage (souring) germs, thereby increasing the food's shelf life. It makes the product stable at room temperature and improves its storage life. Processing parameters, ambient factors, and microorganism and product characteristics are all elements that influence the efficiency of thermal processing (Khan et al., 2017; Tavman, Otles, Glaue, & Gogus, 2019b). It is a mild thermal technique that is commonly used to prolong the shelf life of food from a few days (low-acid products with high water activity) to months (high-acid products and/or with low water activity) at low temperatures, typically 4°C (Augusto et al., 2018).

Pasteurization can be achieved in general by combining time and temperature, such as (i) heating foods to a relatively low temperature and preserving it for a longer period of time, e.g., holding pasteurization, pasteurization by exhaust technique, and (ii) heating foods to a high enough temperature and holding it for a short period of time only. Pasteurization can be done in one of two ways: (i) filling sterile containers with the product first, then pasteurizing, or (ii) pasteurizing the product first, then filling sterile containers. According to the International Dairy Federation (IDF), pasteurization is defined as "a process applied to a food product with the objective of minimizing possible health hazards emerging from microbial pathogens associated with the product while maintaining minimal chemical, physical, and organoleptic changes in the product". Creams, ice cream mixes, eggs, fruit juices, fermented items, soups, and other drinks are all included in this classification (Staal, 1986).

Pasteurization can be done in batches or in real time. A batch pasteurizer is made out of a steam jacket kettle or a tank with steam coils that is used to heat juice or milk to the correct temperature. Pouring the vessel, heating, holding, cooling, emptying the vessel and filling into containers, and washing the vessel are all part of batch processing. and holding durations of up to 30 min. Continuous pasteurization, this is known as the high-temperature-short-time (HTST) technique. Milk, for example, is heated to 71.7°C for 15 s or longer by passing through several heat exchangers. Steam or water is commonly used as a heating medium (Cronshaw & Cronshaw, 1947; Rahman, 2007). Continuous operations have several advantages over batch processes, including rapid heating and cooling rates, shorter holding times, and regeneration, which saves both heating and cooling costs and contributes to the low processing costs associated with thermal processing operations when compared to many novel techniques. A hot water set provides the final heating procedure in the majority of pasteurizers. Steam is utilized to keep the temperature of the hot water constant, between 2°C and 10°C above the desired pasteurization temperature (Lewis & Jun, 2011).

Pasteurization is a decontaminating method which preserves packed liquid foods such as milk, juices, beer, and other fermented beverages by heating them to a certain temperature and duration under atmospheric pressure. Due to the significant food safety problems connected with ready-to-eat (RTE) items, pasteurization is a preferred way of ensuring the safety of prepackaged food products (Qiu, Zhang, Tang, Adhikari, & Cao, 2019). After filling containers, several acidic liquid food products (such as beer and fruit juices) are pasteurized. Pasteurization may be done in a water bath, which is one of the easiest techniques of heating. The water bath might be a square steel tank or a vertical retort with a diameter of 1 m. The product is dipped in the bath and packaged in retort boxes or racks. Cooling can be done in the same tank that was used to heat the containers, or the containers can be transported from the heating tank to the cooling tank. Heating and cooling can also be done in stages. Meats, pickles, sauces, and other acidic foods are all processed in essentially the same way.

If the food is packaged in glass, hot water is usually used to minimize the danger of breaking due to heat shock. For heating, the highest temperature between both the container and the liquid is 20°C, and for cooling, it is 10°C. Steam–air mixes or hot water were used to pasteurize metal and plastic containers. Containers may be sent through a hot water batch immediately by a cold-water bath in a continuous form. Beer and acidic food products are commonly pasteurized using a continuous water spray pasteurizer. The bottles or cans are carried through the pasteurizer by a walking beam or a continuous belt conveyor in this sort of equipment. To achieve optimal performance, it is typical to run the pasteurizer through as many as six distinct temperature zones or sections. First preheat, second preheat, pasteurizing zone, precooling, cooling, and final cooling zone are the sections. Steam tunnels provide the benefit of rapid heating, which leads to a shorter retention period and less equipment. Heat recovery from hot items and recirculating water result in energy and water savings. The heating zone's temperature is progressively raised by lowering the quantity of air in the steam–air combinations, and cooling is accomplished via water sprays or immersion in a water bath. In a retort, acid items like fruit or acidified vegetables like beetroot can be pasteurized (Fellows, 1988; Jackson & Shinn, 1979; Ramesh, 2007; Simpson, 2009).

Pasteurization, on the other hand, does not help in the total removal of all harmful microbes, nor does it destroy spore-forming heat-resistant microorganisms. It is also clear that high-temperature treatment accelerates decomposition processes and reduces the sensory and nutritional value of the treated items. To minimize quality degradation, light heat treatment (40–60°C) is suggested over high-temperature treatment. Thermal pasteurization is achieved by heating/holding/cooling liquid items that may be pumped via tubular pipe lines, such as milk and fruit juices, while they are running through a series of heat exchangers and hold tubes. These systems provide a HTST mechanism that kills specific microorganisms (Bahrami et al., 2020; Liao et al., 2019; Tavman et al., 2019b) (Fig. 1.1).

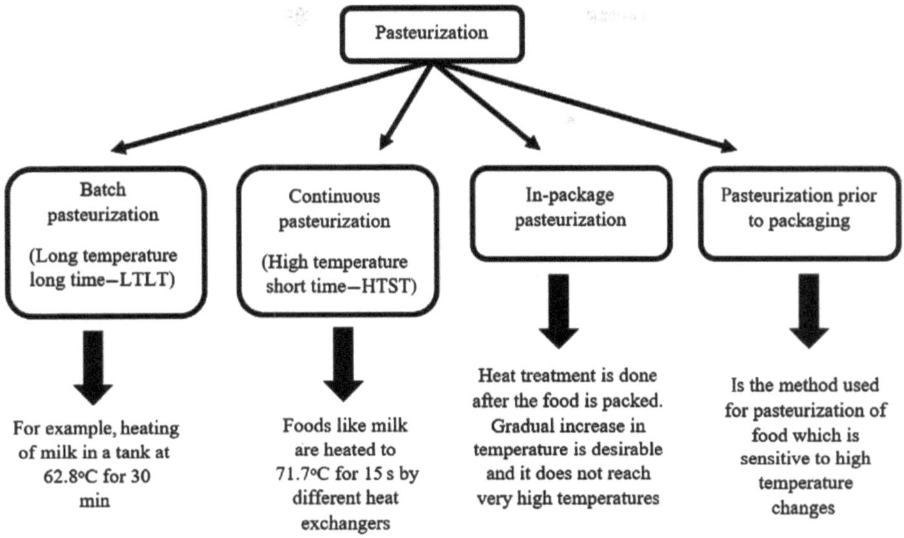

Fig. 1.1 Types of food pasteurization (Rahman, 2007; Tavman et al., 2019b)

1.2.2 Commercial sterilization

Sterilization is described as a regulated heating technique that may destroy/eliminate or inactivate living microorganisms completely (including bacteria, fungi, viruses and spores). The food sterilization methods are divided into two categories: Sterilization by heating (thermal processing) and sterilization without heating (nonthermal processing). Thermal sterilization is performed by applying high intensity heat (usually between 121°C and 140°C) to food items. Sterilization is still a dominant and frequently used technique in the food sector, and it entails that each and every particle of food undergoes appropriate heat treatment. It makes the product stable at room temperature and increases its storage life. Microbes are destroyed by heat in a time/temperature mechanism (Deak, 2014).

There are four phases to the thermal sterilization procedure. To ensure sterilization, the product must first be heated to a temperature of 110–125°C. Second, the product will take a few minutes to reach equilibrium since the surface of the container will be hotter than the center, generating a temperature gradient. The temperature gradient can be reduced during the equilibration stage. Third, the product must be kept at this temperature for a set length of time in order to achieve the F_o value, which is a predefined sterilization value. Finally, the product must be chilled, primarily to prevent overcooking and additional heat treatment (Harlfinger, 1992; Rahman, 2007).

Sterilization may be classified into two parts. The first is full sterilization, which means no living microbes should be present. However, this form of sterilization reduces the quality and nutritional content of the product. The other kind is commercial sterilization, which is described as a product that has been adequately handled so that it will not deteriorate or jeopardize the consumer's health under normal settings while retaining the organoleptic qualities and nutrients. Commercial sterilization is accomplished using a mix of relatively gentle heat treatment, as well as other processing and storage characteristics (Rahman, 2007; Tavman et al., 2019b; Teixeira, 2014). Commercial sterilization, as defined by the US Food and Drug Administration (FDA), is a process that uses heat alone or in combination with other ingredients or treatments to make sure the inactivation of microorganisms, excluding mesophiles and thermophiles, in order to extend shelf life and prevent food spoilage (Awuah et al., 2007; Soni, Smith, Thompson, & Brightwell, 2020). Commercial sterilizing is separated into two categories in the food industry: in-container processing (retorting/bulk canning/in-container) and ultra-high-temperature (UHT) processing (with aseptic filling). Both approaches use the same basic principles for heat sterilization of foods. Batch and continuous retorts are the two types of retorts. The retort is filled with product, closed, and then put through a processing cycle in batch systems. Containers are continually fed into and out of the retort in continuous retorting systems. Static, rotating, steam heated, and hot water with or without air pressure batch retorts are offered in a variety of designs for multiple uses (Featherstone, 2015; Li & Farid, 2016).

Because of its advantages, saturated steam is the most often utilized heating medium for commercial sterilization of packaged goods. When saturated vapor condenses on the container's surface, latent heat is transmitted to the food. When air is trapped inside the retort, it produces an insulating boundary layer around the cans, preventing the steam from condensing and causing the food to be overcooked. The containers are cooled with water after sterilization. In the retort, steam condenses quickly, but the food cools slowly and the pressure in the containers stays high. The overpressure of air is withdrawn once the food has dropped to $<100°C$, and cooling proceeds to around 40°C. Moisture on the container dries to avoid surface corrosion at this temperature, and label adhesives set more quickly. Foods are processed in glass jars or elastic pouches using an overpressure of air in hot water. To offer suitable strength, glass containers are thicker than metal cans, and this, along with glass's weaker thermal conductivity, leads in slower heat penetration and longer processing time than cans, as well as a higher danger of thermal shock to the container. Because of the narrow cross section of the pouch, foods in flexible pouches heat up faster. This allows for energy savings while causing little warming along the container wall. Foods that are liquid or semiliquid are frequently prepared horizontally to guarantee that the thickness of the food remains consistent throughout the pouch. Vertical packs allow for improved hot water circulation in the retort, but specific frames are required to keep the pouches from expanding at the bottom, that would affect the amount of heat penetration and hence the degree of sterilization (Al-Baali & Farid, 2006; Fellows, 1988).

Continuous-flow sterilization involves first heating the product to operating temperature by passing it through a suitable heat exchanger, then passing the product through a holding section for a predetermined time to effect sterilization, and then cooling it to 35°C or less before to aseptic filling. Aseptic packaging of the product has become necessary as a result of the development of HTST processing technologies for sterilizing in a continuous flow. The advantages of HTST treatment can only be completely realized with the use of aseptic packing. When viscous low-acid items are treated, aseptic packing will provide the best quality improvement over traditional canning. Many items may be professionally sterilized prior to packing using continuous methods without compromising their organoleptic or nutritional qualities. Aseptic packaging of foods is a procedure that allows items to be packed and sealed into sterile containers under aseptic conditions after being sterilized in bulk or on steam. It is employed for two reasons: (i) to use containers that aren't suited for in-package sterilization, and (ii) to take advantage of HTST sterilization procedure, which is thermally efficient and produces higher-quality goods than those treated at lower temperatures for longer periods of time (Hersom, 1985; Rahman, 2007).

1.2.3 Blanching

Blanching is a heat treatment that is commonly applied to food before it is dried, frozen, fried, or canned. One of the most essential pretreatment technologies has been studied. The blanching pretreatment is used to (i) inactivate the enzymes that cause off-flavor and odor, (ii) retain freshness color, texture stability, and nutritional quality, (iii) evacuate air between the cells, and (iv) eliminate microorganisms to some amount (Xiao et al., 2017; Xin, Zhang, Xu, Adhikari, & Sun, 2015). Blanching is a relatively light kind of heat treatment from 90°C to 95°C for roughly 1 or 2 min treatment to hot water or steam, and is not a food preservation method. It is generally applied to fresh/raw foods like (fruits and vegetables) prior to further processing. Blanching at a high temperature for a short period of time is referred to as HTST blanching. Nonetheless, over the last 10 years, the usage of blanching temperatures as low as 50°C and blanching periods as long as 40 min has been documented (Reis, 2017).

Blanching is usually achieved in the food processing sector by transporting the prepared raw food item on a belt through an atmospheric hot water bath or steam tunnel. Blanching is a cooking phrase that refers to a precooking procedure in which a product, generally a vegetable or fruit, is heated in steam or hot water for a short period and then cooled by dipping into iced water or water spray to end the cooking process. Blanching is used to soften food by partially or completely cooking it, or to eliminate a sharp flavor (for example, of bacon, cabbage or onions). Blanching is usually done right before heat sterilization, and it can be done before or after filling the containers (cans) with food product. The removal of gas from the raw material's tissues, shrinkage, prevention of enzymatic processes, and elimination of a strong undesired flavor that, if not controlled, may adversely influence the color and nutritional content of the food are all reasons for blanching, and can also kill certain bacteria, depending on how severe it is (Deak, 2014; Teixeira, 2014). In the food sector, hot water and steam are the most widely utilized heating media for blanching. Various hot water and steam blanchers have been developed to improve food product quality, yield, and processing of items with various thermal characteristics and geometries.

1.2.3.1 Hot water blanching

Hot water blanching is still the most common and commercially used blanching process, because of its ease of use, simplicity of equipment, and low investment. Hot water with a high specific heat capacity is used as the heating medium in the hot water blanching process, which results in a high blanching efficiency (Wang et al., 2022; Xiao, Bai, Sun, & Gao, 2014). Products are submerged in hot water (70–100°C) for several minutes in a conventional hot water blanching. The blanched samples are then drained and cooled before moving on to the next step of the procedure. The blanching water, in general, needs to be replaced after a particular length of blanching time since it gets saturated with nutrients leached from the goods. This phase not only uses a lot of water and energy but it also takes a long time. Sodium sulfite and sodium metabisulfite are frequently added to the blanching water to maintain the color of the product and inactivate microbiological activity (Bingol, Wang, Zhang, Pan, & McHugh, 2014).

LTLT blanching, as well as LTLT combined with HTST blanching, have been investigated. Water blanching produces a more homogeneous treatment, allowing for lower processing temperatures. Water blanchers carry the product into the tank, where hot water is supplied, through a screw or chain conveyor. Others submerge and convey the product via a rotating drum. Because water is heated indirectly through a heat exchanger using steam, the steam quality does not need to be "food-grade." Water blanching necessitates longer processing times, increases mineral and nutrient leaching, and resulting in effluents with high biological oxygen demand (BOD). Pipe blanchers are made up of insulated tubes that circulate hot water. The length of the pipe and the velocity of the water regulate the residence period of the food as it is metered into the stream. The blancher-cooker is divided into three sections: preheating, blanching, and cooling. The food is kept on a single conveyor belt during the whole operation (De Corcuera, Cavalieri, & Powers, 2004; Lin & Schyvens, 1995; Rahman, 2007).

1.2.3.2 Steam blanching

One of most basic methods is steam blanching in steam chambers. This treatment deactivates enzymes and eliminates oxygen from intercellular gaps, perhaps protecting the food's color and nutritional content (Di Cesare, Forni, Viscardi, & Nani, 2003). A product is moved through a chamber where "food-grade" steam at roughly 100°C is directly injected in steam blanchers by a chain or belt conveyor. Typically, the temperature in the headspace is monitored and the steam flow rate is adjusted. Because the heat transmission coefficient of condensing steam is larger than that of hot water, steam blanching takes less time than water blanching for sliced and small products. Larger products or parts of product, on the other hand, can be "overblanched" near the surface and "underblanched" at the center due to strong temperature gradients between the surface and the center of the product. Forced convection blanchers were created to improve heat transfer efficiency. These blanchers are built up of nested chambers with a fan that joins both chambers to allow recirculating steam. The steam is forced through a packed bed of goods carried through a mesh belt by the fan. Higher product bed depths and throughput are possible using this technique. Individual quick blanching (IQB) is another method that was created to reduce product treatment inconsistencies. A single layer of product is transported through the steam chamber in IQB, and each "individual" piece of product comes into touch with the steam right away. Steam blanching is more energy efficient than hot water blanching and creates fewer BOD and hydraulic loads. Furthermore, when compared to water blanching, nutrient loss is minimized (De Corcuera et al., 2004).

1.2.4 Hot filling

Filling is important for achieving a high degree of food safety, especially in systems that use food presterilization technology. The drinks are poured into packaging materials at this step, with no post-contamination of the pretreated product. In this instance, it is critical to keep contamination to a minimum in the filling room and on packing materials, as well as to use safe filling technology (Manfredi & Vignali, 2015). Heat-labile spoilage microorganisms such as lactic acid bacteria, yeasts, and some kinds of molds are typically inactivated or killed by the hot-filling process, which is an established and acknowledged thermal method for pasteurization. Product quality is maintained within the specified shelf life till the container is unsealed by the user using this procedure. Juice (fruit and vegetable) is normally filled at a temperature of (70–95°C) (McIntyre, Ikawa, Parkinson, Haglund, & Lee, 1995; Park et al., 2020).

Commercial fruit juice products frequently employ hot-filling technology. This technique assured that the filling of product solution into its package was done in an aseptic manner. Its idea is inserting heated products into material packaging final products at a specific time and temperature. Adjusting the time and temperature is beneficial since it acts as a microbial control to prevent spoiling (Ashurst, 2016; Hariyadi, 2013; Paquin, 2009). After being prepared, the product is pasteurized by continuous heat exchangers (usually tube in tube or plates exchangers) using superheated water supplied from a steam exchange and circulated in a closed circuit by specialized pumps in a hot filling process. The orange juice is pasteurized for 15 s at 95°C, which is maintained by the system. The heated product is subsequently transported at a constant temperature (92°C) from the processing zone to the filling tank. The purpose of the tank, which is normally located adjacent to the filler, is to transfer the product to the filling machine when it is needed. Different filling valves, such as a piston valve or a shatter valve, can be employed. If the filling machine is stopped in the middle of the procedure, the beverage is recirculated and warmed to keep the temperature between 88°C and 92°C. During the early phase of manufacturing, this recirculation system also aids in the removal of air. A maximum 10% of the product is recirculated through the machine. After the containers have been filled, they are sealed and tilted for up to 15 s to ensure that the product is in touch with the surfaces of the packing head area. The containers are then carried to a tunnel, where they are kept at filling temperature for 1–2 min in the first area to assure "commercial sterility," and then chilled with cold water provided by a chiller in the second section. Due to a probable lengthy heat exposure, rapid cooling of the product minimizes sugar build-up and subsequent product color and vitamin content deterioration. The filled container is subsequently palletized, labeled, and packaged in a box or pack (Manfredi & Vignali, 2015).

Application of combined hot filling technique with refrigerator storage temperatures <200°C can be regarded a safe and cost-effective method of preventing spoiling in orange juice products (Spinelli, Sant'Ana, Rodrigues-Junior, & Massaguer, 2009). Similarly, employing the hot-filling procedure to make *aloe vera* drinks can extend the product's shelf life to up to 1 month at room temperature. Depending on this, drink manufacturing processes are modified based on time and temperature parameters to optimize product performance and reduce the danger of spoiling. Meanwhile, it preserves the nutritious product value, which has been the product's key selling point. As a result, the goal of this project is to improve and extend the product's shelf life by process change (Sefrienda, Amri, Ariani, Susanto, & Nurhikmat, 2020).

1.3 Different heat exchangers for thermal processing: A brief overview

The heating and cooling of foods are the most prevalent procedures in a food processing plant. To avoid microbial and enzymatic breakdown, food items must be heated and cooled. Additionally, as food are cooked or chilled, desired sensory properties flavor, color, and texture are imparted. Heat transmission is crucial because it lays the groundwork for knowing how multiple food processes work. Heat exchangers are thermodynamic machines that are commonly utilized in engineering applications like heating and cooling. A heat exchanger is a device designed to efficiently transfer heat between two fluids of different temperatures. To avoid mixing, the media may be separated by a solid wall, or they may be in direct touch.

There are various causes for heat transfer in the food processing industry, including thermal processing such as preheating, pasteurizing, and sterilizing, all of which necessitate the use of high-efficiency heat exchangers. On the other hand, heat exchangers should be cleaned readily and thoroughly as a one-of-a-kind quality assurance device. Heat exchangers are widely employed not just in the food and dairy industries, but also in the pharmaceutical business, alcohol manufacturing, and biochemical processing. They are applied in food processing, dairy processing, biochemical processing, and pharmaceuticals. Heat exchangers are divided into two types: noncontact and contact. The product and the heating or cooling medium are kept physically separated in noncontact heat exchangers, generally by a thin wall. On the other hand, in contact-type heat exchangers, the product and the heating or cooling streams are in direct physical touch (Agarwal, Sikand, & Shanthi, 2014; Başaran, Yılmaz, & Çivi, 2018; Singh & Heldman, 2014). Depending on the necessity, fluid products are heated to a temperature of 62.8°C for 30 min or 71.7°C for 15 s. For the thermal treatment of liquid foods such as whole milk, hazy orange juice, and apple and apricot puree, heat exchangers are commonly utilized (Teke, Ağra, Atayılmaz, & Demir, 2010). In the next subsections, we'll go through some of the most commonly used heat exchangers in the food sector.

1.3.1 Plate heat exchangers (PHEs)

PHEs are widely used in a variety of chemical processes and other industrial applications, with a focus on the food industry for a variety of reasons, including a compact design, a large surface area in a small volume that can be modified as needed by increasing or decreasing the number of plates, and advances in material technology. They are made up of numerous thin plates that are slightly spaced from one another and have a path for fluid flow for heat transfer (Abu-Khader, 2012; Agarwal et a., 2014). Pasteurization and sterilization operations both rely on PHEs. They have a higher overall heat transfer coefficient (OHTC) than tubular ones and are often more compact. However, maintenance costs may be high, as gaskets may have to be replaced at regular intervals, and the integrity of the plates must be inspected on a regular basis, particularly in the regeneration section, where a broken or leakages plate could allow raw product, such as raw milk, to contaminate already pasteurized or sterilized milk (Lewis & Jun, 2011).

PHEs, which were designed over 70 years ago, have a wide range of applications in the dairy and food beverage industries. PHEs are ideal for liquids with a low viscosity (0.5 Pa s). The corresponding diameter of the particles should be <0.3 cm if suspended solids are present. Larger particles can bridge the plate contact sites and cause the heating portion to "burn on." As a result, it is employed for operations like pasteurizing milk, beer, and juices at high temperatures for short periods of time. PHEs are used in the manufacturing of stirred yogurt to chill the yogurt and stop lactic acid fermentation once the required acidity is reached (Fernandes et al., 2006).

1.3.2 Tubular and shell and tube heat exchangers

A double-pipe heat (tubular) exchanger is the simplest noncontact heat exchanger, consisting of a pipe concentrically inside the other pipe. In the annular gap and the inner pipe, the two fluid streams run in opposite directions. The streams may stream in same way or in opposing directions. The product flows in the inner annular space of this heat exchanger, while the heating/cooling medium flows in the inner tube and outer annular space. A shell-and-tube heat exchanger is another typical form of heat exchanger used in the food industry for purposes such as heating liquid food in evaporating processes. One fluid stream flows within the tube, while the other is pushed through the shell and over the tubes. We can attain better heat transfer rates by directing the fluid stream in the shell side to flow over the tubes rather than parallel to them. The cross-flow pattern is enabled by baffles on the shell side. Depending on the design, one or more tube passes can be made (Singh & Heldman, 2014). This design of heat exchanger is made up of a series of tubes that hold the fluid that needs to be heated or cooled, with a second fluid flowing over the tubes which need to be heated or cooled. Due to its toughness, shell and tube heat exchangers are often utilized for high pressure applications (pressures >30 bar and temperatures >260°C) (Hanson, 1989).

It's best to employ a shell and tube heat exchanger with helically corrugated walls. Helically curved heat exchangers are chosen over smooth walls because they improve heat transmission by convection, boosting efficiency. Tubes come in a range of shapes and sizes to accommodate various product properties. Single tubes with an outer jacket, double or many concentric tubes or shells, and tube kinds are examples of these. A multitube design is used in most UHT facilities. Compared to plate heat exchangers, they can endure higher pressures. High pumping pressures can be employed to overcome flow constraints, despite the fact that they are still subject to fouling. With viscous fluids, tubular heat exchangers provide longer processing durations (Söylemez, 2008).

1.3.3 Direct heat exchangers (steam infusion/injection)

A steam-infusion heat exchanger allows steam to come into direct contact with the product. The liquid product is pushed to the top of the heat exchanger and then allowed to flow through the heating chamber in thin sheets. The size of the spreaders is determined by the viscosity of the liquid. Specially constructed spreaders can handle particulate-containing products such chopped vegetables, meat bits, and grains. When steam comes into touch with small particles of food, high rates of heat transmission are achieved. Due to steam condensation, the temperature of the product rapidly rises. The heated goods are discharged from the bottom chamber with condensed steam. To achieve the proper cooking, a specified amount of liquid is held in the chamber's bottom. Concentrated soups, chocolate, processed cheese, custards, fruit, ice cream mixtures, pie fillings, and milk are just a few of the foods that may be cooked and/or sterilized using this sort of heat exchanger (Das, 2005; Heldman, Lund, & Sabliov, 2018).

In direct methods, the product is heated to 75°C, usually by regeneration, before being subjected to cooking steam to reach 140–145°C. This heat treatment moves at a breakneck speed. In a holding tube, the product is retained for a few seconds. Added water is removed via flash cooling, which involves a rapid drop in pressure to lower the product's temperature to between 75°C and 80°C. The elimination of some water vapor is accompanied by a sharp drop in temperature. The amount of water evacuated and the drop in temperature have a direct connection. The pressure (vacuum) in the flash cooling chamber regulates the ultimate temperature and hence the amount of water evaporated. This cooling, like the heating, is a fairly quick process.

Flash chilling not only eliminates the additional water (as vapor), but it also removes other volatile components, however in the situation of UHT milk improves the quality. When compared to an equivalent indirect process with equal holding time and temperature, direct methods use a short, sharp heating profile and cause less chemical damage. They are also less prone to fouling and have longer processing durations, although their regeneration efficiencies are often <50%. Instead of pasteurization, direct methods are most commonly used for UHT. Bringing the steam and the food liquid together can be accomplished in one of two ways. Steam may be injected into a liquid (injection processes) or liquid can be injected into steam (infusion). According to one school of thinking, infusion is less painful than injection since the substance comes into touch with heated surfaces less (Singh & Heldman, 2014; Zhang, Liu, Pain, & Fryer, 1992).

1.4 Applications of thermal processing technologies in food sector

Heat is the most popular method for processing food, due to its capacity to inactivate bacteria and spoilage enzymes. The employment of specific thermal processing procedures, e.g., water immersion, hot air drying, blanching, pasteurization, sterilization, and steaming has been preferred by significant technical advancements over the previous several years, as well as simplified equipment management.

1.4.1 Thermal processing technologies for vegetable's preservation

The heat treatment is used to (i) inactivate the enzymes that cause off-flavor and odor, (ii) retain freshness color, texture stability, and nutritional quality, (iii) evacuate air between the cells, and (iv) eliminate microorganisms to a certain level (Agüero, Ansorena, Roura, & Del Valle, 2008; Garrote, Bertone, Silva, & Avalle, 2001). It was discovered that when soybean pods and seeds were blanched in steam for 10min at 100°C, they preserved soluble sugars (Saldivar, Wang, Chen, & Mauromoustakos, 2010). Thermal pasteurization ensures that microbiological development in basil-bottle gourd juice is controlled during storage. The heat processing temperature was 95°C for 15min. The mixed juice could be stored at room temperature for 6 months and was microbially safe (Gajera, 2017). Various thermal processing processes, such as pasteurization, blanching, autoclaving, and spray drying, can extend the shelf life of beets and beet juice (Prieto-Santiago, Cavia, Alonso-Torre, & Carrillo, 2020). Spray drying treats fruit juice to high temperatures, reducing water activity and water content (to $<5\%$), resulting in a biologically stable powder (Tontul & Topuz, 2017) Likewise, thermal pasteurization of beetroot juice at 96°C for 12min resulted in microbiologically stable juice with the conservation of qualitative characteristics. (Kathiravan, Nadanasabapathi, & Kumar, 2014).

Thermal processing is a widely used technique for extending the shelf life of juices and fluids. The ready to drink ginger bottled drinks were given a 5-min heat treatment at 95°C. The beverage is microbiologically safe for 6 months, with good active component retention (Wadikar, Vasudish, & Premavalli, 2015). Similarly, with spinach juice, heat pasteurization at 80°C for 10min was used. During the 10-week stored at 4°C, no microbiological increase was recorded (Khandpur & Gogate, 2015). Other heat treatment, such as HTLT, might be used on these activities. One example is a 75%–83% drop in PME enzymatic activity. For instance, a broccoli, carrot, red pepper, and tomato smoothie cooked at 80°C for 3min. PPO and PME, whose activity were negligible during storage up to 40 and 58days at 20 and 5°C, accordingly, were completely inactivated by heat processing (Rodríguez-Verástegui et al., 2016).

Before freezing, mushrooms (*Boletus edulis*) are blanched in water or aqueous solutions of citric and L-ascorbic acids/lactic and L-ascorbic acids to minimize adverse texture alterations during freezing and subsequent storage under cold conditions (Jaworska, Bernaś, Biernacka, & Maciejaszek, 2010). Furthermore, blanching can help slow down the breakdown of chlorophylls and carotenoids in frozen dill. (Lisiewska, Kmiecik, & Słupski, 2004). In vegetable systems, peroxidase (POD) is the most heat stable enzyme. As a result, it is commonly employed as a measure of blanching efficiency. Two-phase curves have been seen in studies on the kinetics of heat inactivation for POD, which might be related to the existence of isoenzymes with differing thermal stability (Cruz, Vieira, & Silva, 2006). The inactivation of Peroxidase has been studied on broccoli, green asparagus and carrots (Morales-Blancas & Chandia, 2002), carrots and potatoes (Anthon & Barrett, 2002), tomato juice (Anthon, Sekine, Watanabe, & Barrett, 2002), green beans, green peas and carrots (Güneş & Bayindirli, 1993), pea puree (Icier, Yildiz, & Baysal, 2006), asparagus (Ganthavorn, Nagel, & Powers, 1991), and horseradish (Lemos, Oliveira, & Saraiva, 2000).

1.4.2 Thermal processing technologies for fruits preservation

Fruits are a perishable item that is thermally treated to extend its shelf life. They are used to make juices, nectar, purees, smoothies and other products (Rawson et al., 2011). The intensity of the heat treatment can be used to classify thermal processes. The most frequent method in the processing of juices and drinks is HTLT (80°C and holding durations >30s); it can be categorized as pasteurization (100°C), canning (100°C), or sterilization (>100°C) (Miller & Silva, 2012; Petruzzi et al., 2017). MTLT (mild treatment for long time) pasteurization is a method that uses a temperature < 80°C and a duration >30s. It is a method for making minimally processed foods with an extended shelf life. According to one research, extensive inactivation (about 4.5-log) was discovered after pasteurization operations, but only elevated pasteurization treatments resulted in total removal of microbes in pomegranate juice (Mena et al., 2013).

Pineapple juice was sterilized utilizing developed impingement steam jet technology, with ideal circumstances of two jet nozzles and six steam plates, accordingly, and a steam temperature of 120°C. The development of microorganisms was determined to have stopped, which is compatible with the criterion for sterilized beverages in a sealed container. Moreover, the sterilizing method had no significant effect on the physicochemical properties of color and pH (Sangsom & Inprasit, 2022). Another way of steam extraction is to raise water vapor until it reaches the fruit, where it transfers heat and leaches off the pulp. It is said to inhibit the growth of enzymes and pasteurize juice upon extraction, obviating the need for any further preservation procedure (Bates, Morris, Morris, & Crandall, 2001). Jaboticaba juice, which was also made by steam extraction, included the fruit's phenolic components and had appropriate microbiological purity and sensory acceptability without the need for any further preservation method or chemical preservatives. The use of steam extraction and a hot-filling process resulted in a juice with a storage duration of at least 6 months

(Martins, Canto, Perrone, & Monteiro, 2021). During storage, the nonsweetened acerola tropical fruit juice produced using the hot fill technique retained appropriate stability. During the storage time, the total soluble solid and SO_2 stability were sustained (Freitas, Maia, Sousa, Brasil, & Pinheiro, 2006).

There is no significant differences in physical, physiochemical, or sensory alterations among hot-filling and aseptically treated cashew apple juice over storage time in regarding product stability. (Costa, Maia, Figueiredo, Souza Filho, & Brasil, 2003). The ascorbic acid remained constant through pasteurization in degassed tamarillo nectar at 95°C for 10min (Mertz, Brat, Caris-Veyrat, & Gunata, 2010). Conventional thermal processes guarantee the safety of citrus fruit products while preserving the required bioactive ingredients. The importance of HTST thermal pasteurization (temperatures <80°C and retention periods <30s) is greater (Achir et al., 2016). HTST heat treatment is shown to boost the following: total phenolics, ()-epicatechin, ferulic acid, and *p*-coumaric acid concentration in apricot nectar (Huang et al., 2013), color shade in pomegranate/lemon juice (Mena, Martí, Saura, Valero, & García-Viguera, 2013), and high nutritional value in fermented orange juice (Cerrillo et al., 2015). The effects of HTST treatment on several enzymes were also investigated; it was shown that it could: eliminate PME (95.3%) and PPO (90.9%) in apple juice (Aguilar-Rosas, Ballinas-Casarrubias, Elias-Ogaz, Martin-Belloso, & Ortega-Rivas, 2013); as well as assure total deactivation of PME and POD in grapefruit juice (Gao et al., 2015).

Batch pasteurization is a common method for preserving viscous items, although it takes a long time due to the temperature resistance of liquids and packing. Flow pasteurization, on the other hand, just takes a few seconds to raise the temperature enough to inactivate enzymes and kill bacteria, which should enhance heat-sensitive component retention. Thermal processing had an influence on viscosity; microwave flow-pasteurized apple-based smoothies had greater viscosity and total pectins than those preserved by other means. This shows that the preservation procedure interacts with the microstructure of the smoothie and might change its quality over time. The various heat treatments resulted in items that were acceptable. The various heat treatments resulted in items that were acceptable (Markowski, Celejewska, Rosłonek, & Kosmala, 2017). Furthermore, the juice is heated quickly for about 1min to a temperature that is only 5°C higher than the pasteurization temperature, then poured into bottles that are sealed tightly under steam to disinfect the seal, then chilled. This method may be used to make orange juice, grape juice, apple juice, and other juices (Teixeira, 2014).

Thermal treatment against invasive pests is usually necessary for tropical fruits intended for import, which can be paired with controlled atmosphere (CA) storage to extend storage life and maintain quality of the fruit. The combination of hot water treatment and CA storage provided effective therapy for extending the postharvest life of mangoes without compromising the fruit's nutritional profile (Kim, Brecht, & Talcott, 2007). Carrots were blanched for 4min at 100°C in normal water and acidified water (45g/L citric acid, pH 1.3), and juice was obtained. In blanched samples, total carotenoids, lycopene, and lutein levels increased significantly ($P < 0.05$). The findings imply that a blanching treatment might be used to improve the quality of carrot juice during processing (Jabbar et al., 2014)

1.4.3 *Thermal processing technologies for dairy products*

Thermal processing procedures in milk are characterized as LTLT, HTST, and UHT pasteurization based on treatment duration and heat load applied (UHT). While pasteurization remains the most effective method for increasing shelf life and maintaining microbiological safety of milk till now, it frequently results in negative quality changes. As a result, the dairy industry has expressed an interest in embracing alternative milk processing methods in order to enhance quality attributes (Ziuzina & Misra, 2016). Milk and dairy products are now regarded fundamental aspects in everyone's life and a basic requirement for all humans. Throughout history, human society has always looked for ways to store food for more difficult times. Food preservation technology have been adapted for many years since then. These innovations not only extended the shelf life of the product, but also gave it a new set of physiochemical and sensory properties. Liquid/beverage milk, cheese, milk powders, concentrated milks, fermented milk products, butter, ice cream, newborn formula, creams, and protein-rich goods are all examples of milk products.

One of the most important processing methods in the dairy business is heat treatment. In the dairy sector, heat treatments are widely used. Pathogenic and spoilage microorganisms, as well as several enzymes that might produce taste and texture flaws in heated items, are inactivated by them. Some heat treatments are intended to induce specific effects that improve the processed product's quality or shelf life (Deeth, 2022). Pasteurized (at 72°C for 15 s) When kept at 101°C, buffalo milk has a storage life of up to 10 days (Nasr & Elshaghabee, 2019). Pasteurized buffalo skim milk has a shelf life of up to 12 days at 4–8°C after being pasteurized at 65°C for 30 min. When kept at 4–8°C, sterilization (110°C for 10 min) considerably improved the appearance/color and the aroma of buffalo skim milk and improved its shelf-life up to 28 days (ul Haq et al., 2014). Likewise, Hussain (2011) observed that after 90 days of storage at 10°C and 25°C, commercially processed UHT buffalo skimmed milk had acceptable sensory attributes (Hussain, 2011). Innovative Steam Injection (ISI) heating, developed by Huijs et al., can heat milk at 150–180°C for 0.1 s. The technique proved ability to produce economically sterile milk with minimum chemical alteration, causing just 20%–25% denaturation of -Lg (Huijs, Van Asselt, Verdurmen, & De Jong, 2004).

Heat treatments during cheese production cause denaturation of particular proteins, which improves the texture of the cheeses and the technological output. Thermal treatment of milk and dairy products can also improve their color, taste, and flavor (e.g., cheese browning, cheese melting). The degree of whey protein denaturation is normally controlled by the preheat treatment performed to liquid skim milk before spray drying (Guinee, 2017). Thermal treatment of milk prior to fermentation enhances denaturation of whey proteins (primarily -Lg and -La) and their interaction with -casein on the casein micelle surface through disulfide bonds. Thermally processed milk yogurts have a thin micelle matrix, which reduces hardness and promotes whey retention. Stabilizers and thickeners are added to the milk to maximize the rheological and whey retention qualities, despite the fact that heat treatment enhances hardness and minimizes syneresis (Harte, Luedecke, Swanson, & Barbosa-Cánovas, 2003).

1.4.4 Thermal processing technologies for candies preservation

Candy is a sweet food made from fruits or vegetables that has been impregnated with sugar syrup, then drained of extra syrup and dried to a shelf-stable form. Apples, ginger, mangoes, guava, carrots, and citrus peels have all been utilized in the making of sweets. Blanching unripe mango fruit candy in hot water with 40o Brix honey yields a batter outcome. In terms of general acceptability, the 40oBrix sugar syrup received the highest grade of 8.77 on a 9-point hedonic scale (Mahato, Chakraborty, & Baidya, 2020). The effects of heating (100–140°C/10–180min) on functional component contents, total phenolic content (TPC), and antioxidant activities of six simulated Chinese herb candies containing were investigated, and it was discovered that TPC and antioxidant activity of candies increased significantly ($P < 0.05$) with increasing heating temperature (> 130°C) and time (> 40min) during in the boiling process. During the manufacturing of herb candy, heat processing increases antioxidant activity (Yin et al., 2019). The addition of *C. turgidus*, a microalga, can boost the proximate content of jelly sweets. The ideal jelly candy formulation includes 1.5% biomass powder and 1% water-soluble extract, as well as heat treatment at 80°C (Agustini, Priadi, Hidhayati, Multri, & Rahman, 2021).

For candy making, Indian gooseberries were blanched (at 70–90°C for 5–15min in a pressurized environment), syruped, and then dry (in a cabinet and microwave). Blanching at 80°C for 20min was considered sufficient for the manufacture of Indian gooseberry candy, with 315, 519, 53mg/100g vitamin C retained after blanching. Blanching at 80°C for 20min followed by drying at 70°C can successfully achieve the Recommended Dietary Intake (60mg/day) by retaining more than 42% ascorbic acid (201.75mg/100g) in candy-like additional value products (Yadav, Khanashyam, Wadikar, Patki, & Semwal, 2020). The observed advantage increased with the severity of the heat treatment applied to the food. In comparison to putting menthol to a liquid formulation, introducing l-menthol in -CyD resulted in 86 times improved retention in a hard candy (Reineccius, Reineccius, & Peppard, 2004).

1.5 Conclusion

Traditional thermal processing techniques for shelf-stable foods have been shown to be beneficial in terms of product safety. The use of heat is a critical step in the preservation of foods. Existing commercial thermal processing technology must be enhanced and improved in terms of end product safety and quality, as well as process efficiency improvement. In the food processing industry, heat is used in a variety of unit processes. Many of them have the goal of eliminating the raw state of the material in order to prepare the look and structure of finished goods. The most common finishing process is heat preservation by blanching, pasteurization, and sterilization, as detailed above. Although thermal processing techniques inactivate enzymes, aggregate proteins, and to some extent kill microbes, the major goal of heat preservation is to eradicate microbes to ensure that canned items have a long shelf life without spoiling. Sterilization refers to the use of extreme temperatures (>100°C) to kill all microbes, while pasteurization refers to the use of lower temperatures (usually <100°C) to kill most vegetative harmful bacteria and increase shelf life of the product. Pasteurization is frequently used in combination with another preservation method (concentration, acidification, refrigeration, etc.). Blanching is a unit process that involves heating fruits or vegetables to inactivate enzymes, change texture, preserve color, taste, and nutritional content, and remove trapped air before freezing, canning, or drying. The most popular heating medium for thermal processing are hot water and steam.

References

Aaliya, B., Sunooj, K. V., Navaf, M., Akhila, P. P., Sudheesh, C., Mir, S. A., … George, J. (2021). Recent trends in bacterial decontamination of food products by hurdle technology: A synergistic approach using thermal and non-thermal processing techniques. *Food Research International*, *147*, 110514.

Abu-Khader, M. M. (2012). Plate heat exchangers: Recent advances. *Renewable and Sustainable Energy Reviews*, *16*(4), 1883–1891.

Achir, N., Dhuique-Mayer, C., Hadjal, T., Madani, K., Pain, J.-P., & Dornier, M. (2016). Pasteurization of citrus juices with ohmic heating to preserve the carotenoid profile. *Innovative Food Science & Emerging Technologies*, *33*, 397–404.

Agarwal, P., Sikand, A., & Shanthi, V. (2014). Application of heat exchangers in bioprocess industry: A review. *International Journal of Pharmacy and Pharmaceutical Sciences*, *6* (1), 24–28.

Agüero, M., Ansorena, M., Roura, S., & Del Valle, C. (2008). Thermal inactivation of peroxidase during blanching of butternut squash. *LWT-Food Science and Technology*, *41*(3), 401–407.

Aguilar-Rosas, S., Ballinas-Casarrubias, M., Elias-Ogaz, L., Martin-Belloso, O., & Ortega-Rivas, E. (2013). Enzyme activity and colour changes in apple juice pasteurised thermally and by pulsed electric fields. *Acta Alimentaria*, *42*(1), 45–54.

Agustini, N. W. S., Priadi, D., Hidhayati, N., Multri, H. D., & Rahman, D. Y. (2021). The Chroococcus Turgidus supplements in jelly candy formulation as antioxidant agent. *Journal of Hunan University Natural Sciences*, *48*(11).

Al-Baali, A., & Farid, M. M. (2006). Thermal sterilization of food. In *Sterilization of food in retort pouches* (pp. 1–16). Springer.

Anthon, G. E., & Barrett, D. M. (2002). Kinetic parameters for the thermal inactivation of quality-related enzymes in carrots and potatoes. *Journal of Agricultural and Food Chemistry*, *50*(14), 4119–4125.

Anthon, G. E., Sekine, Y., Watanabe, N., & Barrett, D. M. (2002). Thermal inactivation of pectin methylesterase, polygalacturonase, and peroxidase in tomato juice. *Journal of Agricultural and Food Chemistry*, *50*(21), 6153–6159.

Ashurst, P. R. (2016). *Chemistry and technology of soft drinks and fruit juices*. John Wiley & Sons.

Augusto, P. E., Soares, B. M., & Castanha, N. (2018). Conventional technologies of food preservation. In *Innovative technologies for food preservation* (pp. 3–23). Elsevier.

Awuah, G., Ramaswamy, H., & Economides, A. (2007). Thermal processing and quality: Principles and overview. *Chemical Engineering and Processing: Process Intensification*, *46* (6), 584–602.

Bahrami, A., Baboli, Z. M., Schimmel, K., Jafari, S. M., & Williams, L. (2020). Efficiency of novel processing technologies for the control of listeria monocytogenes in food products. *Trends in Food Science & Technology*, *96*, 61–78.

Başaran, A., Yılmaz, T., & Çivi, C. (2018). Application of inductive forced heating as a new approach to food industry heat exchangers. *Journal of Thermal Analysis and Calorimetry*, *134*(3), 2265–2274.

Bates, R. P., Morris, J. R., Morris, J. R., & Crandall, P. G. (2001). *Principles and practices of small-and medium-scale fruit juice processing*. Food & Agriculture Org.

Bingol, G., Wang, B., Zhang, A., Pan, Z., & McHugh, T. H. (2014). Comparison of water and infrared blanching methods for processing performance and final product quality of French fries. *Journal of Food Engineering*, *121*, 135–142.

Cerrillo, I., Fernández-Pachón, M., Collado-González, J., Escudero-López, B., Berná, G., Herrero-Martín, G., … Gil-Izquierdo, A. (2015). Effect of fermentation and subsequent pasteurization processes on amino acids composition of orange juice. *Plant Foods for Human Nutrition*, *70*(2), 153–159.

Costa, M. C. O., Maia, G. A., Figueiredo, R. W. D., Souza Filho, M. D. S. M. D., & Brasil, I. M. (2003). Storage stability of cashew apple juice preserved by hot fill and aseptic processes. *Food Science and Technology*, *23*, 106–109.

Cronshaw, H. B., & Cronshaw, H. B. (1947). *Dairy information: Dairy industries limited.*

Cruz, R. M., Vieira, M. C., & Silva, C. L. (2006). Effect of heat and thermosonication treatments on peroxidase inactivation kinetics in watercress (Nasturtium officinale). *Journal of Food Engineering*, *72*(1), 8–15.

Das, S. K. (2005). *Process heat transfer*. Alpha Science Int'l Ltd.

De Corcuera, J. I. R., Cavalieri, R. P., & Powers, J. R. (2004). Blanching of foods. In *Encyclopedia of agri, food and biological engineering* (pp. 1–5). New York City, NY, USA: Marcel Dekker.

Deak, T. (2014). Thermal treatment. In *Food safety management* (pp. 423–442). Elsevier.

Deeth, H. C. (2022). Heat treatment of milk: Principles and purpose. In P. L. H. McSweeney, & J. P. McNamara (Eds.), *Encyclopedia of dairy sciences* (3rd Edition, pp. 655–658). Oxford: Academic Press.

Di Cesare, L. F., Forni, E., Viscardi, D., & Nani, R. C. (2003). Changes in the chemical composition of basil caused by different drying procedures. *Journal of Agricultural and Food Chemistry*, *51*(12), 3575–3581.

Featherstone, S. (2015). *A complete course in canning and related processes: Volume 3 processing procedures for canned food products*. Woodhead Publishing.

Fellows, P. (1988). *Processing by application of heat*. In (pp. 221) England: Ellis Horwood.

Fernandes, C. S., Dias, R. P., Nóbrega, J. M., Afonso, I. M., Melo, L. F., & Maia, J. M. (2006). Thermal behaviour of stirred yoghurt during cooling in plate heat exchangers. *Journal of Food Engineering*, *76*(3), 433–439.

Freitas, C. A. S. D., Maia, G. A., Sousa, P. H. M. D., Brasil, I. M., & Pinheiro, A. M. (2006). Storage stability of acerola tropical fruit juice obtained by hot fill method. *International Journal of Food Science & Technology*, *41*(10), 1216–1221.

Gajera, R. (2017). Effect on juice quality of medicinal fruit bottle gourd during storage. *Journal of Medicinal Plants*, *5*(4), 56–60.

Ganthavorn, C., Nagel, C., & Powers, J. (1991). Thermal inactivation of asparagus lipoxygenase and peroxidase. *Journal of Food Science*, *56*(1), 47–49.

Gao, G., Zhao, L., Ma, Y., Wang, Y., Sun, Z., & Liao, X. (2015). Microorganisms and some quality of red grapefruit juice affected by high pressure processing and high temperature short time. *Food and Bioprocess Technology*, *8*(10), 2096–2108.

Garrote, R., Bertone, R., Silva, E., & Avalle, A. (2001). Note. Comparison of two rapid methods of lipoxygenase assay in blanched green peas, green beans and potatoes. *Food science and technology international*, *7*(2), 171–175.

Guinee, T. P. (2017). Pasteurized processed and imitation cheese products. In *Cheese* (pp. 1133–1184). Elsevier.

Güneş, B., & Bayindirli, A. (1993). Peroxidase and lipoxygenase inactivation during blanching of green beans, green peas and carrots. *LWT-Food Science and Technology*, *26*(5), 406–410.

Hanson, F. V. (1989). *Heat exchangers selection, design and construction*. by EAD Saunders, Longman sScientific and Technical, Essex, England, 1988, ISBN 0-47020870-8, pp. 568, price: US $99.95 (co-published in the United States with John Wiley & Sons, Inc., New York). In: Elsevier.

Hariyadi, P. (2013). Hot-fill processing of beverages. *Food Reviews International*, *1*(1), 46–49.

Harlfinger, L. (1992). *Microwave sterilization*. Food technology (USA).

Harte, F., Luedecke, L., Swanson, B., & Barbosa-Cánovas, G. V. (2003). Low-fat set yogurt made from milk subjected to combinations of high hydrostatic pressure and thermal processing. *Journal of Dairy Science*, *86*(4), 1074–1082. https://doi.org/10.3168/jds.S0022-0302(03)73690-X.

Heldman, D. R., Lund, D. B., & Sabliov, C. (2018). *Handbook of food engineering*. CRC press.

Hersom, A. (1985). Aseptic processing and packaging of food. *Food Reviews International*, *1*(2), 215–270.

Huang, W., Bi, X., Zhang, X., Liao, X., Hu, X., & Wu, J. (2013). Comparative study of enzymes, phenolics, carotenoids and color of apricot nectars treated by high hydrostatic pressure and high temperature short time. *Innovative Food Science & Emerging Technologies*, *18*, 74–82.

Huijs, G., Van Asselt, A., Verdurmen, R., & De Jong, P. (2004). High speed milk. *Dairy Industries International*, *69*(11), 30–32.

Hussain, I. (2011). Effect of UHT processing and storage conditions on physico-chemical characteristics of buffalo skim milk. *Journal of the Chemical Society of Pakistan*, *33*(6), 783.

Icier, F., Yildiz, H., & Baysal, T. (2006). Peroxidase inactivation and colour changes during ohmic blanching of pea puree. *Journal of Food Engineering*, *74*(3), 424–429.

Jabbar, S., Abid, M., Hu, B., Wu, T., Hashim, M. M., Lei, S., … Zeng, X. (2014). Quality of carrot juice as influenced by blanching and sonication treatments. *LWT-Food Science and Technology*, *55*(1), 16–21.

Jackson, J. M., & Shinn, B. M. (1979). *Fundamentals of food canning technology*. AVI Pub Co.

Jaworska, G., Bernaś, E., Biernacka, A., & Maciejaszek, I. (2010). Comparison of the texture of fresh and preserved Agaricus bisporus and boletus edulis mushrooms. *International Journal of Food Science & Technology*, *45*(8), 1659–1665.

Karel, M., & Lund, D. B. (2003). *Physical principles of food preservation: Revised and expanded*. CRC Press.

Kathiravan, T., Nadanasabapathi, S., & Kumar, R. (2014). Standardization of process condition in batch thermal pasteurization and its effect on antioxidant, pigment and microbial inactivation of ready to drink (RTD) beetroot (Beta vulgaris L.) juice. *International Food Research Journal*, *21*(4), 1305.

Khan, I., Tango, C. N., Miskeen, S., Lee, B. H., & Oh, D.-H. (2017). Hurdle technology: A novel approach for enhanced food quality and safety–A review. *Food Control*, *73*, 1426–1444.

Khandpur, P., & Gogate, P. R. (2015). Effect of novel ultrasound based processing on the nutrition quality of different fruit and vegetable juices. *Ultrasonics Sonochemistry*, *27*, 125–136.

Kim, Y., Brecht, J. K., & Talcott, S. T. (2007). Antioxidant phytochemical and fruit quality changes in mango (Mangifera indica L.) following hot water immersion and controlled atmosphere storage. *Food Chemistry*, *105*(4), 1327–1334. https://doi.org/10.1016/j.foodchem.2007.03.050.

Lemos, M. A. L., Oliveira, J. C., & Saraiva, J. A. (2000). Influence of pH on the thermal inactivation kinetics of horseradish peroxidase in aqueous solution. *LWT-Food Science and Technology*, *33*(5), 362–368.

Lewis, M., & Jun, S. (2011). Thermal processing. In *Food processing handbook, 2 volume set.*

Lewis, M. J., & Heppell, N. J. (2000). *Continuous thermal processing of foods*. Aspen Publishers.

Li, X., & Farid, M. (2016). A review on recent development in non-conventional food sterilization technologies. *Journal of Food Engineering*, *182*, 33–45.

Liao, X., Bai, Y., Muhammad, A. I., Liu, D., Hu, Y., & Ding, T. (2019). The application of plasma-activated water combined with mild heat for the decontamination of Bacillus cereus spores in rice (Oryza sativa L. ssp. japonica). *Journal of Physics D: Applied Physics*, *53*(6), 064003.

Lin, Z., & Schyvens, E. (1995). Influence of blanching treatments on the texture and color of some processed vegetables and fruits. *Journal of Food Processing and Preservation*, *19* (6), 451–465.

Lisiewska, Z., Kmiecik, W., & Słupski, J. (2004). Contents of chlorophylls and carotenoids in frozen dill: Effect of usable part and pre-treatment on the content of chlorophylls and carotenoids in frozen dill (Anethum graveolens L.), depending on the time and temperature of storage. *Food Chemistry*, *84*(4), 511–518.

Liu, J., Bi, J., McClements, D. J., Liu, X., Yi, J., Lyu, J., … Wu, X. (2020). Impacts of thermal and non-thermal processing on structure and functionality of pectin in fruit-and vegetable-based products: A review. *Carbohydrate Polymers*, *250*, 116890.

Mahato, A., Chakraborty, I., & Baidya, B. K. (2020). Preparation and evaluation of fruit candy from unripe mango. *Int. J. Chem. Stud.*, *8*(1), 2727–2731.

Manfredi, M., & Vignali, G. (2015). Comparative life cycle assessment of hot filling and aseptic packaging systems used for beverages. *Journal of Food Engineering*, *147*, 39–48.

Markowski, J., Celejewska, K., Rosłonek, A., & Kosmala, M. (2017). Impact of different thermal preservation technologies on the quality of apple-based smoothies. *LWT-Food Science and Technology*, *85*, 470–473.

Martins, A. B. N., Canto, M., Perrone, D., & Monteiro, M. (2021). Chemical, microbiological and sensory stability of steam extracted Jaboticaba (Myrciaria Jaboticaba) juice. *Food*, *10* (4), 732.

McIntyre, S., Ikawa, J. Y., Parkinson, N., Haglund, J., & Lee, J. (1995). Characteristics of an acidophilic Bacillus strain isolated from shelf-stable juices. *Journal of Food Protection*, *58* (3), 319–321.

Mena, P., Martí, N., Saura, D., Valero, M., & García-Viguera, C. (2013). Combinatory effect of thermal treatment and blending on the quality of pomegranate juices. *Food and Bioprocess Technology*, *6*(11), 3186–3199.

Mena, P., Vegara, S., Martí, N., García-Viguera, C., Saura, D., & Valero, M. (2013). Changes on indigenous microbiota, colour, bioactive compounds and antioxidant activity of pasteurised pomegranate juice. *Food Chemistry*, *141*(3), 2122–2129.

Mertz, C., Brat, P., Caris-Veyrat, C., & Gunata, Z. (2010). Characterization and thermal lability of carotenoids and vitamin C of tamarillo fruit (Solanum betaceum Cav.). *Food Chemistry*, *119*(2), 653–659.

Miller, F. A., & Silva, C. L. (2012). *Thermal treatment effects in fruit juices.*

Morales-Blancas, E., Chandia, V., & Cisneros-Zevallos, L. (2002). Thermal inactivation kinetics of peroxidase and lipoxygenase from broccoli, green asparagus and carrots. *Journal of Food Science*, *67*(1), 146–154.

Nasr, N., & Elshaghabee, F. (2019). Enhancement of shelf life of pasteurized Milk using NisplinŪ essential oils emulsion. *International Journal of Current Microbiology and Applied Sciences*, *8*(9), 257–266.

Paquin, P. (2009). *Functional and speciality beverage technology*. Elsevier.

Park, S., Thanakkasaranee, S., Shin, H., Ahn, K., Sadeghi, K., Lee, Y., … Seo, J. (2020). Preparation and characterization of heat-resistant PET/bio-based polyester blends for hot-filled bottles. *Polymer Testing*, *91*, 106823. https://doi.org/10.1016/j.polymertesting.2020.106823.

Petruzzi, L., Campaniello, D., Speranza, B., Corbo, M. R., Sinigaglia, M., & Bevilacqua, A. (2017). Thermal treatments for fruit and vegetable juices and beverages: A literature overview. *Comprehensive Reviews in Food Science and Food Safety*, *16*(4), 668–691.

Prieto-Santiago, V., Cavia, M., Alonso-Torre, S. R., & Carrillo, C. (2020). Relationship between color and betalain content in different thermally treated beetroot products. *Journal of Food Science and Technology*, *57*(9), 3305–3313.

Qiu, L., Zhang, M., Tang, J., Adhikari, B., & Cao, P. (2019). Innovative technologies for producing and preserving intermediate moisture foods: A review. *Food Research International*, *116*, 90–102.

Rahman, M. S. (2007). *Handbook of food preservation*. CRC Press.

Ramesh, M. (2007). *Pasteurization and food preservation. In handbook of food preservation* (pp. 589–602). CRC Press.

Rattan, N. S., & Ramaswamy, H. S. (2014). Comparison of free/bi-axial, fixed axial, end-over-end and static thermal processing effects on process lethality and quality changes in canned potatoes. *LWT-Food Science and Technology*, *58*(1), 150–157.

Rawson, A., Patras, A., Tiwari, B. K., Noci, F., Koutchma, T., & Brunton, N. (2011). Effect of thermal and non thermal processing technologies on the bioactive content of exotic fruits and their products: Review of recent advances. *Food Research International*, *44*(7), 1875–1887. https://doi.org/10.1016/j.foodres.2011.02.053.

Reineccius, T., Reineccius, G., & Peppard, T. (2004). Utilization of β-cyclodextrin for improved flavor retention in thermally processed foods. *Journal of Food Science*, *69*(1), FCT58–FCT62.

Reis, F. R. (2017). *New perspectives on food blanching*. Springer.

Rodríguez-Verástegui, L. L., Martínez-Hernández, G. B., Castillejo, N., Gómez, P. A., Artés, F., & Artés-Hernández, F. (2016). Bioactive compounds and enzymatic activity of red vegetable smoothies during storage. *Food and Bioprocess Technology*, *9*(1), 137–146.

Saldivar, X., Wang, Y.-J., Chen, P., & Mauromoustakos, A. (2010). Effects of blanching and storage conditions on soluble sugar contents in vegetable soybean. *LWT-Food Science and Technology*, *43*(9), 1368–1372.

Sangsom, W., & Inprasit, C. (2022). Design and development of innovative steam injection for high-temperature short-time liquid foods. *PRO*, *10*(1), 161.

Sefrienda, A., Amri, A., Ariani, D., Susanto, A., & Nurhikmat, A. (2020). The extended shelf life of aloe vera drink with a modified hot-filling practice for the SME. In *Paper presented at the IOP Conference Series: Earth and Environmental Science*.

Simpson, R. (2009). *Engineering aspects of thermal food processing*. CRC Press.

Singh, R. P., & Heldman, D. R. (2014). Heat transfer in food processing. *Introduction to food engineering*, *5*, 265–417.

Soni, A., Smith, J., Thompson, A., & Brightwell, G. (2020). Microwave-induced thermal sterilization-A review on history, technical progress, advantages and challenges as compared to the conventional methods. *Trends in Food Science & Technology*, *97*, 433–442.

Söylemez, M. (2008). Optimum length of finned pipe for waste heat recovery. *Energy Conversion and Management*, *49*(1), 96–100.

Spinelli, A. C. N., Sant'Ana, A. S., Rodrigues-Junior, S., & Massaguer, P. R. (2009). Influence of different filling, cooling, and storage conditions on the growth of Alicyclobacillus acidoterrestris CRA7152 in orange juice. *Applied and Environmental Microbiology*, *75* (23), 7409–7416.

Staal, P. (1986). IDF and the development of milk pasteurization. *Bulletin, International Dairy Federation(200)*, 4–8.

Tavman, S., Otles, S., Glaue, S., & Gogus, N. (2019a). 4 - food preservation technologies. In C. M. Galanakis (Ed.), *Saving food* (pp. 117–140). Academic Press.

Tavman, S., Otles, S., Glaue, S., & Gogus, N. (2019b). Food preservation technologies. In *Saving food* (pp. 117–140). Elsevier.

Teixeira, A. A. (2014). Thermal food preservation techniques (pasteurization, sterilization, canning and blanching). *Conventional and advanced food processing technologies*, 115–128.

Teke, I., Ağra, Ö., Atayılmaz, Ş.Ö., & Demir, H. (2010). Determining the best type of heat exchangers for heat recovery. *Applied Thermal Engineering*, *30*(6–7), 577–583.

Tontul, I., & Topuz, A. (2017). Spray-drying of fruit and vegetable juices: Effect of drying conditions on the product yield and physical properties. *Trends in Food Science & Technology*, *63*, 91–102.

ul Haq, I., Khaskheli, M., Salman, M., Marri, M. Y., Marri, M. A., Lochi, G. M., … Habib, F. (2014). Effect of heat treatments on sensory characteristics and shelf-life of skimmed milk. *African Journal of Food Science*, *8*(2), 75–79.

Wadikar, D. D., Vasudish, C., & Premavalli, K. (2015). Studies on effect of irradiation on functional properties of two ready-to-drink appetizer beverages and their shelf lives. *Nutrition & Food Science*.

Wang, H., Meng, J.-S., Raghavan, G., Orsat, V., Yu, X.-L., Liu, Z.-L., … Xiao, H.-W. (2022). Vacuum-steam pulsed blanching (VSPB) enhances drying quality, shortens the drying time of gingers by inactivating enzymes, altering texture, microstructure and ultrastructure. *LWT*, *154*, 112714.

Xiao, H.-W., Bai, J.-W., Sun, D.-W., & Gao, Z.-J. (2014). The application of superheated steam impingement blanching (SSIB) in agricultural products processing–A review. *Journal of Food Engineering*, *132*, 39–47.

Xiao, H.-W., Pan, Z., Deng, L.-Z., El-Mashad, H. M., Yang, X.-H., Mujumdar, A. S., … Zhang, Q. (2017). Recent developments and trends in thermal blanching–A comprehensive review. *Information processing in agriculture*, *4*(2), 101–127.

Xin, Y., Zhang, M., Xu, B., Adhikari, B., & Sun, J. (2015). Research trends in selected blanching pretreatments and quick freezing technologies as applied in fruits and vegetables: A review. *International Journal of Refrigeration*, *57*, 11–25.

Yadav, D. K., Khanashyam, A. C., Wadikar, D., Patki, P., & Semwal, A. (2020). The effect of blanching and drying on retention of ascorbic acid in Indian gooseberry (Phyllanthus emblica) candy. In *Paper presented at the Proceedings of the 1st IPB International Conference on Nutrition and Food.*

Yin, Q., Mu, H., Zeng, M., Gao, D., Qin, F., Chen, J., & He, Z. (2019). Effects of heating on the total phenolic content, antioxidant activities and main functional components of simulated

Chinese herb candy during boiling process. *Journal of Food Measurement and Characterization, 13*(1), 476–486.

Zhang, L., Liu, S., Pain, J., & Fryer, P. (1992). Heat transfer and flow in solid-liquid food mixtures. In *Paper presented at the International Chemical Engineering Symposium Series.*

Ziuzina, D., & Misra, N. (2016). Cold plasma for food safety. In *Cold plasma in food and agriculture* (pp. 223–252). Elsevier.

Thermal food process calculations

2

Aman Mohammad Ziaiifar[a] *and Azadeh Ranjbar Nedamani*[b]
[a]Department Food Process, Gorgan University of Agricultural Science and Natural Resources, Gorgan, Iran, [b]Department of Farm Machinery, Faculty of Agricultural Engineering, Sari Agricultural Sciences and Natural Resources University, Sari, Mazandaran, Iran

2.1 Introduction

Thermal processing is a common method to destruct bacterial cells or spores which can lead to human diseases or undesirable properties in food products. While nonthermal technologies are recently involved in food processing, thermal technologies remain yet as the most dominant procedures. They extend the shelf life of foods due to the ability of heat to destroy the microorganisms and inactivate the enzymes. Furthermore, the thermal process leads to cooking and modifying the taste of raw food materials (Ibarz & Augusto, 2017). In food industries, the application of different food domains (engineering, microbiology, chemistry, and technology) leads to improvement in the final food quality (Kutz, 2013; Wang, Dong, Zhang, Yu, & Wang, 2021).

During sterilization process:

- *Engineering aspect*: the rate of heat penetration.
- *Microbiological aspect*: the rate of microbial and enzyme destruction.
- *Chemistry aspect*: the rate of chemical changes in food such as nutrient degradation and changes in color.
- *Technological aspect*: the choice of appropriate sterilization system.

The most common role of food sterilization is decreasing the concentration of bacterial cells or spores at high temperatures. During the sterilization process, we need to have enough knowledge about two important domains: (1) the kinetics of thermal inactivation of target microorganism or enzyme and (2) heat penetration mechanisms based on the process conditions and the food and container characteristics (Heldman & Lund, 2007).

During thermal processing, the food quality is influenced by treatment conditions mainly temperature and duration. Therefore, the selection of a good combination of temperature and time plays a key role. For this, we first need to identify which microorganism or enzyme should be destroyed in a determined food, and then we should select a suitable heating regime to process. Great attention is considered to avoid under-heating or over-cooking food. Many factors should be considered and controlled to achieve a suitable combination of temperature and time during sterilization (Valentas, Rotstein, & Singh, 1997). These factors can be classified as:

- Food characterizations such as thermal properties and pH.
- The type of target microorganism or enzyme and its heat resistance.

Thermal Processing of Food Products by Steam and Hot Water. https://doi.org/10.1016/B978-0-12-818616-9.00005-5

- The heating medium such as type of retort and temperature distribution.
- The container specifications such as geometry and its situation inside the retort.

The bacterial concentration (viable cells or spores) in food material decreases during the heating of food material in which the temperature plays a key role. The temperature changes are dependent on the heat transfer mechanism. The factors affecting heat transfer mechanisms during sterilization are (Valentas et al., 1997):

- *Heating medium (retort)*: temperature and time.
- *Food*: composition, initial temperature, physical and thermal properties, the consistency, presence of particles, the presence of liquid phase inside the container.
- *Package*: location in the retort (Kutz, 2013).

In this chapter, the basic thermal process calculations, factors affecting the heat transfer inside the retorts, and the calculation of sterilization and cook values will be discussed.

2.2 Food sterilization with steam and water

The sterilization process works generally at a temperature > 100°C. The early effort to attain the temperature > 100°C was applying the salt baths. Later, the use of steam was preferred due to its ability to condense on container surfaces and release large amounts of latent heat. Generally, sterilization methods of food materials involves heating of fluid foods in heat exchangers or heating of fluid, semifluid, solid, or a mixture of fluid and solid food materials in packages at a determined temperature and time to eliminate the undesirable microorganisms (Valentas et al., 1997). The retort sterilization of packed foods, high-temperature short time (HTST) and ultra-high temperature (UHT), and aseptic (heating and cooling of food and packaging under aseptic conditions) processes are the most treatment methods used to sterilize the food products. All types of sterilization methods can be categorized into two groups: direct heating and indirect heating.

2.2.1 *Direct heating*

Direct heating of foods (Chapter 7) is based on two main methods of injecting the pressurized steam into the food or injecting the food into the steam (also named infusion). In both, steam contacts with food, and its condensation leads to release the latent heat and quickly increases the food temperature. These methods can be performed by direct or indirect food-steam flow. Foods with low viscosity and homogenous consistency (milk or juices) can be heated by steam injection while for foods that contain small particles steam infusion is recommended as a more gentle process (Richardson, 2001).

2.2.2 *Indirect heating*

The indirect heating of foods relies on the presence of a surface between food and the heating medium. Plate heat exchangers (Chapter 4), tubular heat exchangers (Chapter 5), and scraped surface heat exchangers are three main types of indirect heating systems. The heating of packaged food materials, mainly used in the sterilization of food products, also is considered an indirect heating method.

2.3 Sterilization steps

The most common equipment for the sterilization of foods is the retort (Chapter 13). The sterilization of food inside the container (introduced by Nickolas Appert in the 1800s) should have the ability of heating, and cool with minimum imposed stresses upon the container during the thermal process. This is why a controlled combination of temperature and pressure is used in modern retorts (Richardson, 2001).

Retorts can be designed based on batch or continuous modes. In a continuous retort system, the food containers are automatically transferred. Hydrostatic retorts, hydrolock retorts, and reel and spiral retorts are some common continuous retorts in which the saturated steam is used for sterilization (Richardson, 2001). The batch retort system (Fig. 2.1) is based on separate cycles of heating and cooling periods for every crate loading inside the retort. After each cycle, the lid is opened and the containers will be unloaded from the retort. The rest containers will then load and the process starts again. Batch retorts have more flexibility than continuous retorts and can work easily under over-pressure during the heating and cooling steps. In batch retort, the containers can be fully immersed in a mixture of air/steam and water during heating and cooling, respectively.

Generally, the operation cycles of a retort can be classified as (Richardson, 2001):

1. *Loading the containers*—in batch retorts, the containers are put into the crates (Fig. 2.2) which allow better penetration of heat inside the load and minimize the container damage.
2. *Closing the lid and turning the steam on*—in saturated steam retorts, the presence of entrapped air is an important challenge. The air (especially if the air is entrapped between the containers), has a significant negative effect on the temperature inside the retorts affecting the heating rate. For example, each 10% volume of air inside the closed vessels, can reduce the temperature by about 3°C. Thus, at the start of the thermal process by retorts, the air should be vented from the device through the introduction of steam into the retorts with high velocity. The steam passes through the containers and exits at the opposite side of the retort. The venting process can monitor through a temperature measurement in the different locations inside the retort. The needed time for venting all the air can be experimentally determined. In saturated steam retorts which have a large installation, venting a high amount of steam in the atmosphere leads to an increase in the working environment temperature. Thus, in the retort room, the steam is controlled by a vent manifold to vent steam to the outside of the building.
3. *Closing the valves to increase the temperature*—when all the air inside the retort is vented out, the valve will be closed and the retort pressurizing will be started to reach the desired temperature. Generally, a pressure gauge and safety valve are installed on the retort.
4. *Holding time*—at this stage, the steam is condensed on the outer surface of the container. If the air is properly vented from the retort, a range of temperature less than 0.5°C can be achieved. This is a suitable target for all retorts.
5. *Cooling the containers*—after the holding time, the cooling period of the retort starts. While during the heating process, the pressure inside the container increases and the content

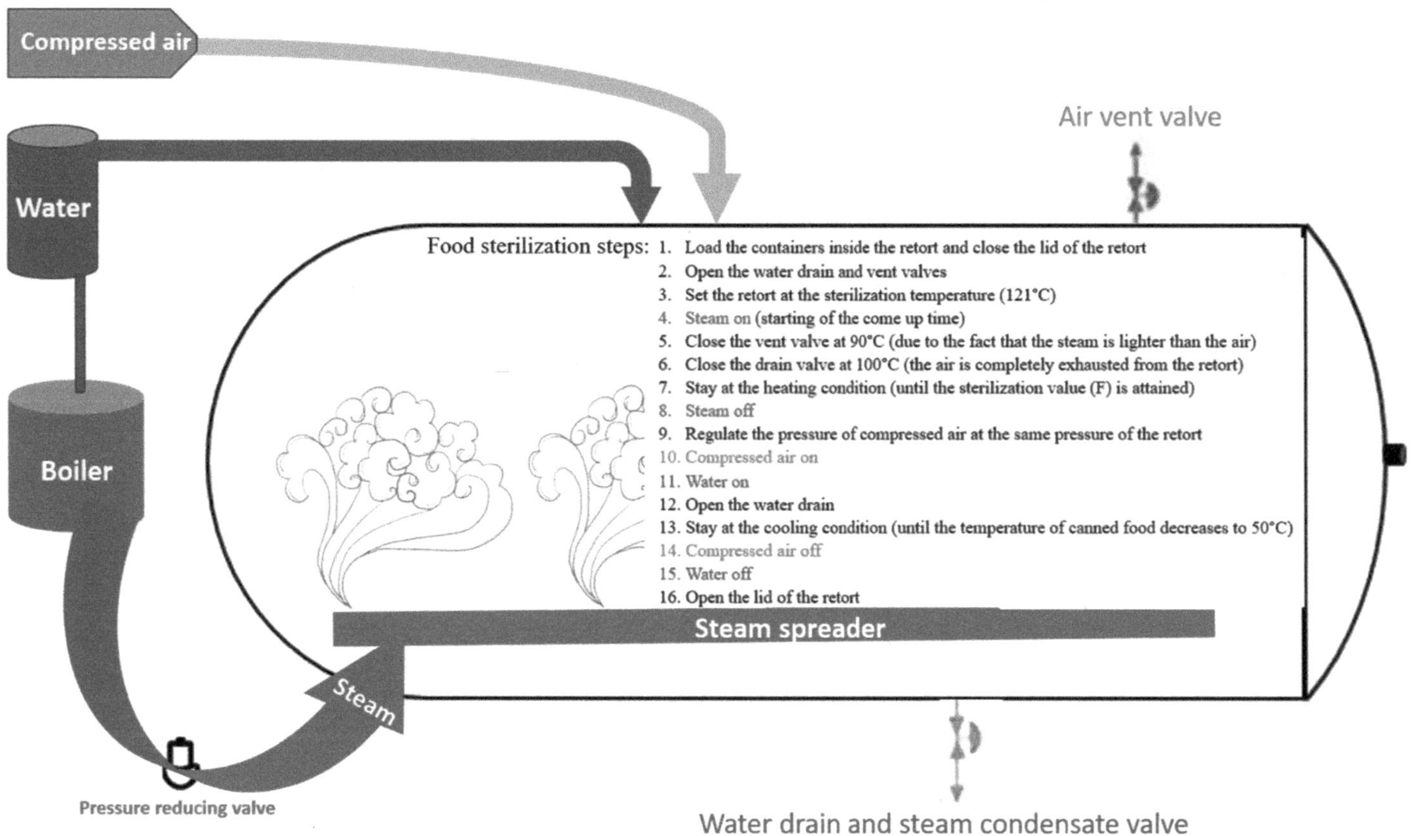

Fig. 2.1 Schematic of a batch retort.

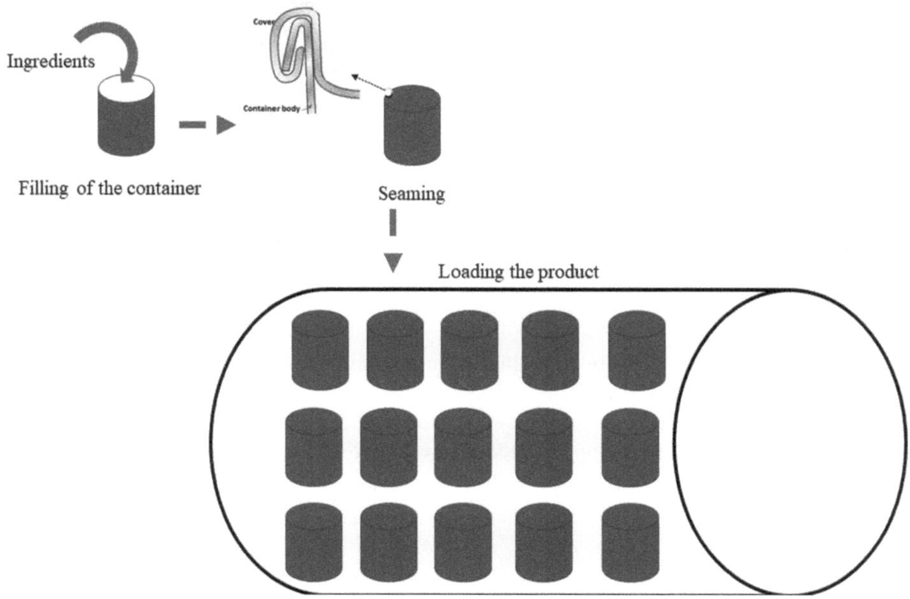

Fig. 2.2 Filling, seaming, and loading inside the retort.

expands, during the cooling period, the pressure inside the can should gradually be reduced. If the retort pressure drops quickly, the internal container pressure can lead to permanent damage to the pack referred "peaking" of the container. During the cooling process, the pressure inside the container should gradually fall or diminish. This is performed using pressured air to compensate for the pressure drop (due to steam condensation) inside the retort during the cooling period. The cold water is used to cooling of containers. The compressed air and steam flow are automatically controlled during the cooling and heating periods, respectively.

6. *Switching off the cooling water*—after reaching the desired cooling temperature (normally 50°C), the residual pressure should release.
7. *Opening the lid and unloading the containers*—the containers finally are conducted to labeling and storage sections.

2.4 Factors affecting the sterilization process

Thermal processing of foods within metal containers is one of the most widespread methods in food preservations. It consists of food heating in a retort (autoclave) with a determined temperature-time regime specified for each food. The time required for the sterilization process is based on the sufficient heating of food to deactivate the undesirable microorganism or enzyme to guarantee public health and food safety (Kutz, 2013).

2.4.1 The heating medium characteristics

Several heating mediums are applied to sterilize the food materials.

2.4.1.1 Saturated steam

Steam at temperatures >100°C is an excellent medium to sterilize food materials (Table 2.1) (Richardson, 2001). The saturated steam is an efficient heating medium because (Berk, 2009):

1. The coefficient of heat transfer is very high in a film layer of condensing vapor on the outer surface of the container.
2. The temperature can be easily controlled by monitoring the pressure of saturated steam.
3. Each unit mass of steam has a high heat content.
4. The deformation and rupturing of the container are avoided. Because the pressure of outside steam and the pressure developed inside the container during the heating process will not have high differences (Berk, 2009).

2.4.1.2 Hot water

The efficiency of hot water is lower than saturated steam. Hot water is preferred for the thermal processing of glass-packaged food or products which are sensitive to heat. To prevent the "cold pockets," a good circulation of hot water is needed (Berk, 2009).

2.4.1.3 Steam-air mixture

This heating medium is fairly used and the efficiency is almost the same as hot water (Berk, 2009). In steam-air retorts, large pockets of air present in retorts, result in under-sterilization of the foods. In this type of retort, after condensation of steam, the air makes an insulation layer around the pack. To eliminate this layer of air

Table 2.1 A typical table of saturated steam properties.

Temperature (°C)	Absolute pressure (kPa)	Enthalpy (kJ/kg)
100	101.3	2676
105	120.8	2683
110	143.3	2691
115	169.1	2698
120	798.5	2706
125	232.1	2713
130	271.2	2719
135	313.2	2726

and also mixing of steam and air, large fans are used to draw the steam-air mixture through the retort and recirculate it to the reverse retort end.

2.4.1.4 Hot gases (combustion gases)

In this system, the food packages are heated by flame resulting from the combustion gases. This method is not usable for canned food sterilization (Berk, 2009).

2.4.2 The container characteristics

The selection of containers for sterilization of food can play a key role in choosing the retort system. Factors such as ease of opening, the overall impact on the product quality, appearance, and microwave ability are the basis of container selection for heat processes (Richardson, 2001). Cans, jars, semirigid plastics or pouches, plastics such as trays and pouches, are common types of packaging materials for food thermal processing. Factors that affect the selection of different types of packaging material for the thermal process are listed below (Richardson, 2001):

- The transport of packaging material before use.
- The weight of packaging material.
- The strength and protectability from some mishandlings before and after filling.
- The need for additional packaging during processing or distribution of final product.
- The filling process condition.
- The sealing process condition.
- The shelf-life of processed food inside the package.
- The effect of light, ultraviolet, oxygen, moisture, and packaging materials on the food characteristics during storage or distribution.

Unlike glass, the metal materials such as tinplate and aluminum have a negligible heat resistance due to higher thermal conductivity and lower thickness of the container wall (Berk, 2009).

The application of spring inside the can (Fig. 2.3) leads to better heat transfer from the head and bottom of the can to the center (Ranjbar Nedamani, Ziaiifar, Parvini, Kashaninejad, & Maghsoudlou, 2015). The spring usage also leads to movement of the cold area from the center to near the wall which this movement generally leads to better and faster heat transfer from the wall and center to the cold area. The spring usage idea in food cans leads to a 47% decrease in processing time.

One of the important factors that affect thermal process safety is pressure. The pressure in combination with temperature and time destroys the microbial loading. The experimental data through measuring the container internal and external pressures are used to determine the temperature profile during thermal processing in retorts. Factors such as the amount of vacuum inside the container during sealing, sealing temperature, the size of headspace, temperature of the thermal process, the formulation of food product, and the headspace gas combination affect the pressure inside the containers (Richardson, 2001). Some overpressure for different types of packages is listed in Table 2.2.

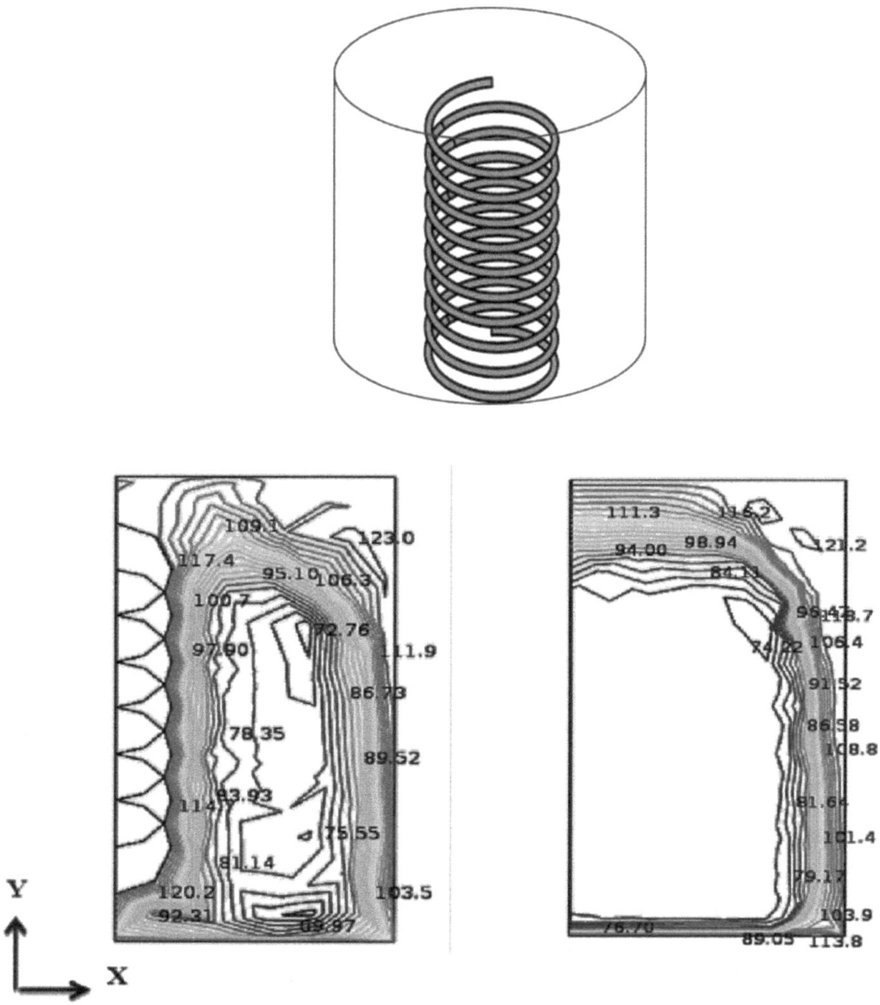

Fig. 2.3 Application of spring inside a can.
Reprinted with permission from Ranjbar Nedamani, A., Ziaiifar, A. M., Parvini, M., Kashaninejad, M., & Maghsoudlou, Y. (2015). *Study of heat penetration coefficients (f and j) using CFD method for canned food based on starch*. PhD, Gorgan University of Agricultural Science and Natural Resources.

Table 2.2 The pressure range can be used for different types of packaging material.

Type of package	Pressure (kg/cm^2)	Tolerance (kg/cm^2)
Plastic trays	2.3–2.9	0.1–0.2
Pouches	1.0	0.1–0.3
Plastic cans	2.9	0.4–0.5

2.5 Heat transfer during sterilization

Understanding the heat transfer mechanisms during sterilization is important to know how the processing time can be determined at a specified lethal temperature. Sterilization is a combination of heating and cooling processes that may be performed at steady-state heat transfer methods such as heat exchangers or unsteady-state heat transfer methods (time-dependent) such as retorts. The ideal form of the healing process is the isothermal form which the temperature of heating or cooling is constant. Methods such as aseptic sterilization, HTST, and UHT are other examples of idealized, isothermal, and steady-state sterilization (Kutz, 2013). The heating or cooling of a food product inside the containers (canned food sterilization) takes place in an unsteady state. In this situation, the product temperature is a function of heat transfer mechanisms. The temperature of the filled product inside the container increases gradually during heating and decreases gradually during cooling. In this case, the lethality is a result of time-temperature history which the product experiences in its coldest heating area inside the package. We can classify three steps for heat transfer in a food container (Berk, 2009):

1. Heat transfer from the heating medium to the surface of the container.
2. Heat transfer inside the container wall.
3. Heat transfer from the inner side of the container to the coldest heating point of the product.

The nature of the food material influences the mechanism of heat transfer. Solid, liquid, solid suspended inside a liquid, and broken heating foods are different types of foods with different heat transfer mechanisms.

2.5.1 Heat transfer in solid food

If the food material inside the package has a solid form, heat transfer occurs by conduction in which the coldest heating area (or slowest heating zone or coldest point) is in its geometric center (Fig. 2.4). To calculate the lethality, the time-temperature history of this area should be therefore considered. These food products (such as squash, canned meat or fish, pet foods, baby foods, and pumpkin) will be sterilized without any mechanical agitation (Heldman & Lund, 2007; Ibarz & Barbosa-Cánovas, 2002; Kutz, 2013). Sometimes, the thermal diffusivity of solid food materials is very low, and when there is no agitation or product movement during the heat process, these products may show a nonuniform distribution of temperature caused by the temperature gradient between the container wall and cold heating point (Heldman & Lund, 2007).

The change in the cold point (the point which receives the lower degree of thermal treatment) temperature of a solid food is calculated as:

$$\frac{T - T_{\mathrm{R}}}{T_{\mathrm{i}} - T_{\mathrm{R}}} = 2.0396 \exp\left[\left(-\frac{2.4048^2}{R^2} - \frac{\pi^2}{4l^2}\right)\alpha t\right] \tag{2.1}$$

where T, T_{R}, and T_{i} are product coldest point, retort, and product initial temperatures (°C), respectively; R radius and l the half-height of the can (m); α thermal diffusivity

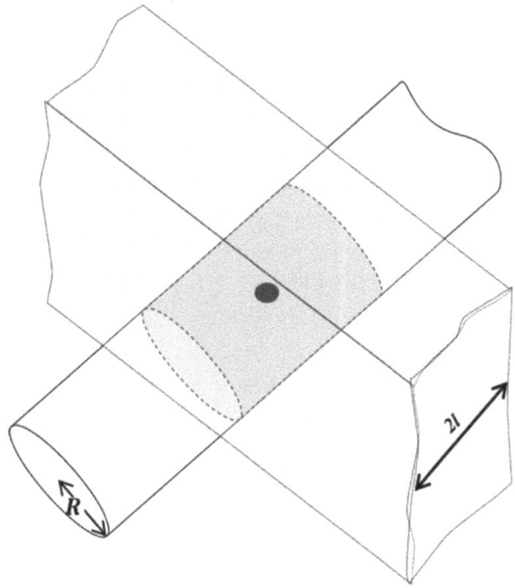

Fig. 2.4 Cold point in solid foods inside a cylindrical container.

(m/s^2); and t is heating time (s). Note that this equation is not valid during early stage of heating.

The processing time can be therefore calculated as:

$$t = \frac{2.303}{\left(\frac{\pi^2}{4l^2} + \frac{2.4048^2}{R^2}\right)\alpha} \log\left(2.0396\frac{T_R - T_i}{T_R - T}\right) \quad (2.2)$$

For example, the temperature of the coldest point in a solid food container ($R = 0.025$ m and $l = 0.0125$ m) with an initial temperature of 40°C and thermal diffusivity of 2×10^{-7} after 10 min of heating in a retort at 121°C can easily be calculated, as:

$$\frac{T - 121}{40 - 121} = 2.0396 \exp\left[\left(-\frac{2.4048^2}{0.025^2} - \frac{\pi^2}{4 \times 0.0125^2}\right)2 \times 10^{-7} \times 600\right] \Rightarrow T = 112^\circ\mathrm{C}$$

A series of charts such as Heisler charts were also developed to find the temperature at any position of the can.

2.5.2 Heat transfer in liquid food

Heat transfer in liquid food material such as soups, milk, and sauces are as below:

- When there is no agitation of the container during sterilization (for example in still cook retort) and heat transfers within the product through natural convection. In this case, the slowest heating zone will be considered at the point below the geometric center

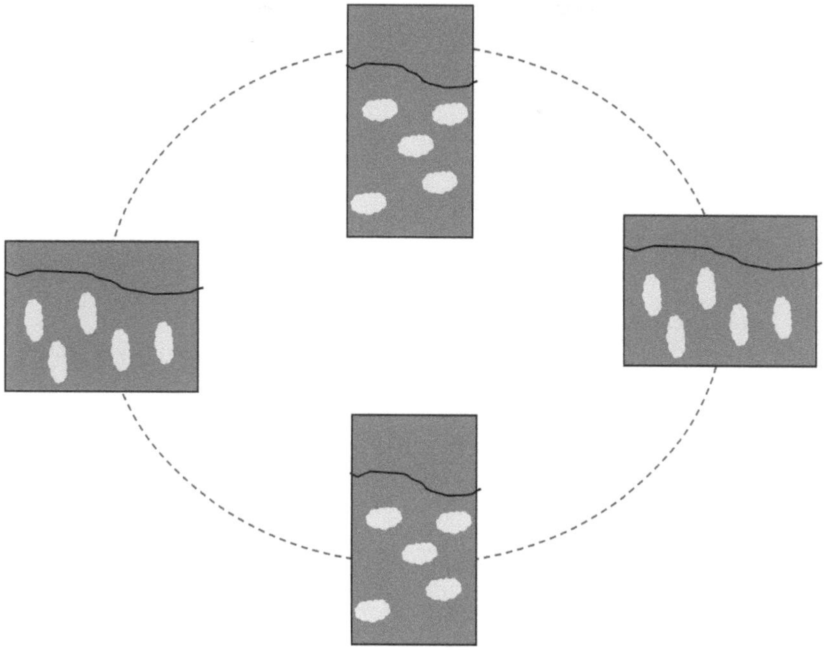

Fig. 2.5 Position of headspace in end-over-end rotation.
Reprinted with permission from Berk, Z. (2009). *Food process engineering and technology*. Elsevier.

(Kutz, 2013). Recently, Ranjbar Nedanami, Ziaiifar, Parvini, Kashaninejad, and Maghsoudlou (2018) showed that the amount of headspace can affect the position and the area of the slowest heating zone in still sterilization. They found that during static sterilization (with no rotation) of a non-Newtonian liquid, the slowest heating zone moves from the "one-tenth height from the bottom" to the interface of the liquid and headspace. Also increasing the volume of headspace can increase the heating rate in broken heating fluids. They reported that higher initial temperature and headspace decrease the heat transfer during the sterilization of 3.5% starch dispersion (Ranjbar Nedanami et al., 2018).

- When there is agitation during sterilization (either axial or end-over-end rotation of can). Because of agitation inside the container and uniform distribution of temperature throughout the product (Heldman & Lund, 2007; Kutz, 2013), there is no slowest heating zone. The agitation leads to forced convection inside the container and the deviation of heating characteristics is comparable with the idealized lumped system. This is referred to as a "well-mixed fluid" situation in heat transfer (Fig. 2.5). Also, the only resistance against heat transfer is at layers of food adjacent to the wall (Valentas et al., 1997). The rotation of the can with an end-over-end mode (Fig. 2.6), provided in batch retorts, is generally preferred. The axial rotation is provided by continuous agitation retorts (Heldman & Lund, 2007; Kutz, 2013).

The change in temperature of the well-mixed fluid is calculated by lumped parameter equation, as:

$$\frac{T - T_R}{T_i - T_R} = \exp\left(-\frac{UAt}{mC_p}\right) \tag{2.3}$$

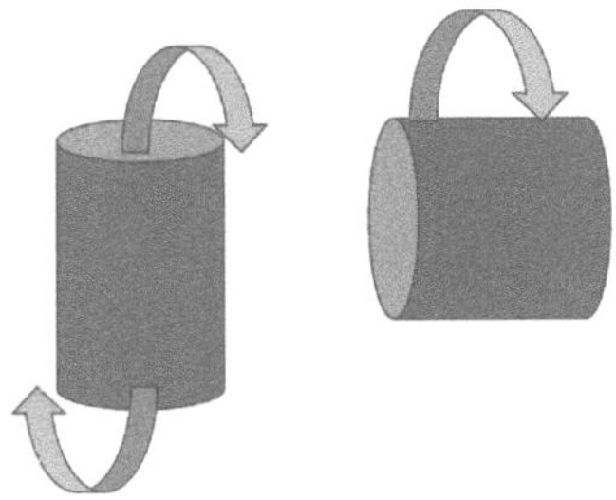

Fig. 2.6 End-over-end and spin agitation.
Reprinted with permission from Berk, Z. (2009). *Food process engineering and technology*. Elsevier.

where U is overall heat transfer coefficient (W/m^{2}°C), A is the surface area of can (m^2), m is mass of food (kg), and C_p is the specific heat of food (kJ/kg°C).

The time is therefore calculated as:

$$t = \frac{2.303mC_p}{UA}\log\left(\frac{T_R - T_i}{T_R - T_R}\right) \quad (2.4)$$

where U overall heat transfer coefficient (W/m^{2}°C), A surface area of heat transfer (m^2), m the mass of fluid food (kg), C_p heat capacity (J/kg°C).

Example 2.1

It is desired to calculate the time required to increase the temperature of concentrated juice from 40 to 110°C. The product is contained in a can (8cm diameter and 10cm height) which is in such a motion that the juice temperature is uniform. The mass and heat capacity of the product is 0.5kg and 3850J/kg°C, respectively. The overall heat transfer coefficient between the heating medium and the product is 400W/m^{2}°C and the heating medium temperature is 125°C.

Solution:

$$A = 2\pi r^2 + 2\pi rL = 0.035\ \text{m}^2$$

$$\frac{110-125}{40-125} = \exp\left(-\frac{400 \times 0.035 \times t}{0.5 \times 3850}\right)$$

$t = 238$ s.

2.5.3 Heat transfer in solid food particles inside the liquid phase

When we have solid food particles inside the liquid phase, a combination of conduction and convection heat transfer is present in food materials inside the container. The coefficient of heat transfer between fluid and particle plays a key role in the heating rate. Since the fluid parameters are not constant, the simplified equations of heat transfer are not applicable. Vidyarthi, Mishra, Dolan, and Muramatsu (2020) applied an inverse method to estimate the lethality of aseptic processing where heat transfer occurs from a fluid to particulate (Vidyarthi et al., 2020).

2.5.4 Heat transfer in broken-heating food materials

Besides, the foods with conduction, convection, or mixed convection/conduction heat transfer, some foods exhibit complex (broken) heating behavior. In these foods, the heat transfer may initially be convection, then due to a physical change in the product, it changes to conduction. For example, during the sterilization of the soups which contain starch, starch gelatinization will occur. This change affects the mechanism of heat transfer. Variations in product ingredients or formulation may cause the transition from convection to conduction to occur at a different temperature and related time. Therefore, the broken-heating foods exhibit more than one mechanism of heat transfer during the heating process. For example, show a broken-heating behavior during the thermal process. The starch solutions firstly show a convection mechanism, and then because of the gelatinization of the starch granules or other thickening agents, the heat transfer mechanism change to conduction (Heldman & Lund, 2007). Some rare products (such as products with solid pieces inside the liquid brine) represent a conduction heat mechanism at the first of the healing process and then the remainder heating of them is based on convection (Heldman & Lund, 2007).

Ranjbar Nedanami, Ziaiifar, Parvini, Kashaninezhad, and Maghsudluo (2016), Ranjbar Nedanami et al. (2018), and Ranjbar Nedanami, Ziaiifar, Parvini, Kashani-Nejad, and Maghsoudlou (2017) studied the effect of headspace and initial temperature on the heat penetration parameters in canned starch dispersion. The starch dispersion (3.5%, w/w) was filled in two cylinders (6×10 cm and 10×6 cm) and one cone (10 cm height and 5 cm radius) with equal volumes. They found in static heating cans, selecting one-third bottom of fulfilled cans as the cold point is mistakable and it is better to select the one-tenth bottom. In the can with headspace, the cold area was near the interface of air food in the food phase. Also, it was clear that in broken heating curve foods which are heated with different headspace in cans, the fulfilled cans have shorter heating time and in cans with headspace, the more headspace helps better heating rate (Ranjbar Nedamani & Hashemi, 2021; Ranjbar Nedanami, 2019; Ranjbar Nedanami et al., 2016, 2017, 2018).

2.6 Temperature measurement during the thermal processing

Measuring and monitoring the temperature of the coldest area in canned food products during the sterilization process is a key step in process control (Ibarz & Barbosa-Cánovas, 2002). Temperature distribution (TD) and heat penetration (HP) tests are used for the validation of a specified heating process. While TD tests are used to identify the slowest heating zone of the retort or container, HP tests are used for measuring

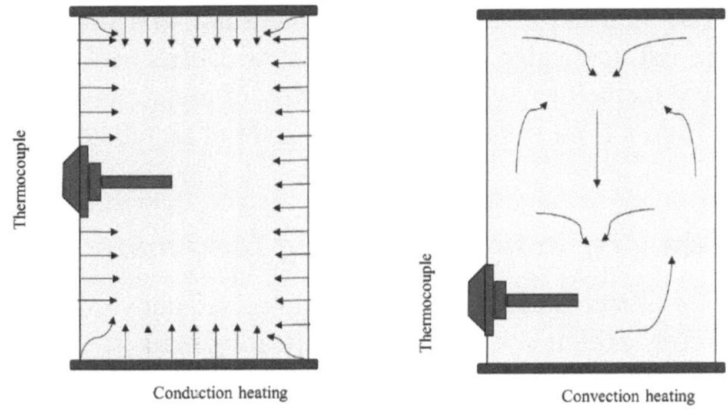

Fig. 2.7 Thermocouples in the container with conduction (in solid food) and convection (in liquid food) heat transfer.
Reprinted with permission from Fellows, P. (2016). *Food processing technology: Principles and practice*. Woodhead Publishing.

the temperature of the slowest heating point of the product (Richardson, 2001). TD test is required when:

- The retort is relocated or another retort is installed that uses the same services.
- Steam, water, or air supply are modified.
- Pumps or valves repaired.
- The retort steam or water circulation system repaired or modified.
- Some doubt about the performance of circulation system is present.
- The size and shape of the container is changed.
- The loading pattern of the container is changed.
- The crate or layer pad design, mode, or use is changed.

On the other hand, the HP tests are used to determine the heating and cooling behavior of specified food materials. This test is used to establish a safe thermal process and provide valid data for analyzing the deviation of the process (Richardson, 2001). To have a valid heat penetration test two factors should be considered: the retort temperature history and the food product temperature history, both in their coldest heating area (Heldman & Lund, 2007; Kutz, 2013).

Monitoring of the temperature changes during the process will be possible with a simple copper-constant thermocouple inserted in the coldest heating area inside the container (Kutz, 2013). The profile of temperature-time can then be obtained. Fig. 2.7 shows where the thermocouples should be inserted to monitor the temperature of canned foods. Fig. 2.8 shows a heat penetration curve. The data can be obtained experimentally using thermocouples and data logger or predicted using the heat transfer equations.

2.7 Effect of thermal processing on microorganisms (*D* and *Z* values)

It is important to know the kinetics of thermal inactivation of undesirable microorganisms or enzymes which lead to food spoilage during storage. This is based on the microbiological history of the product and subsequent storage conditions after

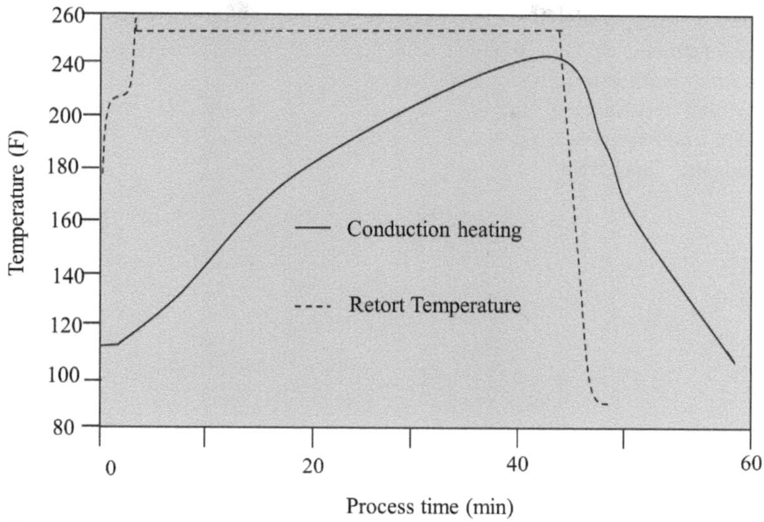

Fig. 2.8 The temperature variation for conduction heating.
Reprinted with permission from Fellows, P. (2016). *Food processing technology: Principles and practice*. Woodhead Publishing.

processing. Some tests should be therefore performed to identify the resistance of the target microorganism/enzyme. In these tests, the product heated uniformly in isothermal conditions to the determined temperature, held for a certain time, finally cooled uniformly to a determined temperature. The TDT curve for a specified microorganism can be used to calculate the time-temperature regime for the thermal process (Kutz, 2013). Fig 2.9 shows the thermal death curve for some biological factors.

The death rate of a microorganism in a determined condition follows the first-order kinetics. Thus, we can show the death rate by (Ibarz & Barbosa-Cánovas, 2002; Valentas et al., 1997):

$$N = N_0 \exp(-kt) \tag{2.5}$$

where N_0 and N are the number of microorganisms at time t and initial number; k is the death rate constant. The units of rate constants are the cycles per time units i.e., t^{-1}. Different temperatures show similar straight lines with different slopes. The k depends on the process temperature, food, and microorganism type and its growth state (vegetative or spore form) (Ibarz & Barbosa-Cánovas, 2002). The constants for the spore form of microorganism are much lower (1/min) than the vegetative form (10^{10}/min), showing that destroying spores is more difficult than vegetative cells (Ibarz & Barbosa-Cánovas, 2002). The logarithmic form of Eq. (2.3) will be a linear equation as below (Ibarz & Barbosa-Cánovas, 2002):

$$\ln\left(\frac{N}{N_0}\right) = -kt \tag{2.6}$$

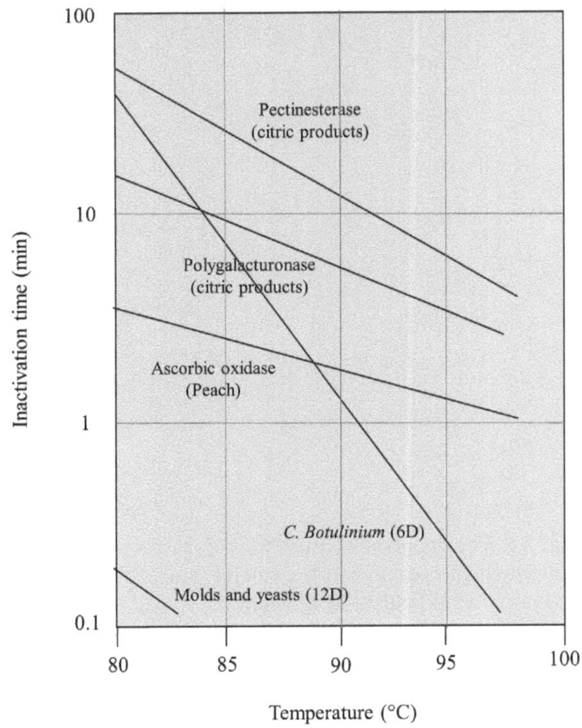

Fig. 2.9 Thermal inactivation of biological factors. Reprinted with permission from Ibarz, A., & Barbosa-Cánovas, G. V. (2002). *Unit operations in food engineering*. CRC Press.

The time needed for decimal reduction is known as D which is the required time for a logarithmic cycle reduction of concentration. In other word, the D shows the time for microorganism reduction to the tenth part (Ibarz & Barbosa-Cánovas, 2002; Valentas et al., 1997). Several factors such as microorganism type and nature of food can affect the D value (Ibarz & Barbosa-Cánovas, 2002).

The D value (decimal reduction time) at a determined temperature is calculated as (Fig. 2.10):

$$D = \frac{t_2 - t_1}{\log(a) - \log(b)} \tag{2.7}$$

where a and b are the number of microorganisms at time t_1 and t_2, respectively.

If a relationship between thermal death time and temperature is established, a semi-logarithmic graph named thermal death time (TDT) or decimal reduction (D_T) is plotted against the temperature (Fig. 2.10).

The Z value (the temperature coefficient of microbial destruction) is the number of degrees the temperature has to be increased to achieve a 10-fold reduction in the D value and is calculated as:

$$Z = \frac{T_2 - T_1}{\log(D_1) - \log(D_2)} \tag{2.8}$$

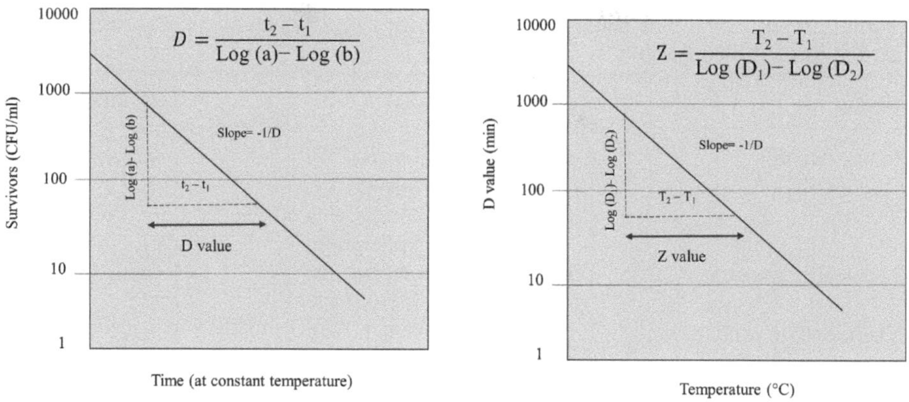

Fig. 2.10 Death rate curve (*left*) and thermal death time (TDT) curve (*right*). Reprinted with permission from Berk, Z. (2009). *Food process engineering and technology*. Elsevier.

where T_1 and T_2 are the temperatures applied to destruct the microorganisms, and D_1 and D_2 are the decimal reduction times at T_1 and T_2.

Each microorganism shows a specified TDT curve. The TDT curve can be drowned in a semilog plot. In sterilization of low acid foods with pH above 4.5, the reference temperature for inactivation of thermophiles spores, which are a relatively high resistant to heat, is 121°C or 250°F. If the pH of the food is low enough that the food is considered as high acid, and the microorganisms are much lower heat resistant than the previous case, a lower reference temperature such as 100 or 66°C for pasteurization can be therefore considered. The reference temperature of inactivation is shown as D_{121} and D_{100}, for example (Heldman & Lund, 2007; Kutz, 2013).

The D value at any other given temperature can be easily obtained:

$$D = D_0 10^{\frac{T_0 - T}{Z}} \tag{2.9}$$

where D is D value at the temperature of T, D_0 is D value at reference temperature i.e., T_0.

Table 2.3 shows the values of N and N_0 to obtain the commercial sterility of some canned foods. The amount of D value is listed for some microorganisms in Table 2.3. Table 2.4 also shows the D and Z values for some microorganisms.

The D and Z values have the same role in thermal inactivation kinetics for spores of microorganisms as the rate constant k and activation energy E_a in Arrhenius equation which shows the effect of temperature on the rate constant of reaction (Heldman & Lund, 2007).

$$k = k_0 \exp\left(\frac{-E_a}{RT}\right) \tag{2.10}$$

where k is the rate constant, k_0 is the factor of frequency, E_a is the energy of activation, R is the constant of gas, and T is the absolute temperature.

Table 2.3 The values of N and N_0 to obtain the target $\ln(N/N_0)$ for commercial sterility of canned foods (Toledo, 2007).

Factor	N	N_0	D_{121} (min)
Public health	10^{-9}	General 10 Meats 10^2 Mushrooms 10^4 Packaging 10^{-5}	0.2
Mesophilic spoilage	10^{-6}	General 10 Meats 10^3	0.5
Thermophilic spoilage	10^{-2}	General 10^2	1.5

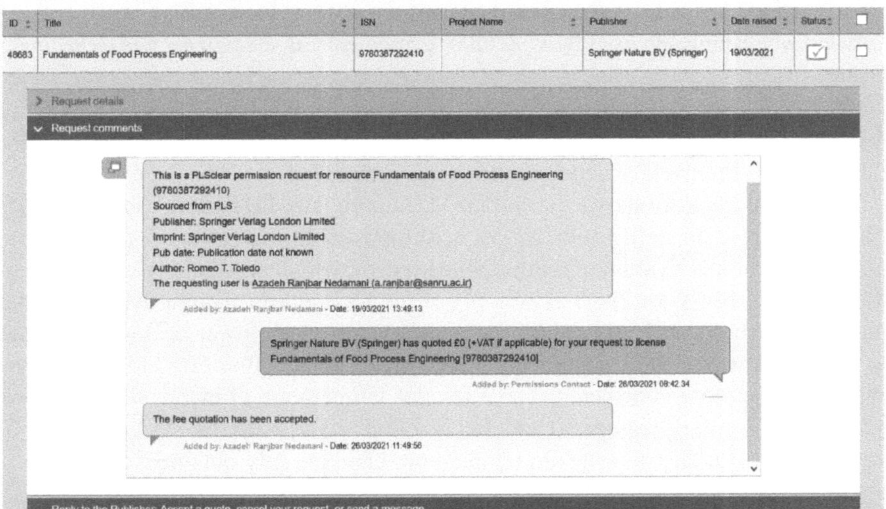

The D_T can be shown as a function of rate constant of microorganism destruction as below (Ibarz & Barbosa-Cánovas, 2002):

$$D_T = \frac{2.303}{k} \tag{2.11}$$

Also, the Z value can be written as a function of activation energy and temperature (Ibarz & Barbosa-Cánovas, 2002), as:

$$Z = \frac{2.303RT^2}{E_a} \tag{2.12}$$

2.8 Process lethality calculations

The purpose of the process lethality determination is to demonstrate the effectiveness of a specific heat process to destroy a microorganism of concern. Eq. (2.5) shows that for obtaining a null concentration of microorganisms or enzymes, an infinite time is

Table 2.4 The resistance of spoilage bacteria against heat in low-acid canned foods.

Organism	Product	D_0 (min)	Z (°C)
C. botulinum 213-B	Green beans	0.22	12
	Peas	0.22	8
C. botulinum 62-A	Green beans	0.22	11
	Corn	0.3	10
	Spinach	0.25	11
Clostridium spp. PA 3679	Asparagus	1.83	13
	Green beans	0.70	9
	Corn	1.20	10
	Peas	2.55	10
	Shrimp	1.68	12
	Spinach	2.33	13
Bacillus stearothermpphilus FS 1518	Asparagus	4.20	11
	Green beans	3.96	10
	Corn	4.32	12
	Peas	6.16	11
	Pumpkin	3.50	13
	Shrimp	3.90	9
	Spinach	4.94	12

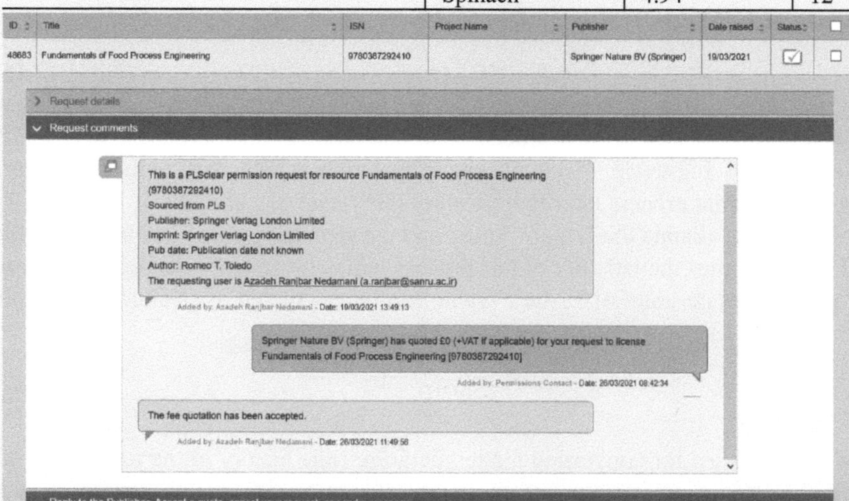

ID	Title	ISN	Project Name	Publisher	Date raised	Status	
48683	Fundamentals of Food Process Engineering	9780387292410		Springer Nature BV (Springer)	19/03/2021		

Request details

Request comments

This is a PLSclear permission request for resource Fundamentals of Food Process Engineering (9780387292410)
Sourced from PLS
Publisher: Springer Verlag London Limited
Imprint: Springer Verlag London Limited
Pub date: Publication date not known
Author: Romeo T. Toledo
The requesting user is Azadeh Ranjbar Nedamani (a.ranjbar@sanru.ac.ir)

Added by: Azadeh Ranjbar Nedamani - Date: 19/03/2021 13:49:13

Springer Nature BV (Springer) has quoted £0 (+VAT if applicable) for your request to license Fundamentals of Food Process Engineering [9780387292410]

Added by: Permissions Contact - Date: 26/03/2021 08:42:34

The fee quotation has been accepted.

Added by: Azadeh Ranjbar Nedamani - Date: 26/03/2021 11:49:56

Reply to the Publisher: Accept a quote, cancel your request, or send a message

Reprinted with permission from Toledo, R. T. (2007). *Fundamentals of food process engineering*. Springer.

required to end the thermal treatment. This implies that complete sterilization of the product is impossible. Thus, for ensuring the product is commercially sterile, a final concentration *b* is defined (Ibarz & Barbosa-Cánovas, 2002). The logarithmic difference between an initial number of microorganisms (*a*) and the final number (*b*) at a given time called reduction degree (Ibarz & Barbosa-Cánovas, 2002):

$$n = \log a - \log b \tag{2.13}$$

where *n* is the numbers of log reduction in microorganism population.

Higher the reduction degree safer the product (Ibarz & Barbosa-Cánovas, 2002). Thus the required treatment time to reach a determined degree of n at a specified T, called thermal death time (F) which is determined by multiplying the reduction degree (n) by decimal reduction time (D) (Ibarz & Barbosa-Cánovas, 2002; Valentas et al., 1997):

$$F = nD \tag{2.14}$$

Determining the sterilizing value is the most critical step in food thermal processing. Process F-value or *process lethality* is expressed in minutes at reference temperature (121°C). The *unit of lethally* is 1 min at reference temperature (121°C) (Heldman & Lund, 2007). Consider the F-value of a process is assigned as 8, this means if we use any time-temperature history for deactivation of bacterial cells or spores, the integrated lethality should be equivalent to lethality that achieved from 8-min exposure to reference temperature (121°C). For specifying an F-value for a thermal process, we need to calculate the minutes required for achieving a specific level of log-cycle reduction of bacterial/spore reduction at the reference temperature. The D_{121} value represents the number of minutes at 121°C for one log-cycle reduction. Thus, the F-value at 121°C can be obtained from multiplying the D_{121} by sterilizing value (required numbers of log cycles for reduction of bacterial cells or spores reduction) (Heldman & Lund, 2007; Kutz, 2013):

$$F = D_{121}(\log a - \log b) \tag{2.15}$$

This F-value is specified for a process. This method of calculation the F-value is derived from a log-cycle in cells/spore population which this reduction is the goal of the thermal process (sterilizing value) (Heldman & Lund, 2007; Kutz, 2013).

For example, assume the $D_{121} = 3\,\text{min}$, and the sterilizing value is 8 log cycle, the $F = 3 \times 8 = 24$, thus the lethality of this process will achieve at $F = 24\,\text{min}$. At a constant process temperature of T, the F-value can be calculated as (Heldman & Lund, 2007; Kutz, 2013):

$$F = 10^{\frac{T-121}{Z}} \times t \tag{2.16}$$

The Z value is used for converting the temperature-time history of the process at reference temperature in minutes and for most calculations of heating processes, it is commonly assumed as 10°C (18°F). F-value calculated with reference temperature and Z of 10°C is F_0.

If the temperature of the product changes during the thermal process (the practical situation during sterilization), the F-value will calculate as an integrated equation (Heldman & Lund, 2007; Kutz, 2013):

$$F = \int_0^t 10^{\frac{T-121}{Z}} \mathrm{d}t \tag{2.17}$$

This method of calculation of F-value is related to a time-temperature history that food is experienced by the thermal process (Kutz, 2013). The Z value can be used also to calculate the efficiency of lethality (L) concerning a reference temperature (Ibarz & Barbosa-Cánovas, 2002). The lethality (L) is therefore defined as the relative effect of each temperature on microbial inactivation in relation to the temperature of reference.

$$L = 10^{\frac{T-121}{z}} \tag{2.18}$$

To guarantee the safety of the process, the F value obtained from the process must be greater than the F value required, as:

$$F_{\text{process}} = \int_0^t 10^{\frac{T-121}{z}} \mathrm{d}t > F_{\text{required}} = D_{121}(\log a - \log b)$$

The most important thing in the thermal process calculation is "when stopping heating?" where the needed sterility is obtained.

2.9 General methods to determine the lethality

The first general method, originally described by Bigelow and Esty (1920), is only of historical attention today and a foundation for subsequent calculation methods (Valentas et al., 1997). This method combined the two laws on microbial destruction with the experimental heat penetration curve, to calculate the optimum time that allowed the attainment of a given sterilization. The general method is the simplest way to calculate the process lethality through a graphical procedure. This method is versatile and can be universally applied for any type of thermal process (Heldman & Lund, 2007; Kutz, 2013). In this method, the lethal rate of the entire heating time to achieve the accomplished value of sterilization is based on the graphical integration (Valentas et al., 1997). The general method uses the actual time-temperature data to analyze by Z-value and the F_0 of a specified user data is a true value. In this method. The calculation by integral, numerical methods, hand, or calculator will lead to similar answers (Valentas et al., 1997).

The Improved General method which uses a graphical approach was suggested by Ball (1923). In this method, the heat penetration data from the time-temperature history of the product can be directly used to evaluate the lethal rate. By referring Fig. 2.11, numerical integration can be written as follow (Heldman & Lund, 2007; Kutz, 2013):

$$F_0 = \int_0^t 10^{\frac{T-T_{\text{ref}}}{z}} dt \approx \sum_0^t 10^{\frac{T-T_{\text{ref}}}{z}} \Delta t \tag{2.19}$$

Fig. 2.11 shows a direct plot of the coldest heating point of food material inside a can during the heat penetration test. When the time is divided into small parts of Δt, the temperature (T_i) at each time interval (Δt) can be extracted from the curve. Finally, by summing all incremental sterilizing values, the total lethality (F_0) will be obtained. The section of the curve should be the shift to the right or left side to determine the required process time to deliver the considered lethality. In this case, the integration should be repeated until the obtained value of sterilizing agree with the value specified for the thermal process (Kutz, 2013).

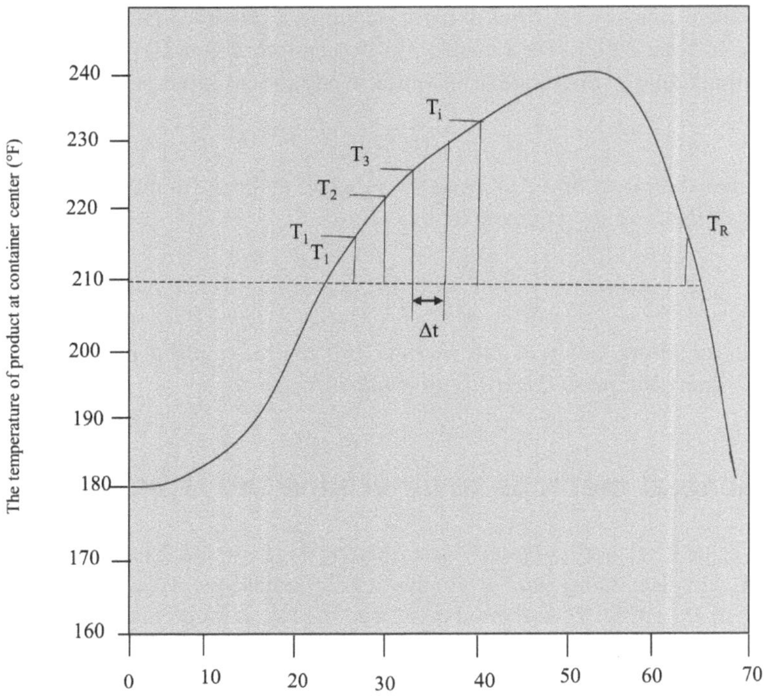

Fig. 2.11 The history of temperature at the container center during the thermal process for general calculation of lethality.
Reprinted with permission from Heldman, D. R., & Lund, D. B. (2007). *Handbook of food engineering*.

Example 2.2

We want to sterilize a low acid solid food filled in a container (110 mm height and 40 mm diameter) with an initial temperature of 30°C in a retort at 121°C. The retort temperature data are shown below. If the effect of come-up time of retort is neglected, find the time at which the $F_0 = 3$ min (as a minimum process sterilization value of the cold point for this product to be safe) is established. The thermal diffusivity of the product is 1.3×10^{-7}. Assume $Z = 10$°C.

Time (min)	Retort temperature (°C)
0	40
1	58
2	92
3	121
⋮	⋮
40	121

Solution:

1. The cold point temperature can be easily calculated from Eq. (2.1), i.e.,

$$\frac{T - T_R}{T_i - T_R} = 2.0396 \ \exp\left[\left(-\frac{2.4048^2}{R^2} - \frac{\pi^2}{4l^2}\right)\alpha t\right]$$

2. The F value is obtained from the cold point temperature (see the table below).
3. After 34 min of heating, the process F value reaches 3.12 min which is > required F value.

$$\frac{T - T_R}{T_i - T_R} = 2.0396 \ \exp\left[(-\frac{2.4048^2}{R^2} - \frac{\pi^2}{4l^2})\alpha t\right]$$

$$F_0 = 10^{\frac{T - Tref}{z}} \times \Delta t$$

Time (min)	Retort temperature or T_R (°C)	Cold point temperature or T (°C)	F_0 (min)	$\sum F_0$ (min)
1	40			
2	58	Where the		
3	92	equation is		
4	121	not valid		
5	121			
6	121	30.18	0.000000	0
7	121	40.38	0.000000	0.000000
8	121	49.43	0.000000	0.000000
9	121	57.47	0.000000	0.000001
10	121	64.61	0.000002	0.000003
11	121	70.94	0.000010	0.000013
12	121	76.56	0.000036	0.000049
13	121	81.55	0.000114	0.000162
14	121	85.98	0.000315	0.000477
15	121	89.91	0.000779	0.001256
16	121	93.41	0.001740	0.002996
17	121	96.50	0.003552	0.006548
18	121	99.26	0.006692	0.013240
19	121	101.70	0.011743	0.024983
20	121	103.87	0.019343	0.044326
21	121	105.79	0.030127	0.074454
22	121	107.50	0.044646	0.119100
23	121	109.01	0.063302	0.182401
24	121	110.36	0.086302	0.268703
25	121	111.56	0.113634	0.382338
26	121	112.62	0.145070	0.527408
27	121	113.56	0.180193	0.707601
28	121	114.39	0.218434	0.926035
29	121	115.14	0.259130	1.185164
30	121	115.79	0.301565	1.486729
31	121	116.38	0.345023	1.831753
32	121	116.90	0.388821	2.220574
33	121	117.36	0.432337	2.652911
34	121	117.77	0.475030	3.127940
35	121	118.13	0.516448	3.644388
36	121	118.45	0.556230	4.200618
37	121	118.74	0.594105	4.794723
38	121	118.99	0.629882	5.424605
39	121	119.22	0.663442	6.088047
40	121	119.42	0.694729	6.782776

Example 2.3

During the production of yogurt, the whole milk is firstly heated from 4 to 95°C in a well-agitated tank and held at this temperature for 15 min. If the mass of milk is 500 kg, the surface area of the tank in contact with milk is 3 m^2, the overall heat transfer coefficient between the heating medium and the milk 450 W/m^2°C, the heat capacity of milk 3900 J/kg°C and heating medium temperature is 125°C, calculate the number of log cycles reduction of the target microorganism with $D_{121} = 0.01$ min and $Z = 10$°C.

Solution:

$F_0 = 0.043$ min.

$F_0 = D \times$ cycles of reduction.

$0.043 = 0.1 \times$ cycles of reduction

Cycles of reduction $= 4.3$.

$$\frac{T - T_R}{T_i - T_R} = \exp\left(-\frac{UAt}{mC_p}\right)$$

$$F_0 = 10^{\frac{T - Tref}{z}} \times \Delta t$$

Time (min)	Milk temperature (°C)	F_0 (min)	$\sum F_0$ (min)
0	4	0.00000	0.00000
1	13	0.00000	0.00000
2	21	0.00000	0.00000
3	28	0.00000	0.00000
4	35	0.00000	0.00000
5	41	0.00000	0.00000
6	47	0.00000	0.00000
7	53	0.00000	0.00000
8	58	0.00000	0.00000
9	63	0.00000	0.00000
10	67	0.00000	0.00001
11	71	0.00001	0.00002
12	75	0.00003	0.00004
13	79	0.00006	0.00010
14	82	0.00012	0.00023
15	85	0.00025	0.00048
16	88	0.00049	0.00097
17	91	0.00090	0.00186
18	93	0.00158	0.00344
19	95	0.00266	0.00610
20	95	0.00251	0.00861
21	95	0.00251	0.01112
22	95	0.00251	0.01364
23	95	0.00251	0.01615
24	95	0.00251	0.01866
25	95	0.00251	0.02117
26	95	0.00251	0.02368
27	95	0.00251	0.02619
28	95	0.00251	0.02871
29	95	0.00251	0.03122
30	95	0.00251	0.03373
31	95	0.00251	0.03624
32	95	0.00251	0.03875
33	95	0.00251	0.04127
34	95	0.00251	0.04378

2.10 Formula method based on heat penetration data

Ball's formula method is a standard method used in the food thermal processing industry, the basis of a number of most recent procedures. The parameters obtained from a heat penetration curve are listed in Table 2.5 (Valentas et al., 1997):

2.10.1 Heat penetration data

The objective of the heat penetration test is to obtain data that can be used to design a sterilization process. The temperature at the cold heating point of the product is an exponential function of time (Fig. 2.8). A linear function is obtained

Table 2.5 Parameters obtained from a heat penetration curve.

Parameter	Description
T_{ih}	Initial food temperature when heating is started
T_{ic}	Food temperature when cooling is started
T_R	Retort temperature
T_w	Cooling water temperature
I_h	Difference between the retort temperature and food temperature at the start of the heating process ($T_R - T_{ih}$)
I_c	Difference between the food temperature and cooling water temperature at the start of the cooling process ($T_{ic} - T_w$)
g	Difference between the retort temperature and food temperature (T) at time t ($T_R - T$)
g_c	The value of g at the end of heating or beginning of cooling ($T_r - T_{ic}$)
f_h	Heating rate index; the time required for the straight-line portion of the heating curve to pass through one log cycle
f_c	Cooling rate index; the time required for the straight-line portion of the cooling curve to pass through one log cycle
l	Come-up period; the time from steam on to the retort reached T_R (the retort requires some time to reach the operating condition)
$0.58L$	Effective beginning of the process; about 42% of this come-up period is generally considered as the time at retort temperature because the product temperature increases even during this period; to accommodate this, the effective beginning of the process is moved left a distance "$0.42L$" from the time the retort reached T_R or moved right $0.58L$ from the time of steam on
T_{pih}	The pseudo-initial temperature during heating; temperature indicated by the intersection of the extension of the heating curve and the vertical line
T_{pic}	The pseudo-initial temperature during cooling; temperature indicated by the intersection of the extension of the cooling curve and the vertical line
j_h	Heating rate lag factor
j_c	Cooling rate lag factor
B	Thermal process time; Ball-corrected for the come-up period (steam on to steam off minus $0.58\ L$)
P_t	Operator process time (time after CUT); $P_t = B - 0.42\ L$

when the plot of "difference between the retort temperature and the cold point temperature," named g, ($g = T_r - T$) is plotted as a function of heat processing time in semilogarithmic scale (Fig. 2.12) (Ibarz & Barbosa-Cánovas, 2002). The slope of the graph is related to the thermal diffusivity of the product and the dimensions of the container (Heldman & Lund, 2007). Based on this graph, the temperature of food product against time can be represented as Fig. 2.13 (Ibarz & Barbosa-Cánovas, 2002).

When cans are placed inside the retort, an induction period exists before the temperature of the food starts to increase. This makes the heat penetration curve in Figs. 2.12 and 2.13 nonlinear at the beginning of the process (Ibarz & Barbosa-Cánovas, 2002). If we extend the curve straight line in a logarithmic plot of temperature-time and make intersect with the ordinate axis, the apparent initial temperature (T_{pih}) can be introduced. The T_{ip}, initial temperature (T_i), and temperature of retort (T_R) are used to calculate one of the heat penetration parameters named j_h or lag factor (Ibarz & Barbosa-Cánovas, 2002). The j_h is the ratio of "the difference between the retort temperature (T_R) and pseudo-initial temperature (T_{pih})" and "the difference between the retort temperature and the initial temperature" (Heldman & Lund, 2007).

Fig. 2.12 Heat penetration curve. Reprinted with permission from Fellows, P. (2016). *Food processing technology: Principles and practice.* Woodhead Publishing.

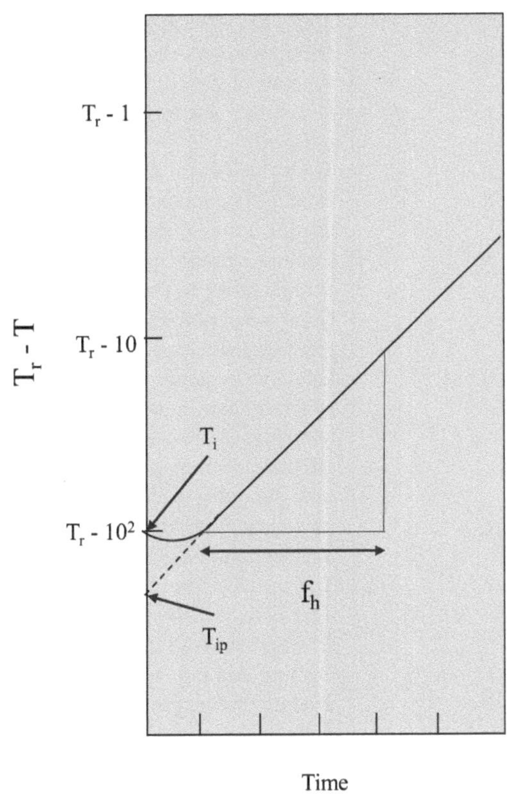

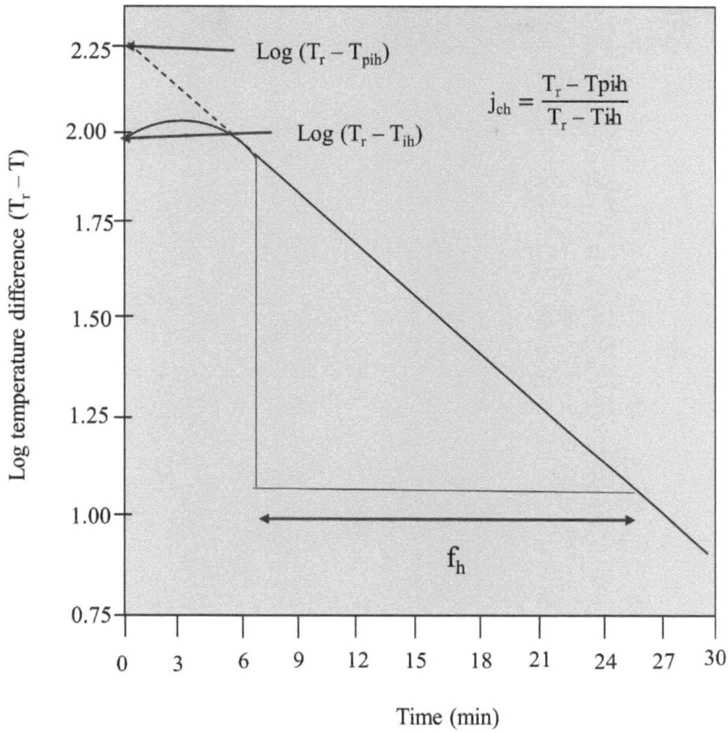

Fig. 2.13 Semi-logarithmic heating curve.
Reprinted with permission from Berk, Z. (2009). *Food process engineering and technology*. Elsevier.

All parameters can be also defined for the cooling stage. But the indices change from h (i.e., heating) to c (i.e., cooling). j_C is therefore similar to j_h but it corresponds to the cooling curve (Ibarz & Barbosa-Cánovas, 2002).

$$j_{\text{h}} = \frac{T_{\text{R}} - T_{\text{pih}}}{T_{\text{R}} - T_{\text{ih}}}$$

$$j_{\text{c}} = \frac{T_{\text{W}} - T_{\text{pic}}}{T_{\text{W}} - T_{\text{ic}}} \tag{2.20}$$

The f_{h} value is also another heat penetration parameter that is calculated for a thermal process setup. The f_{h} is obtained from the slope of a straight section of the heat penetration curve (Heldman & Lund, 2007; Ibarz & Barbosa-Cánovas, 2002). The f_{h} is the time (minutes) to traverse one cycle in the straight-line section of the heat penetration curve (Fig. 2.13) (Heldman & Lund, 2007). It also has a relation with the thermal diffusivity of the product and dimensions of the container (Heldman & Lund, 2007).

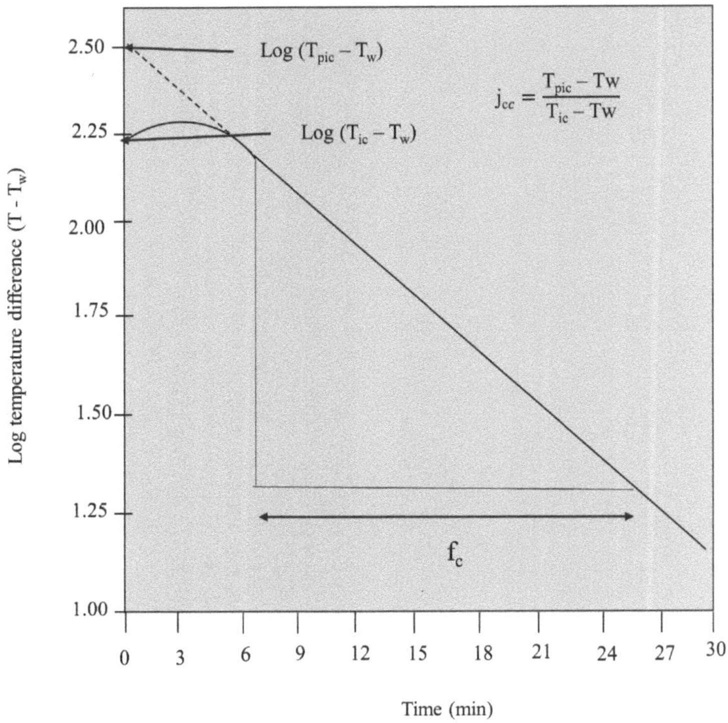

Fig. 2.14 Semi-logarithmic cooling curve.
Reprinted with permission from Valentas, K. J., Rotstein, E., & Singh, R. P. (1997). *Handbook of food engineering practice*. CRC Press.

2.10.2 Ball formula method

Ball formula is:

$$B = f_{\mathrm{h}} \log\left(\frac{j_{\mathrm{h}} I_{\mathrm{h}}}{g}\right) \tag{2.21}$$

As explained earlier, the f_{h} is required minutes to cross on logarithmic cycle in the semilogarithmic heating curve. The j_{h} is the lag factor of heating, and I_{h} is the difference between the temperature of the retort and food at the start of heating. The g is the temperature difference between the treatment device and the temperature at the end of heating at cold point (Ibarz & Barbosa-Cánovas, 2002). Eq. (2.21) includes the lethal effects of the cooling stage. This method can be applied when the semilogarithmic heat penetration curve has a straight line after the initial induction period. Fig. 2.15 shows the Ball formula during the processing based on pure convection and conduction.

Ball's method has some characteristics that make it simple for process time calculation: (1) the possibility of process time if we have the target F-value, (2) calculation

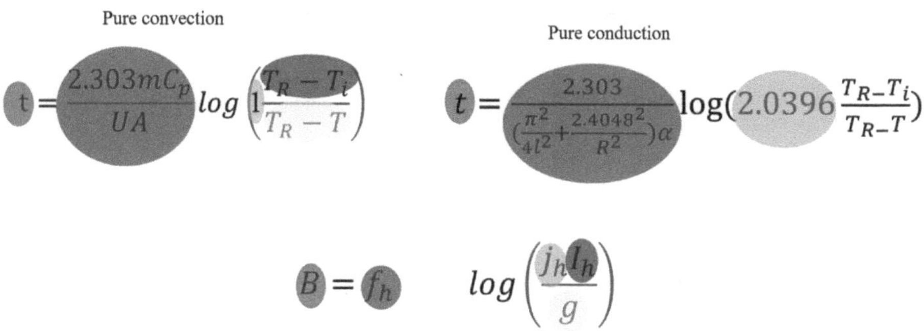

Fig. 2.15 Pure convection vs pure conduction in Ball formula.

of the delivered lethality of a determined heat process, (3) the direct calculations for many sizes of the container through parameter conversion procedures if the heat penetration data for a specific condition was determined, and (4) if the heating medium temperature (T_R) or initial temperature of product (T_i) change, the direct calculation of new process time is possible (Valentas et al., 1997). This method also is faster than general methods.

When a retort is turned on with steam, the temperature does not reach immediately to specified process temperature. A heating time is needed to calculate until the thermal process device reaches operating temperature. This time is called the "come-up period" denoted with l, or CUT, t_{CUT} (Fig. 2.16) (Valentas et al., 1997).

When a food can is placed inside a continuous sterilization system, the latter temperature is T_r. However, in batch thermal system (retort), we have a "come-up" time for reaching the device reaches to determined temperature. In this case, we should consider the come-up time in lethality. Balls proposed that 40% of come-up time involve in the thermal process (Ibarz & Barbosa-Cánovas, 2002; Valentas et al., 1997):

$$B = t_p + 0.42 t_{CUT} \tag{2.22}$$

where t_p is the operator process time (Fig. 2.16) and B is resulted from the calculation of process to affect the target F-value in Ball method.

However, in the case of thin containers or other containers that heat rapidly, this consideration should be under the standards of industry (Valentas et al., 1997). Come-up time is an important parameter that is used at a later time to find the correct test zero. In the best case, the come-up time should be very short. In thermal

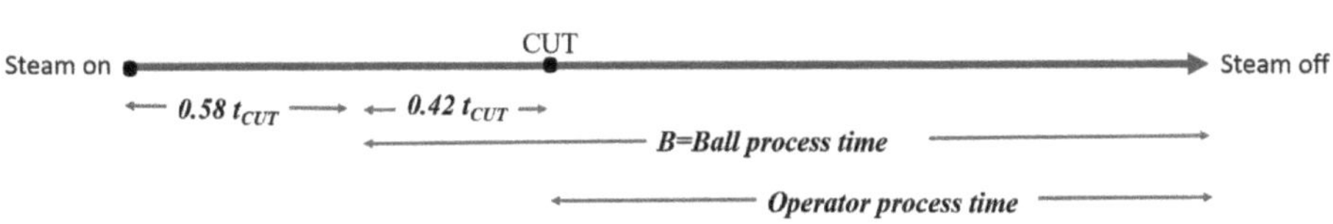

Fig. 2.16 The times in formula methods.

processing with water as the heating medium, the long CUT and small container size lead to the mean values of f_h and j cannot be obtained (Valentas et al., 1997). If we want to have a meaningful value for f_h and j, the CUT should be less than 0.5 f_h (Valentas et al., 1997).

Also, we should notice that in the general method, the length of CUT is included in the lethality calculations because all temperatures we used in the general method, are a reflection of the heating effect inside the product during CUT (Valentas et al., 1997).

To calculate Ball process time (B):

1. f_h (the slope of the heating line) is calculated from Fig. 2.15 or by experiments.
2. I_h (retort temperature minus initial product temperature) i.e., $T_R - T_{ih}$ is defined.
3. j_h ($\frac{T_R - T_{pih}}{T_R - T_{ih}}$) can be easily calculated from Fig. 2.14.
4. $g(T_R - T)$, i.e., the difference between the retort temperature and the temperature of the cold point of the product at the end of heating is not as easily obtained because the final temperature to ensure an adequate food sterility level cannot be known in advance. The g is actually the only parameter that determines the end of the heating process.

How to find *g*?

The end of the process is determined by g. g itself depends on (Fig. 2.17):

- The slope of the heating curve (f_h).
- F value (TDT) of the target microorganism at retort temperature (process time).
- Z value of target microorganism.
- The cooling driving force ($T_R - T_W$), i.e., the difference between the retort temperature and cooling water temperature.

There is a parameter U which is equivalent to the F value for thermal process time to reach a determined microorganism inactivation (the time required to accomplish a certain degree of microbial inactivation at the temperature of the treatment equipment, equivalent to the value F of the process) (Heldman & Lund, 2007; Ibarz & Barbosa-Cánovas, 2002):

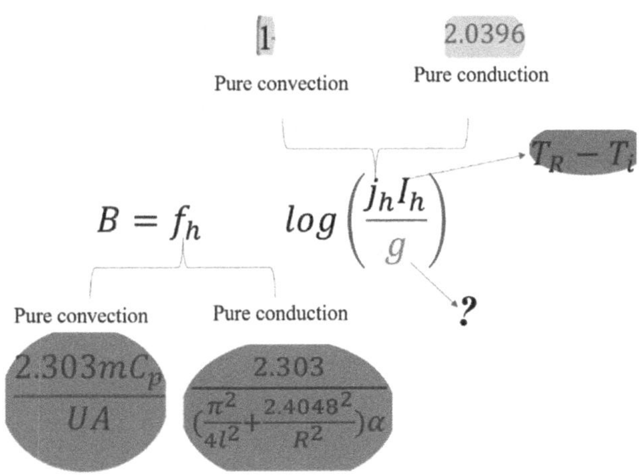

Fig. 2.17 Ball formula calculation steps.

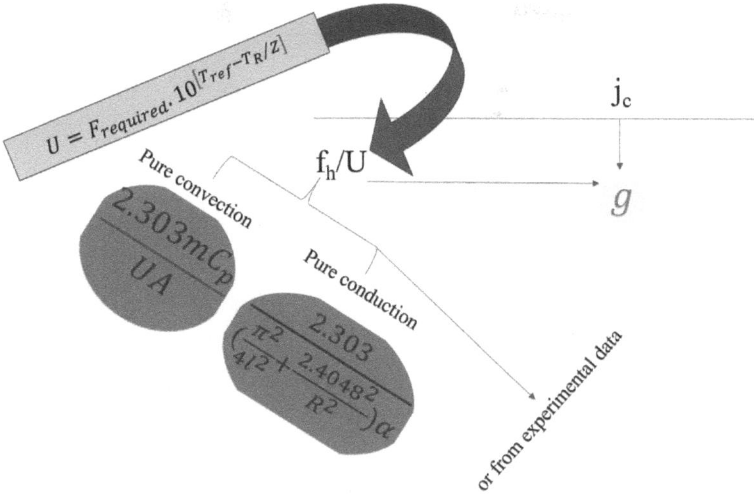

Fig. 2.18 The way to find g: schematic of table of f_h/U:g correlation for various j_c at different Z values.

$$U = F_0 \times F_i \tag{2.23}$$

where $F_0 = F^{Z}_{250} = F_{\text{Required}}$ and $F_i = 10^{\frac{T_{\text{ref}}-T_R}{Z}}$, i.e., the time at temperature T_R, equivalent to 1 min at 121°C.

Therefore,

$$U = F_{\text{required}} \times 10^{\frac{T_{\text{ref}}-T_R}{Z}} \tag{2.24}$$

Ball suggested the f_h/U and then found a relationship between f_h/U and g (Valentas et al., 1997) (Fig. 2.18).

Table 2.6 shows the amount of f_h/U as a function of g at a specific Z-value (Ibarz & Barbosa-Cánovas, 2002; Valentas et al., 1997). This table is based on conduction heat transfer and the assumptions of shape and temperature profile at the cooling stage are nonarbitrary. Fig. 2.19 shows a schematic of calculation in Ball method.

Stumbo method is similar to the method introduced by Ball, but it can be applied when the j_c is different from 1.41 (Valentas et al., 1997). Several tables for the evaluation of the process were published by Stumbo and Longley (1966) which considered the j_c variability (Valentas et al., 1997). These 57 Stumbo's tables result to answers which are in high agreement with general methods. Some tables are available for Z-value of 10°F for inactivation of vegetative bacteria during the pasteurization process, other tables for other Z-values can be found in references for sterilization process based on *C. botulinum*, and for nutrients (Valentas et al., 1997).

Pham (1987) developed two algebraic equations, one for $U/f_h > 1$ and another for $U/f_h < 1$, and simplified tables (Table 2.7). Pham's method is useful for mass-average

Table 2.6 The values of f_h/U vs log g for Ball's formula method.

f_h/U	log g	f_h/U	log g
0.350	−2.147	4.000	0.655
0.400	−1.790	4.500	0.702
0.450	−1.512	5.000	0.742
0.500	−1.290	5.500	0.776
0.550	−1.108	6.000	0.805
0.600	−0.949	7.000	0.854
0.650	−0.843	8.000	0.894
0.700	−0.736	9.000	0.927
0.750	−0.635	10.00	0.955
0.800	−0.544	15.00	1.052
0.850	−0.463	20.00	1.112
0.900	−0.392	25.00	1.155
0.950	0–0.328	30.00	1.187
1.000	−0.273	35.00	1.214
1.100	−0.173	40.00	1.235
1.200	−0.090	45.00	1.254
1.300	−0.019	50.00	1.270
1.400	0.042	60.00	1.296
1.500	0.097	70.00	1.318
1.600	0.146	80.00	1.336
1.700	0.183	90.00	1.352
1.800	0.229	100.0	1.365
1.900	0.265	120.0	1.388
2.000	0.298	140.0	1.406
2.500	0.430	160.0	1.422
3.000	0.525	180.0	1.435
3.500	0.598	200.0	1.447

Reprinted with permission from Valentas, K. J., Rotstein, E., & Singh, R. P. (1997). *Handbook of food engineering practice*. CRC Press.

lethality calculations. Also, it can be applied when the f_h is not equal to f_c (Valentas et al., 1997).

Hayakawa developed tables for heating and cooling sections of the thermal process to calculate the lethality of these sections. These tables are based on Z-value at 20°F. Also, he introduced a parameter of g/K_s with $K_s = Z/20$ which is tabulated against U/f_h (the reciprocal term of Stumbo's f_h/U) (Toledo, 2007).

In the lethality table for the cooling phase, the U is based on T_g, and before adding to heating U, it should be converted to U based on T_r through (Toledo, 2007):

$$U = 10^{\left(\frac{-g}{Z}\right)} \tag{2.25}$$

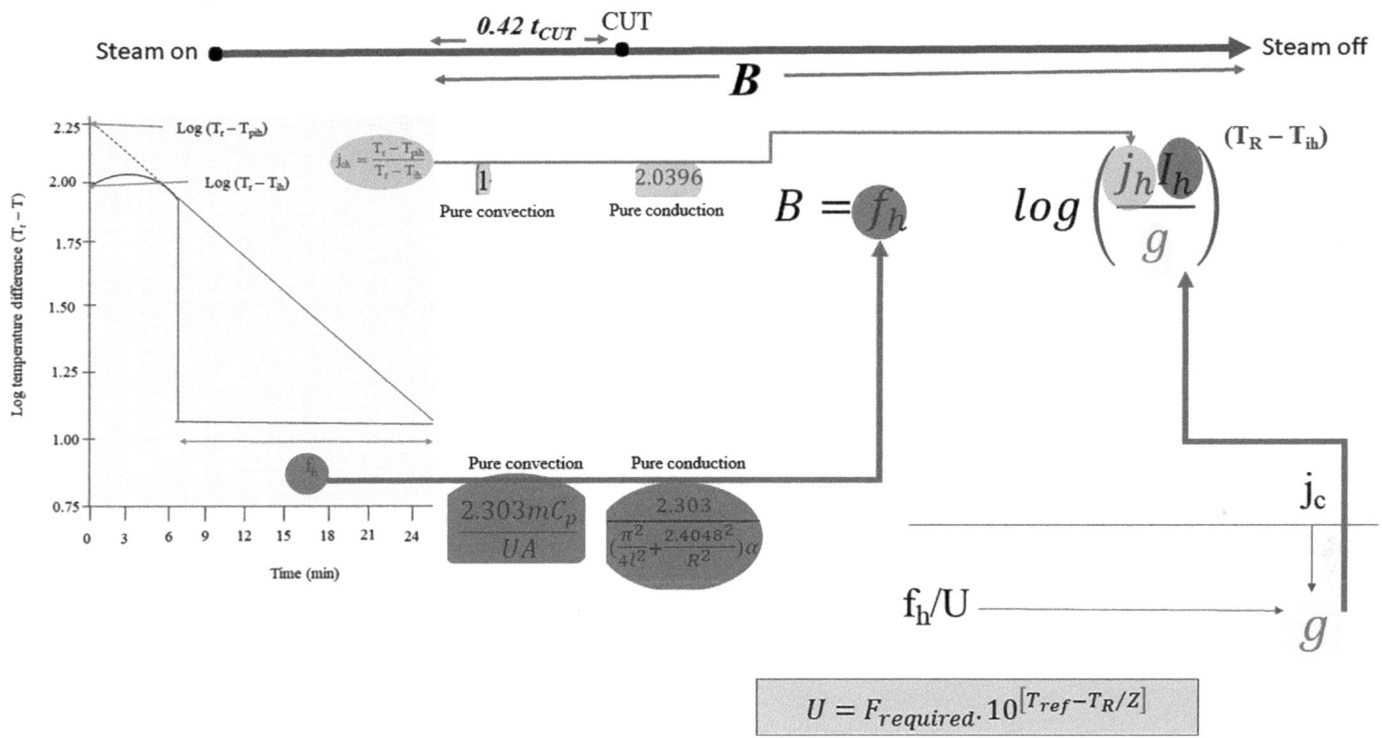

Fig. 2.19 A schematic of formula method.

Table 2.7 The equations and coefficients in Pham's method.

For high sterilizing values: W > 1
$\log(g/z) = -W + A_j - B/j + C$
$A = 0.088 + 0.107N_2$
$B = 0.102N_1$
$C = 0.074N_1 + 0.177N_2 - 0.653$
$W = U/f_h$
$N_1 = z/(T_r - T_i)$
$N_2 = z/(T_r - T_w)$
***For low sterilizing values:* 0.04 <*W* < 1.5**
$\log(a) = -W + A_1 + A_2 \exp.(-2.7W^{0.5})$
$\log(b) = -W + B_1 + B_2 \exp.(-2.7W^{0.5})$
$g = z(a + bj_{cc})$
$W = U/f_h$
$A_1 = -0.71 - 0.41(N_1/N_2) \exp.(-0.58/N_2)$
$A_2 = 2.14(N_2)^2 + 0.6(N_2)^2/N_1 - 0.26(N_1)^2 - 1.24N_1 + 1.02$
$B_1 = 0.31(N_2/N_1)^{0.5} + 0.55(N_2)^{0.5} + 0.61(N_1)^{0.5} - 1.86$
$B_2 = (N_2/N_1)^{0.5}(0.91(N_1)^2 - 3.18N_1 - 0.755) - 1.38(N_1)^2 + 2.55N_1 + 1.52$

2.11 Numerical calculations of heat transfer during sterilization

The data relating to heat penetration curves can be used in numerical calculations for estimating the heat transfer parameters during sterilization (Heldman & Lund, 2007). Numerical calculations are useful to predict the thermal process after simulation with specified software. This leads to simulating the considered process conditions on heat penetration of determining food material in a specified thermal process.

Datta and Teixeira (1988) numerically predicted the velocity and temperature of canned water in static retort and suggested that the cold area is donate-like and is near one-tenth length from the bottom. Kumar, Bhattacharya, and Blaylock (1990) also used the finite element method to simulate non-Newtonian food heating in vertical metal cans which were heated from the top at 121°C (Kumar et al., 1990). They realized that in natural convection, the cold area is near the bottom of the can. Wiese and Wiese (1992) used different numerical methods to determine the breaking point (Wiese & Wiese, 1992). Ghani, Farid, Chen, and Richards (1999) numerically simulated the natural convection of canned water and CMC with PHOENICS and predicted the fluid flow and internal temperature of this model fluid. They found cold area moves to the bottom during the heating process and its size and shape are different in water and CMC (Ghani et al., 1999). Ghani, Farid, and Chen (2002) also studied the sterilization of canned orange soup. They found that the temperature of the cold area reaches a maximum temperature of 107°C and SHZ moves to the bottom due to natural convection and stays at 20%–25% (Ghani et al., 2002). Farid and Ghani (2004) used a new computational technique to estimate the sterilization time of canned foods. To predict the temperature profile of the cold area and natural convection in the can, they used computational fluid dynamics with PHOENIX. They suggested that due to little vertical dimension, the effect of natural convection in a horizontal can is lower than vertical can (Farid & Ghani, 2004). Dimou and Yanniotis (2011) predicted a temperature and velocity profile of 7.6 × 10.9 cm cans containing 4% brine and 0, 1, 8, and 19 asparaguses. They found that the cold area is near the %13.5 from the bottom and is not dependent on the asparagus numbers (Dimou & Yanniotis, 2011). Dimou and Yanniotis (2011) studied the effect of process variables on temperature profile, flow patterns, and location of the cold area in static cans with computational fluid dynamics. They used asparagus, olive in %4 brine, and peach in %20 sucrose in 76 × 109 mm cans and have found the flow and temperature are affected by the shape, type, and arrangement of particles inside the can, and properties of the filling fluid (Dimou & Yanniotis, 2011). Sarghini and Erdogdu (2016) investigated the effect of rotation rate (end-over-end) and viscosity (water and 1.5% CMC) on heat transfer characteristics of canned food with headspace (Sarghini & Erdogdu, 2016).

Ranjbar Nedanami et al. (2018) studied the amount of heat penetration parameters and temperature of the slowest heating zone in 3.5% starch dispersion. The computational fluid dynamics software was used and governing equations for energy, momentum, and continuity were computed using a finite volume method. A 3.5% starch dispersion was used. The effect of 50 and 75°C initial temperatures, and 20% and 10% headspace were investigated. It was determined that in static sterilization with headspace, the SHZ is near the interface of the air-product. An increase in initial temperature and a decrease in headspace leads to an increase in j_h. The f_h of 20% headspace with 50°C initial temperature is lower in all samples. Increasing the initial temperature leads to f_h increase and f_2 decrease. Finally, it was demonstrated that instead of the wall surface or the volume of food, the ratio of total heat surface to the volume of food is effective on its temperature pattern (Ranjbar Nedanami et al., 2018). In this study, the effect of container geometry on heat penetration parameter (j) and microbial lethality (L) in the slowest heating point were also investigated (Tables 2.8 and 2.9). The temperature of the slowest heating zone was monitored by a thermocouple and then compared with the predicted temperature by software. It was determined that the cone-shaped container had the fastest heat transfer during sterilization (Fig. 2.20). Also, container geometry has a significant effect on the slowest heating zone shape, position, final temperature, j, L, and F-value (Ranjbar Nedanami, 2019).

2.12 Cook value calculations

Thus, a *cook value C* is defined for thermal processes which are applicable for determining the degradation of sensory characteristics (Ibarz & Barbosa-Cánovas, 2002). In a simplified form of cooking value with a reference temperature of 100°C (water

Table 2.8 The heat penetration parameter j and Lethality factor at 15 min (900 s) after reaching the container wall to 121°C for different container geometry filled with IT = 50°C of starch dispersions.

Can characteristics	j_h	The temperature of SHZ (°C)	*L*	*F* (min)
Cone 0 degree	0.4999	85.5	0.0195	0.293
Cone 180 degree	0.731	69.07	0.00017	0.0025
Cylinder1 (6 × 10 cm)	0.803	63.9	0.00008	0.001
Cylinder2 (10 × 6 cm)	0.652	74.7	0.00096	0.014

Table 2.9 The heat penetration parameter *j* and lethality factor at 20 min (the 1200s) after reaching the container wall to 121°C for different container geometry filled with IT = 50°C of starch dispersions.

Can characteristics	j_h	The temperature of SHZ (°C)	*L*	*F* (min)
Cone 0 degree	0.496	85.72	0.0323	0.646
Cone 180 degree	0.608	77.78	0.00086	0.0172
Cylinder1 (6 × 10 cm)	0.735	68.8	0.0003	0.006
Cylinder2 (10 × 6 cm)	0.613	77.45	0.0024	0.048

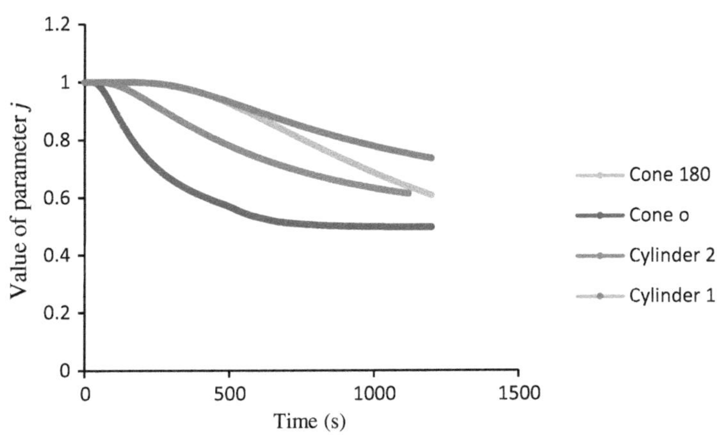

Fig. 2.20 Values of parameter *j* for different container geometries.

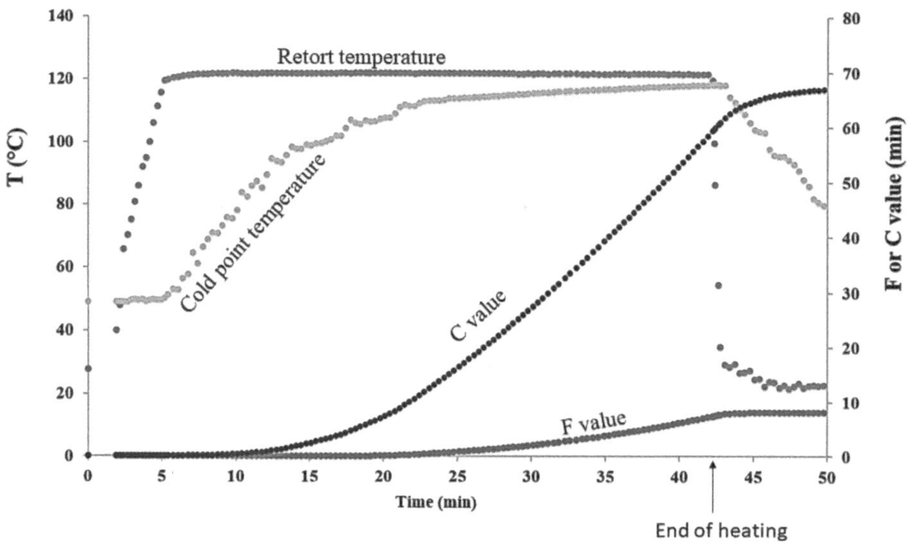

Fig. 2.21 A typical graph of F and cook values.

boiling temperature) and Z in the range of 20–40°C, inducing higher values for cook value when compared to F value (Fig. 2.21). We can write (Ibarz & Barbosa-Cánovas, 2002):

$$C = \int_0^t 10^{\frac{T-100}{Z}} dt \tag{2.26}$$

2.13 Degradation of quality factors

Sterilization is a thermal process that can affect food quality and characteristics. To select a specified thermal process to destroy the unwanted microorganisms or enzymes, we should also consider the characteristics and nutrients of the food. A high-temperature processing results in the destruction of microorganisms over the food quality factors, because the decimal reduction time and Z-value of microorganisms are lower than most food quality factors (Heldman & Lund, 2007). If we use a high temperature with a short exposure time, the potential of quality optimization by the thermal process will be high. But we should remember that the high temperatures may have undesirable effects on sensitive food nutrition or quality factors. If we know the quality factors and their degradation kinetics during the thermal process, we can find the optimum combination of retort temperature and process time to maximize the food product quality (Heldman & Lund, 2007). Also, some interactions between proteins, sugars, and lipids may occur during the thermal processing of foods, which are important due to nutritional and qualitative viewpoints (Zhang et al., 2021).

In a thermal process in which heat transfers through a conduction mechanism in a food product, the processing time may be high and higher temperatures near the surface of the product can lead to a higher degradation of food quality than the cold area of the product center (Heldman & Lund, 2007). In this case, we can calculate the volume integration of the different levels of quality factors at different locations of a food product through a finite element feature by a numerical simulation model (Heldman & Lund, 2007).

If the degradation parameters of *D*- and *Z*-values for nutrients is determined, the nutrient loss calculations are the same as microorganism deactivation. Nutrient retention during a constant temperature process in holding tubes of aseptic processing systems is estimated as (Toledo, 2007):

$$\log\left(\frac{C}{C_0}\right) = -\left[\frac{D_{\mathrm{m}}}{D_{\mathrm{n}}} S_{\mathrm{v}}\right] 10^{(T_0 - T)\left(\frac{1}{Z_{\mathrm{m}}} - \frac{1}{Z_{\mathrm{n}}}\right)} \tag{2.27}$$

where the subscripts n and m are the nutrient degradation and microorganism inactivation parameters, respectively, and S_v is $\log(N_0/N)$.

For this, the degradation of quality factors should be determined through the integrated lethality of food inside the container volume. The nutrients are substantially retained in the same regions in the container. If we consider the container as separated incremental cylindrical shells, calculate each shell temperature at specified time increments, determine the nutrient degradation extended, and finally by summing these degradations extend, we can calculate the quality degradation throughout the container. The residual nutrient concentration also can be calculated at the end of the process by integrating the concentration of residuals at each incremental cylindrical shell in the container volume. Based on two observations, Stumbo (1973) suggested an equation for residual nutrients in a container with conduction heating of food (Toledo, 2007):

1. When the value of j at a point of the container (j_v) is 0.5_j, the g at any time in that point (g_v) is 0.5 g, and the volume is 19% of the total volume of the container, the condition of that point is isotherm.
2. The isotherm volume is v, F-value of isotherm and F-value at the critical point are F_v and F, respectively. Thus the $(F_v - F)$ is proportional to $(1 - v)$.

$$\overline{F} = F + D\log\left(\frac{D + 10.92(F_v - F)}{D}\right) \tag{2.28}$$

where the $\overline{F}$ is integrated lethal effect of the heating process on nutrients based on lethality at the critical point of F.

2.14 Conclusion

In food sterilization, a series of information about different science domains such as the-bacteriology, heat transfer, and modeling are needed. Knowledge about the sterilization system and their characteristics are also required. To estimate sterilization time (required F value), it is first important to identify the target microorganism and its behavior to the heat via the parameters D and Z values. Then, by determining the changes in cold point temperature (by performing the experiments or by modeling the process) the process F value can be calculated which needs to be higher than the required one.

References

Ball, C. O. (1923). *Thermal process time for canned food (Vol. 7)*. National Research Council of the National Academy of Sciences.

Berk, Z. (2009). *Food process engineering and technology*. Elsevier.

Bigelow, W. D., & Esty, J. (1920). The thermal death point in relation to time of typical thermophilic organisms. *The Journal of Infectious Diseases*, 602–617.

Datta, A., & Teixeira, A. (1988). Numerically predicted transient temperature and velocity profiles during natural convection heating of canned liquid foods. *Journal of Food Science, 53* (1), 191–195.

Dimou, A., & Yanniotis, S. (2011). 3D numerical simulation of asparagus sterilization in a still can using computational fluid dynamics. *Journal of Food Engineering, 104*(3), 394–403.

Farid, M., & Ghani, A. A. (2004). A new computational technique for the estimation of sterilization time in canned food. *Chemical Engineering and Processing: Process Intensification, 43*(4), 523–531.

Ghani, A. A., Farid, M., & Chen, X. (2002). Numerical simulation of transient temperature and velocity profiles in a horizontal can during sterilization using computational fluid dynamics. *Journal of Food Engineering, 51*(1), 77–83.

Ghani, A. A., Farid, M., Chen, X., & Richards, P. (1999). Numerical simulation of natural convection heating of canned food by computational fluid dynamics. *Journal of Food Engineering, 41*(1), 55–64.

Heldman, D. R., & Lund, D. B. (2007). *Handbook of food engineering*. CRC Press.

Ibarz, A., & Augusto, P. (2017). *Principles of unit operations in food processing*. Chichester, UK: John Wiley & Sons, Ltd.

Ibarz, A., & Barbosa-Cánovas, G. V. (2002). *Unit operations in food engineering*. CRC Press.

Kumar, A., Bhattacharya, M., & Blaylock, J. (1990). Numerical simulation of natural convection heating of canned thick viscous liquid food products. *Journal of Food Science, 55*(5), 1403–1411.

Kutz, M. (2013). *Handbook of farm, dairy and food machinery engineering*. Academic Press.

Ranjbar Nedanami, A. (2019). Numerical calculation F-value and lethality of non-Newtonian food fluid during sterilization based on can geometry. *Iranian Food Science and Technology Research Journal, 14*(6), 113–125.

Pham, Q. (1987). Calculation of bound water in frozen food. *Journal of Food Science, 52*(1), 210–212.

Ranjbar Nedamani, A., & Hashemi, S. J. (2021). RSM-CFD modeling for optimizing the apricot water evaporation. *Journal of Food and Bioprocess Engineering*, *4*(2), 112–119.

Ranjbar Nedamani, A., Ziaiifar, A. M., Parvini, M., Kashaninejad, M., & Maghsoudlou, Y. (2015). *Study of heat penetration coefficients (f and j) using CFD method for canned food based on starch*. PhD Gorgan University of Agricultural Science and Natural Resources.

Ranjbar Nedanami, A., Ziaiifar, A. M., Parvini, M., Kashaninejad, M., & Maghsoudlou, Y. (2018). Numerical calculation of sterilization heat penetration parameters based on initial temperature and headspace in canned non-Newtonian fluid. *Journal of Food Processing and Preservation*, *42*(10).

Ranjbar Nedanami, A., Ziaiifar, A. M., Parvini, M., Kashani-Nejad, M., & Maghsoudlou, Y. (2017). Numerical study of starch concentration and filling initial temperature on cold area changes in canned starch-based food. *Iranian Food Science and Technology Research Journal*, *13*(1), 129–140.

Ranjbar Nedanami, A., Ziaiifar, A. M., Parvini, M., Kashaninezhad, M., & Maghsudluo, Y. (2016). Numerical study of the effects of can fill level on the heat transfer pattern in starch dispersion. *Innovative Food Technologies*, *3*(3), 1–11.

Richardson, P. (2001). *Thermal technologies in food processing*. Taylor & Francis.

Sarghini, F., & Erdogdu, F. (2016). A computational study on heat transfer characteristics of particulate canned foods during end-over-end rotational agitation: Effect of rotation rate and viscosity. *Food and Bioproducts Processing*, *100*, 496–511.

Stumbo, C. (1973). *Thermobacteriology in food processing*. New York, NY: Academic, Inc.

Stumbo, C., & Longley, R. E. (1966). New parameters for process calculation. *Food Technology*, *20*(3), 341.

Toledo, R. T. (2007). *Fundamentals of food process engineering*. Springer.

Valentas, K. J., Rotstein, E., & Singh, R. P. (1997). *Handbook of food engineering practice*. CRC Press.

Vidyarthi, S. K., Mishra, D. K., Dolan, K. D., & Muramatsu, Y. (2020). Inverse estimation of fluid-to-particle heat transfer coefficient in aseptic processing of particulate foods. *Biosystems Engineering*, *198*, 210–222.

Wang, W., Dong, L., Zhang, Y., Yu, H., & Wang, S. (2021). Reduction of the heterocyclic amines in grilled beef patties through the combination of thermal food processing techniques without destroying the grilling quality characteristics. *Foods*, *10*(7), 1490.

Wiese, K. L., & Wiese, K. F. (1992). A comparison of numerical techniques to calculate broken line heating factors of a thermal process. *Journal of Food Processing and Preservation*, *16* (5), 301–312.

Zhang, Y., Dong, L., Zhang, J., Shi, J., Wang, Y., & Wang, S. (2021). Adverse effects of thermal food processing on the structural, nutritional, and biological properties of proteins. *Annual Review of Food Science and Technology*, *12*, 259–286.

Design and simulation of heat exchangers for the food industry

3

Gokcem Tonyali Karsli and Deniz Cekmecelioglu
Department of Food Engineering, Middle East Technical University, Ankara, Turkey

3.1 Introduction to heat exchangers

Heat exchangers are devices, which are used to accomplish the transfer of heat between two fluids. The heat exchange is commonly facilitated without mixing the fluids in food processing due to product quality concerns. Heat is transferred by convection from the hot fluid stream to a separating wall, by conduction through the wall, and again by convection from the separating wall to the cold fluid stream (Fig. 3.1). The wall could be a tube wall, a flat or curved surface based on the type of the heat exchanger (Geankoplis, 1993).

The overall rate of heat transfer from the hot fluid to cold fluid becomes:

$$\dot{Q} = UA\Delta T_m \tag{3.1}$$

where U is the overall heat transfer coefficient, A is the effective heating area, and ΔT_m is the mean temperature difference between the two fluids and this term varies according to the type of the heat exchanger (Cengel, 2003). The two convective terms and conduction are embedded in the coefficient of U, as shown in Eq. (3.2), which is written according to the network of resistances depicted in Fig. 3.1:

$$U = \frac{1}{A\sum R} = \frac{1}{A(R_{\text{conv1}} + R_{\text{wall}} + R_{\text{conv2}})} \tag{3.2}$$

Thus, the rate of heat transfer between the two fluids at any location depends on the temperature difference, which changes along the heat exchanger, contact area, and the overall heat transfer coefficient U.

Heat exchangers are widely used in food industry for conditioning of various streams of diary, beverage, brewery, oil, and water processing. It is also a common practice to use heat exchangers for generation of steam, which finds application in heating of feeds for distillation, evaporation, and extraction units.

Thermal Processing of Food Products by Steam and Hot Water. https://doi.org/10.1016/B978-0-12-818616-9.00012-2

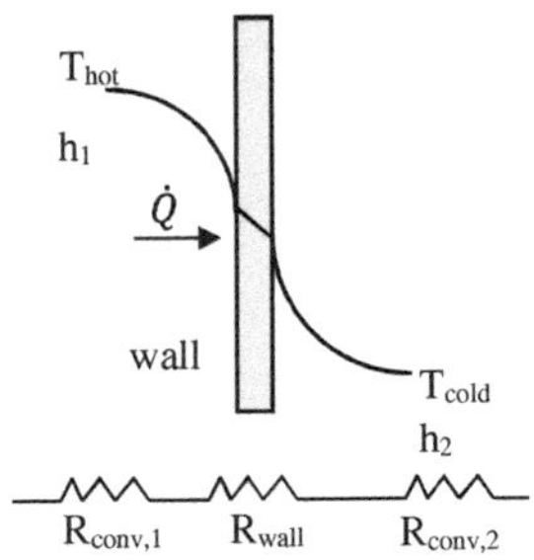

Fig. 3.1 Mechanisms of heat transfer and network of thermal resistance for two fluids separated by a wall of a heat exchanger.

3.2 Types of heat exchangers

The hardware and configuration of heat exchangers depend upon field of application and heat transfer needs. In food processing, fluid properties and quality requirements are additional constraints to the choice of the equipment. All this has led to numerous types of heat exchanger designs:

3.2.1 Double-pipe heat exchanger

The simplest heat exchanger is the double-pipe or concentric-pipe heat exchanger, where one fluid flows inside the small pipe and the other fluid flows through the annular space between the inner and outer pipes, as depicted in Fig. 3.2. Two types of flow arrangements, named as parallel (co-current) flow and counter flow, are used in this type of heat exchanger. In the parallel flow, the hot and cold fluids enter the heat exchanger at the same end and flow in the same direction, whereas in the counter flow the fluids enter at the opposite ends and flow in the opposite directions (see Fig. 3.2 for counter flow).

3.2.2 Shell-and-tube heat exchanger

A widely used heat exchanger type in industrial applications is the shell-and-tube heat exchanger (Chapter 6). This equipment contains a large number of tubes packed parallel in a shell to enable either hot or cold fluid to flow inside the tubes while the other flows through the shell (Fig. 3.3). The shell-side fluid can flow in normal direction to

Fig. 3.2 Schematic drawing of a double-pipe heat exchanger (single-pass counter flow).

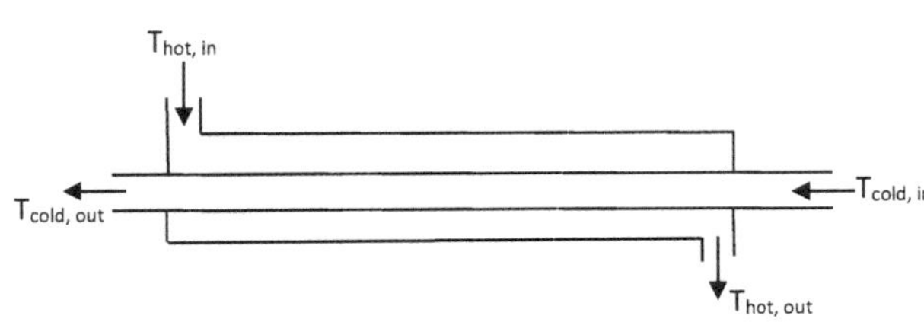

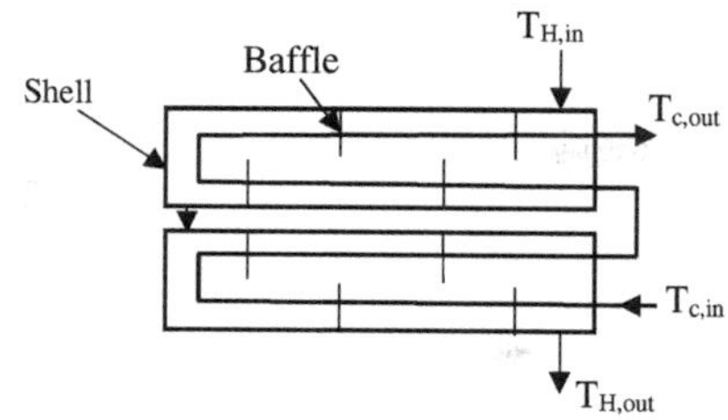

Fig. 3.3 Schematic for two-shell-four-tube-pass heat exchanger.

tubes or across the tube bank by the baffles so that the rate of heat transfer is improved. The shell-and-tube heat exchangers are preferred for large liquid flows and classified further based on the number of shell and tube passes. The simplest shell-and-tube heat exchanger is 1 shell pass and 1 tube pass, or also known as double-pipe heat exchanger. Fig. 3.3 shows 2-shell pass-and-4-tube-passes heat exchanger.

3.2.3 Compact (or cross-flow) heat exchanger

The most common device for heating or cooling of gases, e.g., air, is the compact heat exchanger. Generally used in gas-gas and gas-liquid configurations and overcomes the low heat transfer coefficient with gases by accommodating high ratio of surface area to volume. In this equipment, the liquid, for instance, flows inside the tubes while the gas is forced to flow perpendicularly in the tube bundle. Thus, it is also known as cross-flow heat exchanger. The gas flowing across is called a *mixed stream* because it can move freely as heat is being exchanged and the liquid is called *unmixed stream* having confined and separate channel in the exchanger, where it cannot mix with itself (Fig. 3.4). Compact heat exchangers provide high rates of heat transfer within a small volume and are appropriate when strict constraints on weight and volume exist. The

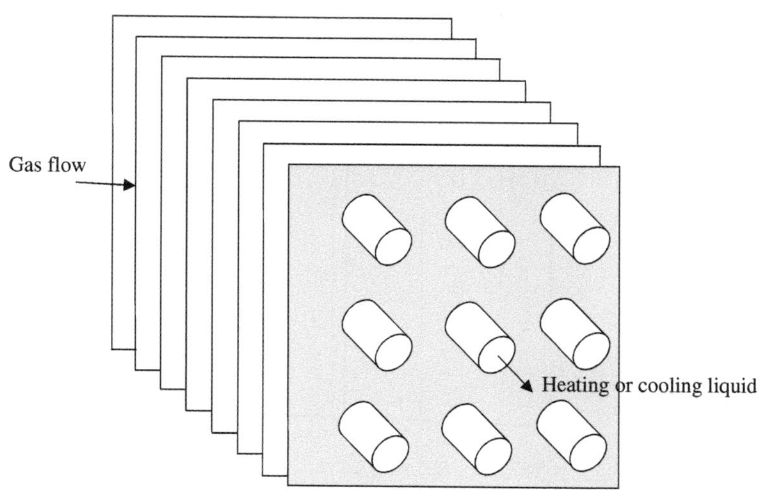

Fig. 3.4 Cross-flow heat exchanger.

area of these exchangers are enlarged by attaching thin plates or corrugated fins to the walls of the heat exchanger.

3.2.4 Plate heat exchangers

Plate heat exchangers (Chapter 4) are comprised of a pack of gasketed corrugated metal plates, which are pressed together with a frame to form the required flow as shown in Fig. 3.5. The gaskets seal around the plates to avoid fluid mixing and also used to create the desired flow configurations by closing or opening the ports at the plate corners, e.g., series, parallel, and multipass arrangements. The hot and cold fluids flow through the formed channels and alternate passages so that heat is transferred to cold fluid by two hot fluid streams through the thin corrugated metal plates (Fig. 3.5). The heating area can be increased by assembling more plates according to the required heat load. Because the plates provide each fluid high rates of heat transfer, the plate heat exchangers are considered more compact than the shell-and-tube heat exchangers with the same heat transfer capacity. Plate heat exchangers are of primary choice for many engineering applications involving liquid-liquid heat transfer and widely used in breweries, dairy, and beverage industries due to easy maintenance and cleaning, and effective heat-transfer characteristics.

3.3 Heat transfer fundamentals

Two mechanisms of heat transfer occur between fluids and the separating wall of the heat exchangers, e.g., plate heat exchangers, shell-and-tube heat exchangers, etc.; conduction within the wall and convection from fluid toward the wall and vice versa or inside the channels in case of plate heat exchangers or tubes of the double-pipe heat exchangers, or alike. Continuity and momentum equations are used to develop the flow expressions within exchangers, i.e., 3D or else transient laminar flow equations.

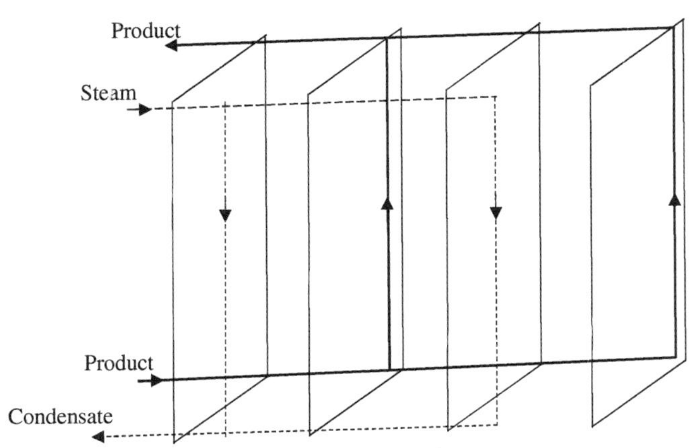

Fig. 3.5 A counter flow diagram of a plate heat exchanger.

When the flow is incompressible and the density does not change with time and space variables, the equations read as, for instance on cartesian coordinates:

$$\nabla \cdot v = 0 \tag{3.3}$$

$$\frac{\partial v}{\partial t} + v \cdot \nabla v = -\frac{1}{\rho}\nabla p + \nu \cdot \nabla^2 \cdot v \tag{3.4}$$

Eq. (3.4) is known as the Navier-Stokes equation that is used to describe the laminar flow of the fluids. Similarly, the transient energy equation is expressed as:

$$\frac{\partial T}{\partial t} + v \cdot \nabla T = \frac{k}{\rho C_p}\nabla^2 T \tag{3.5}$$

where ρ is the density, v is the velocity, p is the pressure, ν is the kinematic viscosity, T is temperature, k is thermal conductivity, and C_p is the heat capacity.

The continuity, momentum, and energy Eqs. (3.3)–(3.5) are used to calculate the flow and temperature field of the fluid material in question when solved simultaneously. The overall heat transfer coefficient U is considered constant. This is reasonable for single-phase compact heat exchangers, however in larger ones such as shell-and-tube exchangers, U varies with respect to position and local temperature. If U is not dependent on T, the changing flow configuration and the variation of temperature still lead to variations of U. When U is dependent on position, an average U defined as in Eq. (3.6) must be used. For cases of constant U, Eq. (3.1) is valid to deal with the varying temperatures of the fluids by finding the appropriate mean temperature difference.

$$U_{\text{ave}} = \int_0^A U \frac{dA}{A} \tag{3.6}$$

The equations of U now read for plate heat exchangers and tubular heat exchangers as in Eqs. (3.7) and (3.8), respectively:

$$U = \frac{1}{\frac{1}{h_1 A} + \frac{\Delta x}{kA} + \frac{1}{h_2 A}} \tag{3.7}$$

$$U = \frac{1}{\frac{1}{h_i A_i} + \frac{ln(r_0/r_i)}{2\pi k L} + \frac{1}{h_0 A_0}} \tag{3.8}$$

The local convective heat transfer coefficient h_x or the average value of h is calculated using the standard *Nusselt number* expressions defined as a function of *Reynolds* and *Prandtl numbers*. The local heat flux between the fluid and the wall at any location in the heat exchanger is written as:

$$\dot{q} = h_x(T_{\text{fluid}} - T_{\text{wall}}) \tag{3.9}$$

If we make simply an energy balance on each of the fluids in a heat exchanger, regardless of type and flow arrangement, at steady state and constant pressure conditions, we get:

$$\dot{Q} = \dot{m}_c C_{p,c}(T_{c,\text{out}} - T_{c,\text{in}}) = \dot{m}_h C_{p,h}(T_{h,\text{in}} - T_{h,\text{out}}) \tag{3.10}$$

or in differential forms:

$$d\dot{Q} = C_c dT_c = -C_H dT_H \tag{3.11}$$

where $C_c = \dot{m}_c C_{p,c}$ and $C_H = \dot{m}_h C_{p,h}$ denote capacitance terms of the cold and hot fluids, respectively.

Rewriting Eq. (3.1) in differential form for the two fluids and varying area term:

$$d\dot{Q} = U\, dA\, (T_H - T_C) \tag{3.12}$$

which is useful to determine the temperature change with respect to position or area of the heat exchanger, when combined with Eq. (3.11). A number of authors have used this approach, i.e., the steady-state thermal model, for various flow configurations.

3.3.1 Logarithmic mean temperature difference

For complete design of heat exchangers, the flow configuration need to be specified. The most widely used flow configurations in plate heat exchanger and double-pipe heat exchanger are parallel and counter flows. It is quite useful to sketch simply the temperature variation observed in both fluids when we consider either of the flow arrangements such as those presented in Fig. 3.6. There is a quite noticeable temperature change in both hot and cold fluids along the heat exchanger, i.e., with respect to length or area, under parallel or counter-current flows (Figs. 3.3–3.6A and B). This yields varying ΔT with position and needs to be quantified reasonably, e.g., as average value, for use in Eq. (3.1).

In parallel flow, the temperature difference between hot and cold fluids at the inlet is large but decreases exponentially with length. If capacitance of both fluids are equal, the outlet temperatures can reach halfway values of the inlet temperatures. By examination of Fig. 3.6A, it is evident that $C_H < C_C$ and if they were equal the two temperature lines would be parallel straight lines. On the other hand, more complete heat exchange can be achieved with counter-flow arrangement (Fig. 3.6B). In other words, it is possible for the hot fluid to leave the heat exchanger at a lower temperature than the exit cold fluid. This implies the greater heat transfer per unit area than the parallel flow configuration. If the area of heat transfer is increased infinitely, the outlet temperature of hot fluid will approach inlet temperature of cold fluid for $C_H < C_C$, but outlet of cold stream cannot reach the inlet temperature of hot stream. This case implies the theoretical maximum rate of heat transfer between fluids defined as:

$$\dot{Q}_{\max} = C_{\min}(T_{Hi} - T_{Ci}) \tag{3.13}$$

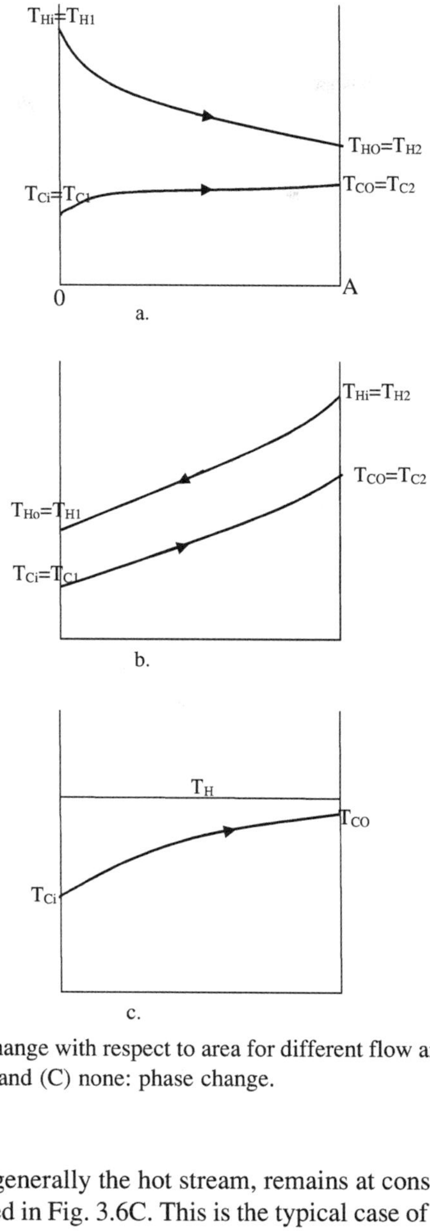

Fig. 3.6 Temperature change with respect to area for different flow arrangements. (A) Parallel flow, (B) counter flow, and (C) none: phase change.

where $C_{min} = C_H$.

One of the fluids, generally the hot stream, remains at constant temperature while the other fluid is heated in Fig. 3.6C. This is the typical case of a phase change, which occurs in either the evaporator or the condenser. The direction of the fluid at constant temperature is not noted because it has no effect on the analysis. As a result, counter flow arrangement is the most commonly used flow arrangement.

To determine ΔT_m for such arrangements, say counter flow; we write from Eq. (3.11):

$$dT_H = -\frac{d\dot{Q}}{C_H} \tag{3.14}$$

$$dT_C = \frac{d\dot{Q}}{C_c} \tag{3.15}$$

Taking their difference and substituting each expression, we obtain

$$dT_H - dT_C = d(T_H - T_C) = -d\dot{Q}\left(\frac{1}{C_H} + \frac{1}{C_C}\right) \tag{3.16}$$

Combining with Eq. (3.12),

$$\frac{d(T_H - T_C)}{T_H - T_C} = -U\left(\frac{1}{C_H} + \frac{1}{C_C}\right)dA \tag{3.17}$$

Integrating both sides of Eq. (3.17) for $T_C: 1 \rightarrow 2$ and $T_H: 1 \rightarrow 2$, where 1 and 2 denote inlet and outlet points as indicated in Fig. 3.6 and replacing C_H and C_C terms from Eqs. (3.14) and (3.15):

$$\ln \frac{T_{H2} - T_{C2}}{T_{H1} - T_{C1}} = \frac{UA}{\dot{Q}}(T_{H2} - T_{H1} - T_{C2} + T_{C1}) \tag{3.18}$$

or

$$\dot{Q} = UA\frac{\Delta T_2 - \Delta T_1}{ln\frac{\Delta T_2}{\Delta T_1}} = UA\Delta T_{LM} \tag{3.19}$$

where $\Delta T_1 = T_{H1} - T_{C1} = T_{HO} - T_{Ci}$, $\Delta T_2 = T_{H2} - T_{C2} = T_{Hi} - T_{CO}$ for counter flow, and ΔT_{LM} is the logarithmic-mean temperature difference and its value changes according to the flow arrangement, however the final expression (3.19) remains the same in case of parallel flow or phase-change flow depicted in Fig. 3.6. Comparing to Eq. (3.1), we conclude that $\Delta T_m = \Delta T_{LM}$ and this is valid only for a single pass double-pipe heat exchanger with parallel or counter flow. For larger heat exchangers:

$$\Delta T_m = F\Delta T_{LM} \tag{3.20}$$

where F is the correction factor. The charts of the correction factor F are available in Cengel (2003) for different types of heat exchangers (see fig. 13.18, p. 684).

Example 3.1

A shell-and-tube heat exchanger (2-shells and 8-tube passes) is used to heat 1.8 kg/s of water ($c_p = 4180$ J/kg K) from 20 to 78°C with ethylene glycol ($c_p = 2760$ J/kg K). Ethylene glycol cools down from 95 to 55°C at the shell side. If the heat transfer area is 15 m^2, calculate the heat transfer rate and overall heat transfer coefficient.

Solution:

Assumptions: Steady-state conditions; the heat exchanger is well-insulated, so that heat loss to the surroundings is negligible; changes in the kinetic and potential energies are negligible; fluids have constant properties; there is no fouling.

The heat transfer rate gained by water stream:

$$\dot{Q} = \dot{m}_c \cdot C_{p,c} \cdot (T_{c,\text{out}} - T_{c,\text{in}})$$

$$\dot{Q} = 1.8\,\text{kg/s} \times 4.18\,\text{kJ/kg K} \times (78 - 20)^{\circ}\text{C} = 436.4\,\text{kW}$$

To find out the logarithmic mean temperature difference for counterflow configuration, the temperature difference terms are:

$$\Delta T_1 = T_{Hi} - T_{Co} = 95 - 78 = 17^{\circ}\text{C}$$

$$\Delta T_2 = T_{Ho} - T_{Ci} = 55 - 20 = 35^{\circ}\text{C}.$$

Thus

$$\Delta T_{LM} = \frac{\Delta T_1 - \Delta T_2}{ln\,\frac{\Delta T_1}{\Delta T_2}} = 25^{\circ}\text{C}$$

F, the correction factor should be read from the graph using the relevant components:

$$P = \frac{t_2 - t_1}{T_1 - t_1} = \frac{T_{CO} - T_{Ci}}{T_{Hi} - T_{Ci}} = 0.77$$

$$R = \frac{T_1 - T_2}{t_2 - t_1} = \frac{T_{Hi} - T_{HO}}{T_{CO} - T_{Ci}} = 0.69$$

So, $F = 0.89$. Finally, the overall heat transfer coefficient is calculated from:

$$436.4\,\text{kW} = U \times 15\,\text{m}^2 \times 0.89 \times 25^{\circ}\text{C}.$$

$$U = 1300\,\text{W/m}^2\,\text{K}.$$

3.3.2 Fouling

The surfaces of heat exchangers are coated with various deposits from food commodity; e.g., denaturated milk protein, or corroded due to interaction of fluids and the construction material of the heat exchangers. The deposits create an additional resistance to heat transfer, lowering remarkably the performance of the equipment and may block the flow area as well. This effect is known as fouling and represented by R_f.

The fouling factor R_f can be calculated according to Eq. (3.21) by determining the experimental values of U for clean and dirty conditions as (Welty, Wicks, Wilson, & Rorrer, 2008):

$$R_f = \frac{1}{U_{dirty}} - \frac{1}{U_{clean}}\text{m}^{2\circ}\text{C/W} \tag{3.21}$$

As a result, the relations in Eqs. (3.7) and (3.8) are true for new equipment, with fouling factor of zero at new state. For an operating heat exchanger, e.g., shell-and-tube heat exchanger, the expression of overall heat transfer coefficient is now written as:

$$\frac{1}{UA}=\frac{1}{h_iA_i}+\frac{R_{f,i}}{A_i}+\frac{\ln(r_0/r_i)}{2\pi kL}+\frac{R_{f,o}}{A_0}+\frac{1}{h_0A_0} \tag{3.22}$$

Milk fouling, for instance, causes an increase in energy consumption, maintenance, and labor costs during milk-pasteurization in plate heat exchangers. The dairy industry was reported to spend 80% of the product costs on cleaning-in-place process (Choi et al., 2013).

Example 3.2

In a milk processing plant, a plate heat exchanger is designed with 40 plates made of 304 stainless steel (k=16.2 W/m K) to heat milk (c_p=3850 J/kg K) from 45 to 72°C at a mass flow rate of 20 ton/h. The heating is accomplished by water (c_p=4208 J/kg K) available at 90°C at a flow rate of 40 ton/h. The heat transfer coefficient on the milk side is 8000 W/m^2 K, and on the hot water side is 7100 W/m^2 K. If the thickness of the plates is 4 mm, calculate

(a) Area of plates for a counter-flow configuration.

(b) How many more identical plates will be needed to maintain the same heat transfer rate when fouling factors of 0.0002 m^2 K/W on the milk side and 0.0001 m^2 K/W on the water side are considered?

Solution:

Assumptions: Steady-state conditions, negligible heat loss, negligible changes in the kinetic and potential energies, constant fluid properties.

(a) The rate of heat from milk to water:

$$\dot{m}_{\text{milk}} = 20\ \text{ton/h} = 5.6\,\text{kg/s}$$

$$\dot{Q} = \dot{m}_c.C_{p,c}.(T_{c,\text{out}} - T_{c,\text{in}})$$

$$\dot{Q} = 5.6\,\text{kg/s} \times 3.850\,\text{kJ/kg K} \times (72-45)\,\text{K} = 582.1\,\text{kW}$$

The outlet temperature of water:

$$\dot{m}_{\text{water}} = 38\,\text{ton/h} = 10.6\,\text{kg/s}$$

$$\dot{Q} = \dot{m}_h.C_{p,h}.(T_{h,\text{in}} - T_{h,\text{out}})$$

$$582.1\,\text{kW} = 10.6\,\text{kg/s} \times 4.208\,\text{kJ/kg K} \times (90-T)^{\circ}\text{C}$$

$$T_{h,\text{out}} = 77^{\circ}\text{C}.$$

The logarithmic mean temperature difference for counterflow configuration:

$$\Delta T_{LM} = \frac{\Delta T_1 - \Delta T_2}{\ln\frac{\Delta T_1}{\Delta T_2}} = \frac{18-32}{\ln\frac{18}{32}} = 24.1^{\circ}\text{C}$$

And the overall heat transfer coefficient:

$$U = \frac{1}{\frac{1}{h_{\text{water}}} + \frac{1}{h_{\text{milk}}} + \frac{\Delta x}{k_{\text{steel}}}} = \frac{1}{\frac{1}{7100} + \frac{1}{8000} + \frac{0.004}{16.2}} = 1950\ \text{W/m}^2$$

Area of heat transfer is then calculated by:

$$\dot{Q} = UA_s\Delta T_{lm}, \text{where } A_s = A_p \times N\ (N: \text{number of plates})$$

$$582.1\ \text{kW} = \left(1.950\ \text{kW/m}^2\text{K}\right)\left(A_s\ \text{m}^2\right)(24.1)^{\circ}\text{C}$$

$$A_s = 12.4\ \text{m}^2$$

Finally, the area of each plate is calculated as:

$A_s = A_p \times N$ where $N = 40 \Rightarrow A_p = 0.3\ \text{m}^2$

(b) Now, we need to update U:

$$U = \frac{1}{\frac{1}{h_{\text{water}}} + \frac{1}{h_{\text{milk}}} + \frac{\Delta x}{k_{\text{steel}}} + R_{f,\text{water}} + R_{f,\text{milk}}}$$

$$U = \frac{1}{\frac{1}{7100} + \frac{1}{8000} + \frac{0.004}{16.2} + 0.0001 + 0.0002} = 1230$$

Heat transfer area is calculated from:

$$\dot{Q} = UA_s\Delta T_{lm}$$

$$582.1 = (1.230)(A_s)(24.1) \Rightarrow A_s = 19.6\ \text{m}^2$$

The number plates is found by:

$$N = \frac{A_s}{A_p} = \frac{19.6}{0.3} \cong 65\ \text{plates}$$

As a result, 25 plates are needed.

3.4 Selection of heat exchangers

Heat exchangers are used in many heating or cooling applications, e.g., in boilers, furnaces, refrigerators, etc. For a given application, there may be only one suitable equipment type. Frequently heat exchangers are purchased as off-the-shelf item in process industries and thus the required equipment is selected on the basis of cost and a number of specifications furnished by the manufacturers. Generally, suitability of the heat exchanger to application is evaluated and this step is of critical importance. A typical heat exchanger selection process or selection request requires the user to submit a number of different specifications to the manufacturer in order to design the product or when sourcing a standard heat exchanger, these specifications should be understood in order to properly select the right design for use:

Heat transfer rate: Also known as the heat load or heat transfer requirement, is a measure of the heat transferred in the heat exchanger per unit time. This is the most fundamental specification for describing heat exchanger performance. The steady-state heat transfer Eq. (3.10) can be used to calculate the heat load.

Type of duty: Examples of duties are heating/cooling of liquids, gases or slurries, boiling or condensation. Certain equipment are less suitable than others for some of these duties. For example, plate heat exchangers are not good for gases.

Operating temperatures/pressures: The operating temperatures or pressures of the exchanger are usually set by process conditions. However, in certain cases, the user or designer will specify the operating temperatures and pressures including startup, shutdown, normal and process upset conditions. In fact, these limits are also dictated by material considerations, safety, and economics. On a heat exchanger design/selection sheet, the design pressure or temperature indicates the highest pressure/temperature the system will reach at a normal operation. Often manufacturers will request both inlet and outlet temperatures for each side of the heat exchanger.

Fouling factors: Liquid foods such as milk, fruit juices, oil, etc., have a tendency to foul the heat transfer surfaces. For instance, particulate fouling occurs in low velocity zones around the baffles in shell-and-tube heat exchangers. Accounting for the effect of fouling on pressure drop requires an estimate of the fouling layer thickness. Arbitrarily increasing the heat transfer area by 15%–50% would also be a safety approach to fouling.

Safety: Safety is the most important factor to be considered in heat exchanger selection. On nonhazardous duties it may be acceptable to have some risk of equipment failure. With a plate heat exchanger, there is always a possibility of gasket failure, although with correct installation and maintenance this should be low.

Repair: Corrosion or fatigue may limit the working life of an exchanger. Some designs of exchanger can be repaired in an acceptable and economic fashion. The likelihood and consequences of failure should be assessed at the selection stage.

Dimensions and weight: Installed cost of an item is greater than the item cost due to the need for pipework, instruments, foundations, lagging, etc. Many of these are related to dimensions and weight of the item. Considerable savings can be made by the use of a compact design.

Cost: Cost is the total installed cost mention above. There may be extra running costs such as power requirement for pumping and maintenance. Up-to-date information on costs can be obtained by consulting with manufacturer.

Pressure drop: The pressure drop through an exchanger result from three losses: the frictional loss due to flow, the losses due to changes in direction of flow and losses due to expansion and contraction into and out of nozzles and tubes. In gas-flow heat exchangers, the pressure drop on both fluid sides is as important as the thermal performance. However, it is less critical in liquid heat exchangers. Increasing pressure drop provides improved heat transfer, while it requires higher pumping power and cost.

Flow rate: It is a measure of the rate at which the fluids flow through the heat exchanger, commonly expressed in gallons per minute (gpm) for liquids and cubic feet per minute (cfm) for gases. Flow is limited by the size of the pump and the maximum allowable pressure drop in the system. To simplify, the typical fluid velocities range at 0.7–7 m/s for liquids and at 3.30 m/s for gases. Minimizing the pressure drop and mass flow rate can reduce the operating costs, but will increase the size of the equipment.

Materials of construction: Materials are selected on the basis of corrosion resistance, mechanical strength, thermal conductivity, and cost (Tables 3.1 and 3.2). The engineer making the materials selection must also know all the aspects involved in the construction, operation and maintenance of the heat exchanger.

The minimum cost strategy would imply use of less expensive materials and rectifying the problems as they show up. Optimum material properties include high thermal conductivity for efficient heat transfer, high strength to withstand high operating pressures, high corrosion resistance to corrosive fluids over time, and low cost. Common materials for heat exchangers include copper, brass, aluminum, and stainless steel (Table 3.2).

The most widely used heat exchangers in process industries are air-cooled heat exchangers, plate heat exchangers, and shell-and-tube heat exchangers. The users could be overwhelmed by the large variety of these heat exchangers (Larowski & Taylor, 1983). Detailed information on these heat exchangers is also elsewhere (GBH Enterprises Ltd, n.d.). In food processing, the major heat exchangers are shell-and-tube heat exchangers and plate heat exchangers. The most typical type is clearly the shell-and-tube heat exchangers. An example of practical hints on selection of shell-and-tube heat exchangers, is given here in Tables 3.3 and 3.4.

3.5 Design variables of heat exchangers

Selection and design are inter-related tasks. Most of the factors discussed for selection also apply to the design stage of the equipment. An illustration of basic steps of a heat exchanger design is given in Fig. 3.7. Different applications require different design

Table 3.1 Special considerations which are taken into account during materials selection.

Physical properties	High heat transfer coefficient, high *k*-value, low thermal expansion constant, compatible with tube-sheet, tube support, and shell
Mechanical properties	Good tensile and creep properties (high creep rupture strength and adequate creep ductibility), good fatigue, corrosion fatigue and creep fatigue behavior, high fracture toughness
Corrosion resistance	Low corrosion rate, resistance to corrosion from leaks and failure, tolerance to fluids mixing
Manufacture	Ease of fabrication
Operating experience	Deep knowledge from experience, lessons from failures

Table 3.2 Common materials used for construction of heat exchangers.

	Advantages	Disadvantages
Copper	Best performance of heat transfer, high thermal conductivity	Poor corrosion resistance
Brass (copper +zinc)	Stronger than copper, high conductivity	Low corrosion resistance
Aluminum	Light-weight, high heat transfer efficiency, low cost	Alloyed with bronze and nickel
Stainless steel	Excellent corrosion resistance, lower heat transfer efficiency	Expensive

Table 3.3 Description of shell-and-tube heat exchangers with respect to shell type.

Shell type	Description	Shell type	Description
E	One-pass shell Counter-or-co-current flow	J	Divided flow Full support plate under center nozzle
F	Two-pass shell Longitudinal baffle	K	Kettle reboiler or vaporizer Liquid disengages from vapor in dome
G	Split flow Longitudinal baffle Full support plate under nozzle	X	Crossflow Multiple nozzles typical for flow distribution
H	Double split flow Two longitudinal baffles Full support plate under nozzles and at shell midpoint		

Table 3.4 Determination of number of tube passes based on shell ID.

Shell ID		Recommended maximum number of tube passes
in.	mm	
<10	<254	4
10–20	254–508	6
20–30	508–762	8
30–40	762–1016	10
40–50	1016–1270	12
50–60	1270–1524	14

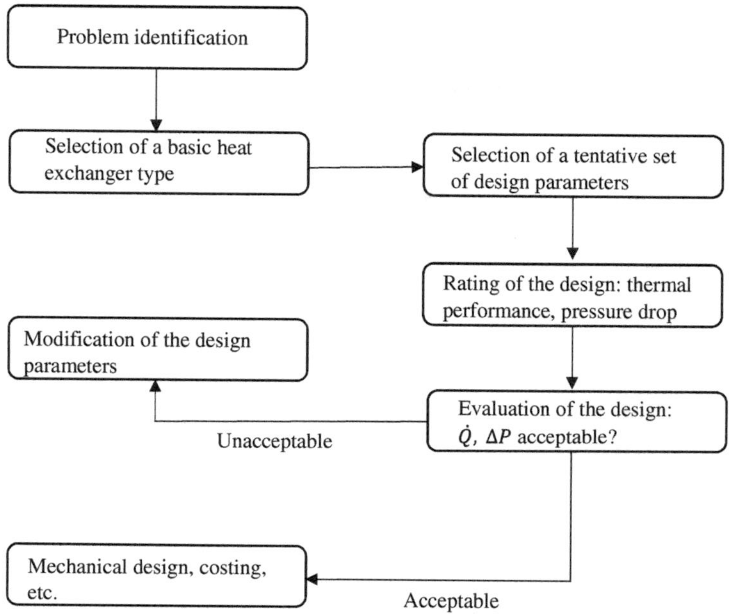

Fig. 3.7 Basic logic for a heat exchanger design.
Adapted from Özden, E. (2007). Detailed design of shell-and-tube heat exchangers using CFD. A Master Thesis in Department of Mechanical Engineering, Middle East Technical University, Ankara, Turkey, pp. 4, 14, 17.

criteria. The design is based on maximizing performance while minimizing cost, or could be based on minimizing pressure drop, size, and weight.

The heat transfer requirement, cost, size, and pressure drop characteristics are simultaneously considered in the design of the heat exchangers. Pumping the fluids at high velocities leads to high values of overall heat transfer coefficient but causes a large pressure drop through the heat exchanger and higher pumping costs. If area increases, the overall heat transfer coefficient and pressure drop decrease, however, the size and hence the cost may be a limitation.

The pumping power for each fluid stream is calculated by Eq. (3.23):

$$\text{Pumping power} = (\dot{m}\ \text{kg/s})\left(\frac{\Delta P}{\rho}\ \frac{\text{N/m}^2}{\text{kg/m}^3}\right) = \frac{\dot{m}\Delta P}{\rho}(W) \tag{3.23}$$

where $\dot{m}$ is the mass flow rate, ΔP is the pressure drop, and ρ is the density.

The pressure drop for a typical straight flow in a pipe is given by:

$$\Delta P = f\left(\frac{L}{D_h}\right)\frac{\rho {v_{\text{ave}}}^2}{2} \tag{3.24}$$

where L is the length of the pipe, D_h is the hydraulic pipe diameter, v_{ave} is the average fluid velocity, and f is the Darcy-Weisbach friction factor.

For tubes with any cross-sectional shape or a bank tubes within the heat exchanger, the following equation is recommended for determination of the pressure drop:

$$P_1 - P_2 = \frac{G^2}{\alpha}\left[\frac{1}{\rho_2} - \frac{1}{\rho_1}\right] + 4f\frac{L}{D}\frac{G^2}{2\rho_m} + g\rho_m(z_2 - z_1) \quad (3.25)$$

where the subscripts 1 and 2 denote inlet and outlet points, $G = \frac{\dot{m}}{A_c} = \rho v$, K_c and K_e are loss coefficients for contraction and expansion, respectively.

There are two major different designs in food processing; shell and tube, and plate heat exchangers and three types of flow arrangements; counter flow, parallel flow, and cross flow. Among all, the counter flow is the most efficient design, leading to the largest log-mean temperature difference and the greatest transfer of heat. Shell and tube heat exchanger can be built with bare tubes or finned tubes and generally designed according to the standards of the Tubular Exchanger Manufacturers Association (TEMA).

Example 3.3

A counter-flow shell-and tube heat exchanger is to be used to heat air from 20 to 80°C by condensation of steam at 100°C in the shell as schematically drawn below. The air flows at 3 kg/s and has a mass velocity of 10 kg/s m^2. The internal diameter of the steel tubes is 6.35 cm and length is 8.5 m. Thus, what is the air-side pressure drop $P_o - P_3$?

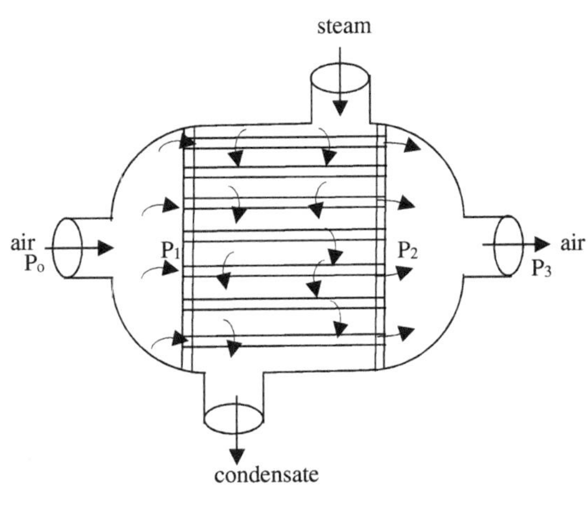

Solution:

The mean temperature of air is

$$T_b = \frac{20 + 80}{2} = 50^{\circ}\text{C}$$

At $T_b = 50°C$, the viscosity of air is $\mu = 1.95 \times 10^{-5}$ kg/m s. Then, the Reynolds number is:

$$Re = \frac{Dv\rho}{\mu} = \frac{DG}{\mu} = \frac{(10)(0.0635)}{1.95x10^{-5}} = 32{,}564 \text{ (Turbulent)}$$

For a turbulent flow through smooth pipe, the friction factor (f) is:

$$f = \frac{0.046}{Re^{0.2}} = 0.0057$$

The pressure drop $P_1 - P_2$:

$$P_1 - P_2 = \frac{G^2}{\alpha}\left[\frac{1}{\rho_2} - \frac{1}{\rho_1}\right] + 4f\frac{L}{D}\frac{G^2}{2\rho_m} + g\rho_m(z_2 - z_1)$$

For a horizontal tube, $\alpha = 1$ and $z_2 = z_1$, and $\rho_1 = 1.207\,\text{kg/m}^3$, $\rho_2 = 1.002\,\text{kg/m}^3$, and $\rho_m = \frac{1.207 + 1.002}{2} = 1.1045\ \text{kg/m}^3$. Then

$$P_1 - P_2 = \frac{10^2}{1}\left[\frac{1}{1.002} - \frac{1}{1.207}\right] + 4(0.0057)\frac{8.5}{0.0635}\frac{10^2}{2(1.1045)}$$

$$P_1 - P_2 = 155.1\ \text{N/m}^2 \tag{3.26}$$

Assuming the area of the header to be twice as large as the area of tubes, we can compute the corresponding velocities:

$$v_1 = \frac{10}{1.207} = 8.3\ \text{m/s}$$

$$v_2 = \frac{10}{1.002} = 10\ \text{m/s}$$

$$v_0 = \frac{v_1}{2} = 4.15\ \text{m/s}$$

$$v_3 = \frac{v_2}{2} = 5\ \text{m/s}$$

and also $K_e = 0.21$ and $K_c = 0.31$. Then applying the equations:

$$\frac{P_0 - P_1}{\rho_1} + \frac{v_0^2 - v_1^2}{2} = K_c\frac{v_1^2}{2}$$

$$\frac{P_2 - P_3}{\rho_2} + \frac{v_2^2 - v_3^2}{2} = K_e\frac{v_2^2}{2}$$

Inserting values and solving for pressure difference terms:

$$P_0 - P_1 = 44.07\ \text{N/m}^2 \tag{3.27}$$

$$P_2 - P_3 = -27.05\ \text{N/m}^2 \tag{3.28}$$

And finally summing Eqs. (3.26)–(3.28), we obtain:

$$P_0 - P_3 = (P_0 - P_1) + (P_1 - P_2) + (P_2 - P_3)$$

$$P_0 - P_3 = 44.07 + 155.1 - 27.05$$

$$P_0 - P_3 = 172.1\ \mathrm{N/m^2}$$

3.6 Analysis of heat exchangers

3.6.1 Simple approach (U constant and outlet fluid temperatures are known/calculable)

Basic analysis of heat exchangers starts with taking the overall heat transfer coefficient U constant, a reasonable case for compact single-phase heat exchangers, to deal with rate of heat transfer, heating area, or the mean temperature difference, according to Eqs. (3.1), (3.21), and (3.22):

$$\dot{Q} = UA\Delta T_m = UA(F\Delta T_{LM}) \tag{3.29}$$

where $F = 1$ for double pipe heat exchangers only, i.e., a limited case to single-pass parallel and counter flow configurations. The value of F needs to be determined for multi-pass and cross flows. This approach is useful when the inlet and outlet temperatures of hot and cold fluids are known or when they can be calculated using Eq. (3.10).

In combination with (3.10), Eq. (3.29) also serve for selection of heat exchangers by providing the heat load or enabling to find out the temperature change in the fluid streams or area of a specified heat exchanger.

A second type of problem is to determine the heat load and outlet fluid temperatures when the type and size (e.g., area) of the heat exchanger are specified. In this case, an iterative approach is employed as the followings:

1. Assume the outlet temperatures
2. Calculate $\dot{Q}$ using Eq. (3.29)
3. Calculate the outlet temperatures using Eq. (3.10)
4. Compare the results with the assumed values and repeat 1–3 until convergence.

Although the recommended iterative method is clear, it is time-consuming and even could be tedious if the assumed values are far to the point. For such cases, a more practical method known as **the effectiveness-NTU method** is the solution to the second type of problem.

3.6.2 The effectiveness-NTU method

The effectiveness of heat exchangers is defined as:

$$\varepsilon = \frac{\text{actual rate of heat transfer}}{\text{theoratical maximum rate of heat transfer}} = \frac{\dot{Q}}{\dot{Q}_{\max}} \tag{3.30}$$

Substituting the relevant expression from Eqs. (3.10) and (3.13), we have:

$$\varepsilon = \frac{C_H(T_{H,\text{in}} - T_{H,\text{out}})}{C_{\min}(T_{H,\text{in}} - T_{C,\text{in}})} = \frac{C_C(T_{C,\text{out}} - T_{C,\text{in}})}{C_{\min}(T_{H,\text{in}} - T_{C,\text{in}})} \tag{3.31}$$

where maximum temperature difference is always the difference between inlet temperatures of the hot and cold fluids, $T_{H,\text{in}} - T_{C,\text{in}}$ for parallel and counter flows.

Kays and London (1964) first developed the effectiveness method in detail and provided effectiveness relations for various heat exchanger flow arrangements in tabular and graphical forms, which have been cited by others (Cengel, 2003; Holman, 2010), and partially repeated here (Table 3.5).

The graphical and tabular relations of the effectiveness method contribute to quick estimation of size of the selected heat exchanger. Thus, a comparison between different types of the heat exchangers can be made and an appropriate one is selected. A rearranged form of Table 3.5 to solve for the NTU is available for easier calculation (see Table 3.6).

Table 3.5 Effectiveness relations for heat exchangers (Cengel, 2003).

Type	Effectiveness (ε) relation
Double pipe Parallel flow	$\varepsilon = \frac{1-\exp[-\text{NTU}(1+c)]}{1+c}$
Counter flow	$\varepsilon = \frac{1-\exp[-\text{NTU}(1-c)]}{1-c\exp[-\text{NTU}(1-c)]}$
Shell and tube One-shell pass 2, 4, …, tube passes	$\varepsilon = 2\left\{1 + c + \sqrt{1+c^2}\frac{1+\exp\left[-\text{NTU}\sqrt{1+c^2}\right]}{1-\exp\left[-\text{NTU}\sqrt{1+c^2}\right]}\right\}^{-1}$
Cross-flow (single pass) Both fluids unmixed	$\varepsilon = 1 - \exp\left\{\frac{\text{NTU}^{0.22}}{c}\left[\exp\left(-c\text{NTU}^{0.78}\right) - 1\right]\right\}$
$C_{\max}$ mixed, $C_{\min}$ unmixed	$\varepsilon = \frac{1}{c}(1 - \exp\{1 - c[1 - \exp(-\text{NTU})]\})$
$C_{\min}$ mixed, $C_{\max}$ unmixed	$\varepsilon = 1 - \exp\{-\frac{1}{c}[1 - \exp(-c\text{NTU})]\}$
All heat exchangers with $c = 0$	$\varepsilon = 1 - \exp(-\text{NTU})$

NTU = $UA/C_{\min}$ (number of transfer units); $c = C_{\min}/C_{\max} = (\dot{m}C_p)_{\min}/(\dot{m}C_p)_{\max}$ (capacity ratio).

Table 3.6 NTU relations for heat exchangers (Cengel, 2003).

Type	Effectiveness (ε) relation
Double pipe Parallel flow	$\text{NTU} = -\frac{\ln[1-\varepsilon(1+c)]}{1+c}$
Counter flow	$\text{NTU} = \frac{1}{c-1}\ln\left(\frac{\varepsilon-1}{\varepsilon c-1}\right)$
Shell and tube One-shell pass 2,4, …, tube passes	$\text{NTU} = -\frac{1}{\sqrt{1+c^2}}\ln\left(\frac{\frac{2}{\varepsilon}-1-c-\sqrt{1+c^2}}{\frac{2}{\varepsilon}-1-c+\sqrt{1+c^2}}\right)$
Cross-flow (single pass) C_{max} mixed, C_{min} unmixed	$\text{NTU} = -\ln\left[1+\frac{\ln(1-\varepsilon c)}{c}\right]$
C_{min} mixed, C_{max} unmixed	$\text{NTU} = -\frac{\ln[c\ln(1-\varepsilon)+1]}{c}$
All heat exchangers with $c=0$	$\text{NTU} = -\ln(1-\varepsilon)$

Example 3.4

A shell and tube heat exchanger with one shell and eight tube passes is used to heat 0.4 kg/s of ethanol ($c_p = 2570$ J/kgK) at 20°C. The heating is accomplished by 0.8 kg/s of water ($c_p = 4206$ J/kg K) available at 90°C. The heat exchanger consists of thin-walled tubes of 6 m length and 1.1 cm in diameter. If the overall heat transfer coefficient is 800 W/m^2 K, calculate the outlet temperatures of the fluids by using the effectiveness-NTU method.

Solution:

Assumptions: Steady-state conditions, negligible heat loss, tubes are thin-walled, so their thickness is negligible, changes in the kinetic and potential energies are negligible, and fluids have constant properties.

We start with calculating thermal capacitance of hot and cold fluids:

$$C_c = \dot{m}_c \times c_{p,c} = (0.4)(2570) = 1028\ \text{W/K}$$

$$C_h = \dot{m}_h \times c_{p,h} = (0.8)(4206) = 3365\ \text{W/K}$$

Cold fluid has the smaller heat capacity rate. This leads to the maximum possible heat transfer rate:

$$\dot{Q}_{\max} = C_{\min}\left(T_{h,\text{in}} - T_{c,\text{in}}\right) = 1028 \times (90 - 20)$$

$$\dot{Q}_{\max} = 71.9\ \text{kW}$$

From Table 3.6, the NTU relation for the shell and tube heat exchanger:

$$\text{NTU} = -\frac{1}{\sqrt{1+c^2}}\ln\left(\frac{\frac{2}{\varepsilon_1} - 1 - c - \sqrt{1+c^2}}{\frac{2}{\varepsilon_1} - 1 - c + \sqrt{1+c^2}}\right)$$

where c is the capacity ratio, and defined as:

$$c = \frac{C_{\min}}{C_{\max}} = \frac{1028}{3365} = 0.31$$

And

$$A_s = n(\pi DL) = 8(\pi)(0.011)(6) = 1.66\ \text{m}^2$$

Inserting into NTU formula:

$$\text{NTU} = \frac{UA_s}{C_{\min}} = \frac{(800)(1.66)}{1028} = 1.30$$

Then we calculate ε using NTU and c values:

$$\varepsilon = 0.65$$

The actual heat transfer rate:

$$\dot{Q} = \varepsilon \times \dot{Q}_{\max}$$

$$\dot{Q} = 46.7\ \text{kW}$$

Finally, the outlet temperatures of the fluids can be calculated:
The outlet temperature of water:

$$\dot{Q} = C_h \cdot (T_{Hi} - T_{Ho})$$

$$46{,}700 = 3365(90 - T_{Ho})$$

$$T_{Ho} = 76.1^{\circ}\text{C}$$

Similarly, the outlet temperature of ethanol:

$$\dot{Q} = C_c \cdot (T_{Co} - T_{Ci})$$

$$46{,}700 = 1028(T_{Co} - 20)$$

$$T_{Co} = 65.4^{\circ}\text{C}$$

3.6.3 Steady state energy balance (U is constant)

For any number of tubes in parallel with any of shape, we can use the area A as the independent variable rather than length L and make energy balance for the control volume shown in Fig. 3.8:

$$\partial \dot{q} = U \partial A (T_H - T_C) \tag{3.32}$$

Rewriting the rate of heat transfer for cold fluid as:

$$\partial \dot{q} = C_C \frac{dT_C}{dA} \partial A \tag{3.33}$$

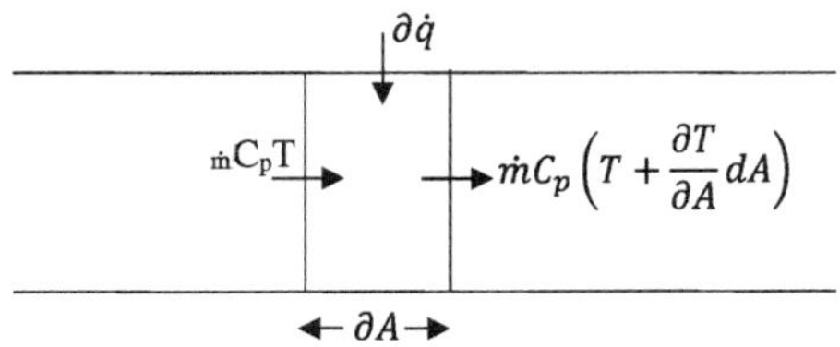

Fig. 3.8 Control volume for heat exchanger.

where $C_C = \dot{m}_c C_{p,c}$.

Combining the two equations:

$$C_C \frac{dT_C}{dA} \partial A = U \partial A (T_H - T_C) \tag{3.34}$$

Rearranging and integrating for T_c: in $\rightarrow$ out and considering T_H constant to mimic a phase change condition:

$$\int_{T_{c,i}}^{T_{c,o}} \frac{dT_c}{T_H - T_C} = \int_0^A \frac{UdA}{C_C} \tag{3.35}$$

$$-\ln\left(\frac{T_H - T_{C,o}}{T_H - T_{C,i}}\right) = \frac{UA}{C_C} \tag{3.36}$$

Taking $C_C = C_{\min}$:

$$-\ln\left(\frac{T_H - T_{C,o}}{T_H - T_{C,i}}\right) = \mathrm{NTU}_{\max} \tag{3.37}$$

Or

$$\frac{T_H - T_{C,o}}{T_H - T_{C,i}} = e^{-\mathrm{NTU}_{\max}} \tag{3.38}$$

Finally rearranging to write the effectiveness term:

$$\frac{T_{C,o} - T_{C,i}}{T_H - T_{C,i}} = 1 - e^{-\mathrm{NTU}_{\max}} \tag{3.39}$$

and finally,

$$\varepsilon = 1 - e^{-\mathrm{NTU}_{\max}} \tag{3.40}$$

3.6.4 Variable U approach

The convective heat transfer coefficients, h_i and h_o, depend on fluid properties such as velocity, density, viscosity, and temperature. Hence, the overall heat transfer coefficient, U, varies along the heat exchangers, particularly when the aforementioned properties are temperature-dependent. To illustrate this, let us consider the double-pipe heat exchanger case with parallel flow (Holman, 2010). The heat exchanger is divided into segments of surface are ΔA_i and the corresponding hot and cold temperature points are called T_{Hi} and T_{Ci}, respectively. As a result, the overall heat transfer coefficient U becomes dependent of these temperatures:

$$U_i = U_i(T_{Hi}, T_{Ci}) \tag{3.41}$$

The rate of heat transfer for ΔA_i is then:

$$\Delta \dot{Q}_i = -\left(\dot{m}_H C_{p,H}\right)_i (T_{H,i+1} - T_{H,i}) = \left(\dot{m}_C C_{p,C}\right)_i (T_{C,i+1} - T_{C,i}) \tag{3.42}$$

And also

$$\Delta \dot{Q}_i = U_i \Delta A_i (T_H - T_C)_i \tag{3.43}$$

Noting $(dT_H - dT_C)_i = d(T_H - T_C)_i = -d\dot{Q}\left[\frac{1}{\dot{m}_H C_{pH}} + \frac{1}{\dot{m}_C C_{pC}}\right]$ and $d\dot{Q} = U_i (T_H - T_C)_i \Delta A_i$.

We obtain:

$$\frac{d(T_H - T_C)_i}{(T_H - T_C)_i} = -U_i\left[\frac{1}{\dot{m}_H C_{pH}} + \frac{1}{\dot{m}_C C_{pC}}\right]\Delta A_i \tag{3.44}$$

or the finite difference equation:

$$\frac{(T_H - T_C)_{i+1} - (T_H - T_C)_i}{(T_H - T_C)_i} = -U_i\left[\frac{1}{\dot{m}_H C_{pH}} + \frac{1}{\dot{m}_C C_{pC}}\right]\Delta A_i = -K_i \Delta A_i \tag{3.45}$$

Finally rearranging:

$$\frac{(T_H - T_C)_{i+1}}{(T_H - T_C)_i} = 1 - K_i \Delta A_i \tag{3.46}$$

where K_i is dependent on T_H and T_C.

The total heat-transfer rate is then calculated from

$$\dot{Q}_{\text{total}} = \sum_{i=1}^{n} \Delta \dot{Q}_i \tag{3.47}$$

where n is the number of segments.

The numerical solution of Eq. (3.46) when the inlet temperatures and flows are given involves the following steps:

1. choose a convenient ΔA_i;
2. compute the value of U for the inlet conditions and initial ΔA increment;
3. calculate $\Delta \dot{Q}_i$;
4. calculate the values of T_H, T_C, and $T_H - T_C$ for the next increment;
5. repeat steps 2–4 until all increments are employed.

Example 3.5

An oil at 371.9 K is being cooled in a heat exchanger as flowing at a mass flow rate of 3630 kg/h. Water flows at 288.6 K and 1450 kg/h as coolant stream. Therefore, compute the overall heat transfer coefficient U along the tube for parallel flow and exit temperatures of oil and water if the heat transfer area is 3 m^2. Take $C_p = 4.187$ kJ/kg K for water and $C_p = 2.30$ kJ/kg K for oil.

Solution:

We choose $\Delta A = 0.5\ m^2$, and divide the total area into six segments to use the approach represented by Eq. (3.46) (see figure below). Once the value of K_i is known, the exit temperatures of water and oil for each segment can be calculated to finally compute U for each segment.

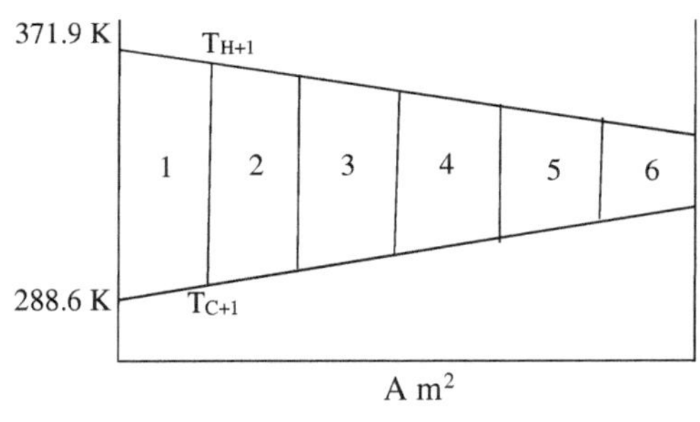

Considering slight changes in density of oil and water over the temperature change, we also take K_i constant (note that the mass flow rates of oil and water are in kg/s):

$$K_i = U_i \left[\frac{1}{\dot{m}_{\text{oil}} C_{p,\text{oil}}} + \frac{1}{\dot{m}_w C_{p,w}} \right] = U_i \left[\frac{1}{(1)(2300} + \frac{1}{(0.403)(4187)} \right]$$

$$K_i = 0.001 U_i.$$

Because no initial U is given or calculable, the values of U_i, T_{C+1}, and T_{H+1} through initial ΔA have been found by iterating Eqs. (3.42) and (3.43) using the inlet conditions:

$$U_i(0.5)(83.3) = (0.403)(4187)(T_{C+1} - 288.6) = (1)(2300)(371.9 - T_{H+1})$$

The results are:

$$U_i = 206\ \text{W/m}^2\text{K}, T_{H+1} = 368.2\ \text{K, and } T_{C+1} = 293.7\ \text{K}.$$

Knowing the inlet and outlet temperatures of both fluids in the first increment, the value of U_i can be recalculated to check on the accuracy of our iteration by the equation:

$$\Delta\dot{Q} = \dot{m}C_p\Delta T = U_i\Delta A_i\Delta T_{LM}$$

$$(1)(2300)(371.9 - 368.2) = U_i(0.5)(78.8)$$

$$U_i = 216\ \text{W/m}^2\ \text{K}$$

Or using the cold fluid:

$$(0.403)(4187)(293.7 - 288.6) = U_i(0.5)(78.8)$$

$$U_i = 218\ \text{W/m}^2\text{K}.$$

So, this makes an average value of 217 W/m^2 K. Inserting the known values into (3.41) and new value of U_i into (3.45),

$$(1)(2300)(371.9 - T_{H+1}) = (0.403)(4187)(T_{C+1} - 288.6)$$

$$T_{H+1} - T_{C+1} = 74.3$$

And solving simultaneously:

$$T_{H+1} = 368.1\ \text{K, and}\ T_{c+1} = 293.8\ \text{K}$$

which are almost the same as those found in the iteration, which make the iteration acceptable. The values of T_H, T_C, and $T_H - T_C$ for the next increments were calculated repeating similar work through Eqs. (3.42) and (3.46). The results for all increments are tabulated as:

Increment no.	T_{H+1}	T_{C+1}	ΔT_{LM}	U_i	U_{ave}
1	368.2	293.7	78.8	216/218	217
2	365.0	298.2	70.6	209/215	212
3	362.0	302.2	63.2	218/214	216
4	359.2	305.9	56.5	228/221	225
5	356.6	309.4	50.2	238/235	237
6	**354.2**	**312.6**	44.3	249/244	247

Finally, we conclude that the exit temperatures of oil and water are 354.2 and 312.6 K, respectively.

3.7 Mathematical modeling of heat exchangers

Modeling is a powerful tool that enables process control and optimization. A major challenge in modeling and analysis of heat exchangers is due to geometry and physical phenomena of the heat exchange between the two fluids. Different approaches have been used to model heat exchangers, which are described by hyperbolic partial differential equations (PDEs), or ordinary differential equations (ODEs). The PDEs are helpful for detailed description of internal behavior of the system while the energy balance equation leads to ODEs under uniform conditions, e.g., the same temperature within wall, and the resulting equations describe the dynamic behavior of a heat exchanger.

Generally, the models are obtained from fundamental laws of energy and fluid mechanics, which together describe thermal behavior of fluids involved in the heat exchanger. For a co-current (or parallel) flow of fluids around a plate of a plate heat exchanger (Bastida, Ugalde-Loo, Abeysekera, & Qadrdan, 2017), the energy balance equations defined as $\dot{Q}_{\text{stored}} = \dot{Q}_{\text{in}} - \dot{Q}_t$ for each stream are developed as:

$$m_h C_{p,h} \frac{dT_{h2}}{dt} = \dot{m}_h C_{p,h}(T_{h1} - T_{h2}) - UA(T_{h2} - T_{c2}) \tag{3.48}$$

$$m_c C_{p,c} \frac{dT_{c2}}{dt} = \dot{m}_c C_{p,c}(T_{c1} - T_{c2}) - UA(T_{c2} - T_{h2}) \tag{3.49}$$

where m_h and m_c are mass of hot and cold streams, subscripts 1 and 2 indicate inlet and outlet streams, and the total heat transferred between the plates is $\dot{Q}_t = UA\Delta T$.

The obtained equations are typical non-linear ODEs describing behavior of co-current heat exchanger. The overall heat transfer coefficient U depends on thermo-physical properties of fluids, operating mean velocities, and type of heat exchanger. The assumptions supporting the above equations are (1) negligible heat loss, (2) negligible changes in kinetic and potential energies, and (3) thermal properties are constant. This approach can easily be applied to double pipe heat exchangers once the plate wall is replaced with tube-wall.

Another approach is to consider a differential element in a heat exchanger, say plate heat exchanger. Based on the law of the energy conservation, the rate of heat transfer between process fluid or heating medium in a channel i and adjacent plates p_i and p_{i+1} (Georgiadis & Macchietto, 2000):

$$A_x p_i C_{p,i}\left(\frac{\partial T_i}{\partial t} + n_i v_x \frac{\partial T_i}{\partial x}\right) = U_i A_i\left(T_{pi-1} - T_i\right) + U_i A_i\left(T_{pi} - T_i\right) \tag{3.50}$$

and for plate i:

$$\rho_i C_{p,p} A_{xp} \frac{\partial T_{pi}}{\partial t} = U_i A_i\left(T_i - T_{pi}\right) + U_{i+1} A_i\left(T_{i+1} - T_{pi}\right) \tag{3.51}$$

where x is axial position, $T_i(x,t)$ is the temperature in channel i, v_x is average fluid velocity, n_j is $+1$ or -1 dependent on the direction of the flow. The advantage of such model is that it can be used for counter flow, parallel flow or multi-channel flows of several types of plate heat exchangers.

A simplified model in a steady state gives a linear system of first-order ODEs, which are comprised of energy balance for each channel and the corresponding boundary conditions under assumptions of plug flow, uniform flow distribution, no heat loss, and no heat exchange in the flow direction (Gut & Pinto, 2003; McKillop & Dunkley, 1960):

$$m_i C_{pi} \frac{dT_i}{dx} = n_i w\, \phi\, U_{i-1}(T_{i-1} - T_i) + n_i w\, \phi\, U_i (T_{i+1} - T_i) \tag{3.52}$$

and

$$\frac{1}{U_i} = \frac{1}{h_i} + \frac{1}{h_{i+1}} + \frac{x_p}{k} + R_h + R_c \tag{3.53}$$

where i shows channel number, w is plate width, ϕ is the area enlargement factor, x_p is plate thickness, k is thermal conductivity of the plate, and R_h and R_c are fouling factors of hot and cold fluids. The overall heat transfer coefficient U becomes constant when all fluid properties are independent of temperature.

In case heat loss is taken into account, for instance around a pipe heater, the energy equation is written as:

$$\dot{m}_1 C_{p1} \frac{dT_1}{dx} = h_1\, p\, (T_1 - T_w) \tag{3.54}$$

and

$$h_1(T_1 - T_w) = h_2(T_w - T_2) + \varepsilon\sigma\left(T_w{}^4 - T_2{}^4\right) \tag{3.55}$$

where 1 and 2 indicate the two fluids flowing inside and outside of the tube, respectively, T_w is the pipe wall temperature.

Determination of fouling or antifouling performance of plate heat exchangers is another interest of modeling concept. There are various modeling studies based on heat and mass transfer equations and fouling kinetics in literature including three dimensional computational fluid dynamics model (Choi et al., 2013), dynamic thermal model using PDE (Georgiadis & Macchietto, 2000; Narataruksa, Triratana, Pana-Suppamassadu, Heggs, & Tia, 2008), and thermal steady-state model (Gut & Pinto, 2003; Ribeiro Jr. & Andrade, 2002). Empirical models relying on steady-state prediction of temperatures are also developed to predict the dry mass of deposits in each channel (Delplace & Leuliet, 1995). Empirical models are efficient when theoretical approach is complicated due to presence of many factors (Wan Mokhtar, 2012). The mathematical models commonly used to predict the *Nusselt number* as a function of geometric and operating parameters are empirical but widely used to determine convective heat transfer coefficients and then the overall heat transfer coefficient is calculated.

3.7.1 Empirical modeling

Empirical modeling is a simple but widely used method that involves regression of experimental data to obtain some pre-decided quantities. One can work with empirical models to describe the relationship between the process inputs and outputs. The fundamental physical laws, e.g., the mass and heat transfer in heat exchangers, are not considered. Disadvantages of empirical models are such that they are only supported by experimental data, i.e., they are applicable only under a particular operating condition and they cannot explain the physical reasons for the behavior of the process. Therefore, empirical models are not able to predict beyond the range of the designed experiment. However, empirical models can be used for prediction of experimental data to achieve process optimization and control of a specified system. Regression equations, artificial neural network (ANN) and design of experiments (DoE) toolbox models such as response surface model are examples of empirical models. The ANN models are shown to build useful nonlinear equations for prediction of fast and accurate fouling with less computational load (Biyanto et al., 2015; Thirumarimurugan & Kannadasan, 2009), and prediction of overall heat transfer coefficient, effectiveness, and efficiencies of hot and cold side fluids (Thirumarimurugan & Kannadasan, 2009).

The empirical modeling approach can be used for quick simulation of processes in combination with theoretical approach to ensure process control. For instance, in pasteurization process, controlling of outlet product temperature from heating section of the heat exchanger is a key element to provide uniform product quality or to satisfy some industrial standard. An accurate dynamic process model to achieve such a process control strategy can be developed by fitting empirically the experimental data to a general transfer function model or to a more specific transfer function model, which can be derived by use of fundamental conservation laws.

Transfer functions are process models used in control system design and relate process inputs and outputs. Generally, first-order plus time delay (FOPTD) model is adequate to describe many dynamic industrial processes (see example below) and its structure is represented by the transfer function:

$$G(s) = \frac{Y(s)}{X(s)} = \frac{K_p\, e^{-\theta s}}{\tau s + 1} \tag{3.56}$$

where $Y(s)$ and $X(s)$ denote the input and output variables, K_p is the process gain, θ is the time delay, and τ is the time constant. The response of an output to a step change in an input is recorded to prepare curves such as process reaction curve to determine the parameters of the transfer function above.

Example 3.6

Coconut milk at 65°C and flowing at 1.5 m/s is to be heated furthermore with hot water in a plate heat exchanger. The transfer function model relating milk temperature to hot water temperature is described by Eq. (3.56):

$$G(s) = \frac{K_p e^{-\theta s}}{\tau s + 1}$$

Thus, predict milk temperature for the first 30 min using the transfer function model.

Solution:

In order to obtain the parameters of the process model, (i) the hot water temperature is changed from 70 to 80°C after 5 min and (ii) the unknowns K_p, θ, and τ are determined by the process reaction curve method (see Wan Mokhtar, 2012), that is:

$$G(s) = \frac{0.7e^{-1.17s}}{7.2s + 1}$$

The response equation after the described change in hot water temperature becomes:

$$T(t) = T(0) + 7\left[1 - e^{-\left(\frac{t-1.17}{7.2}\right)}\right]$$

where $T(0)$ is the initial milk temperature.

3.7.2 CFD modeling

Computational fluid dynamics (CFD) is the numerical modeling of systems that exhibit fluid flow, heat transfer and mass transfer. The basis for the analysis is the derivation of governing equations involving the conservation of the fluid flow, heat transfer and mass transfer over a given domain of interest. To numerically solve these equations, they are transformed into a finite algebraic form and then solved using a computer software following the specification of the physical properties and boundary conditions.

The solution of system equations is achieved using the finite volume or control volume technique. The control volume technique involves dividing the domain (geometry) of interest into a number of finite portions, referred to as a grid or mesh. The governing transport equations are integrated over the finite mesh volumes to obtain a set of algebraic equations, which can be solved numerically using the commercial package, for instance to examine the behavior of transport properties such as the velocity and temperature on a local scale.

A typical CFD software consists of three main parts: a pre-processor, a solver and a postprocessor. In preprocessor, input data of the model is processed and computational grid is created. In order to balance the accuracy of the simulations and CPU time, an optimum size of mesh need to be chosen. In solver execution, the discretized physical equations are solved according to boundary conditions of the model, until convergence criteria is met. In postprocessing the results are examined using data visualization tools.

Following the solution of the numerical equation, the CFD solutions are capable of returning a complete picture of the behavior of fluid flow and the other phenomenon under given conditions. CFD techniques are used in rating, and iteratively in sizing of heat exchangers. In CFD simulations, by modeling the geometry as accurately as possible, the flow structure and the temperature distribution inside the shell can be obtained. This detailed data can be used for calculating global parameters like heat transfer coefficient and pressure drop.

This numerical approach delivers detail that is unable to be obtained using the traditional experimental approach. Using CFD, together with experiments, may speed up the shell-and-tube heat exchanger design process and may improve the quality of the final design. Hence, CFD provides a relatively cheap technique that is completely nonintrusive and capable of detailing the interactions of parameters on a local scale.

3.7.2.1 CFD governing equations

The general transport equation for a property is shown by Eq. (3.57):

$$\frac{\partial \varphi}{\partial t} = \nabla.(\delta \nabla \varphi) - \nabla(\varphi.v) + \dot{\varphi}_G \tag{3.57}$$

where the property φ is either mass (ρ), momentum ($\rho.v$) or energy ($\rho.e$) and $\dot{\varphi}_G = -\Delta p + \rho g$. The governing transport Eq. (3.57) represents the laminar form of the transport equation.

Turbulent flows are characterized by fluctuating velocity and pressure fields. In practice, the turbulent flows contain small eddies with high frequency fluctuations. It is computationally intensive to solve the Navier-Stokes equation directly to determine such flow structures. To make it feasible, the governing equation is modified introducing additional variables by a method known as *time-averaging*. This step involves breaking the velocity into two components; one represents the mean value and the second is the representation of changes that are due to turbulent flow. Thus, the equation for turbulent flow becomes:

$$\frac{\partial}{\partial t}\rho v + v \cdot \nabla(\rho v) = \nabla \cdot \left[\mu\left(\nabla v + (\nabla v)^T\right)\right] + \nabla \cdot \left(-\rho \overline{u'u'}\right) - \nabla p \tag{3.58}$$

The obtained expression is also called Reynolds-averaged Navier-Stokes (RANS) equation. The RANS model is the most widely used model for industrial flows (see next example). A major difference between Eqs. (3.57) and (3.58) is the second term on the right hand side of Eq. (3.58) referred to as the Reynolds stresses.

According to the theory of Boussinesq in 1877 (Walker, 2005), the Reynolds stress is related to the rate of deformation but utilizes a value of the turbulent viscosity as the proportionality constant as:

$$\tau_{ij} = -\rho \overline{u'_i u'_j} = \mu_T \left(\frac{\partial u_i}{\partial x_j} + \frac{\partial u_j}{\partial x_i}\right) \tag{3.59}$$

The task now is to derive expressions for calculation of the turbulent viscosity, μ_T. The $k - \varepsilon$ model provides an algebraic definition:

$$\mu_T = \rho C_\mu \frac{k^2}{\varepsilon} \tag{3.60}$$

The values of k and ε are computed directly from the differential transport equations for the turbulence kinetic energy k and turbulence dissipation rate ε:

$$\frac{\partial \rho k}{\partial t} + \nabla \cdot (\rho v k) = \nabla \cdot \left[\left(\mu + \frac{\mu_T}{\sigma_k} \right) \nabla k \right] + P_k - \rho \varepsilon \tag{3.61}$$

$$\frac{\partial \rho \varepsilon}{\partial t} + \nabla \cdot (\rho v \varepsilon) = \nabla \cdot \left[\left(\mu + \frac{\mu_T}{\sigma_\varepsilon} \right) \nabla \varepsilon \right] + \frac{\varepsilon}{k} (C_{\varepsilon 1} P_k - C_{\varepsilon 2} \rho \varepsilon) \tag{3.62}$$

$$P_k = \mu_T \nabla v \cdot \left(\nabla v + \nabla v^T \right) - \frac{2}{3} \nabla \cdot \overline{v} (3 \mu_T \nabla \cdot v + \rho k) + P_{kb} \tag{3.63}$$

Example 3.7

A shell-and-tube heat exchanger with plain tube bundle was used for heating of water flowing at 2 kg/s and 25°C by hot oil at 100°C and at flow rates of 2–10 kg/s (Ahmed, Bhattacharjee, & Saha, 2017). The specifications of the heat exchanger are given in Table E.3.7.1. Thus:

(a) Simulate in 3D velocity profile and pressure using CFD method.
(b) What are the pressure drop and heat transfer coefficients at the shell side?

Solution:

Mass, momentum, energy, turbulent kinetic energy k, and turbulent dissipation rate ε equations (or so-called RANS equations) were used as governing equations and the heat exchanger system was simulated using finite element method (FEM). The entire domain discretized with unstructured mesh elements (mesh size: 995,441) is presented as in Fig. E.3.7.1:

Table E.3.7.1 Physical specifications of the shell-and-tube heat exchanger.

Tube	ID: $d_i = 11$ mm OD: $d_0 = 15$ mm Effective length: $l = 500$ mm Tube pitch: 25 mm Layout pattern: triangular (30 degree) Quantity: 37 Material: structural steel
Shell	ID: 200 mm OD:204 mm Material: structural steel
Baffle	Cut ratio: 25% Thickness: 2 mm Number: 4 Baffle pitch: 100 mm Material: structural steel

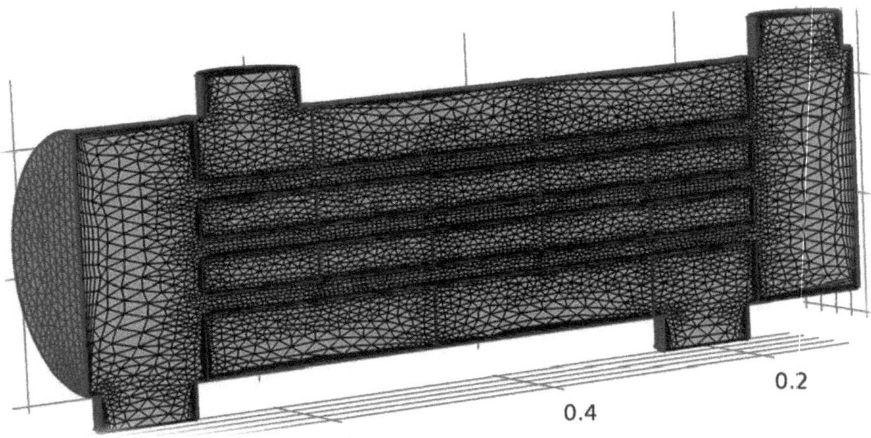

Fig. E.3.7.1 Meshed view of the shell-and-tube heat exchanger (Ahmed et al., 2017).

Fig. E.3.7.2 3D visualization of streamline of shell-side fluid flow for shell-and-tube heat exchanger (Ahmed et al., 2017).

The simulation results were expressed in terms of Nusselt number vs mass flow rate of oil and the results were reported to be compatible with experimental work of Peng (2005).

The velocity profile is depicted in Fig. E.3.7.2, where the variation of color of streamline represents change of pressure within the heat exchanger as well. In case of a turbulent flow, the flow streams bounce back due to baffled sections and creates dead zones at the corners (Fig. E.3.7.2). When the shell-side fluid flows parallel to the tubes between the baffled section, part of the flow is blocked by the annular fins and redirects in the radial direction within the shell as an improved cross flow (this alternative design not shown here).

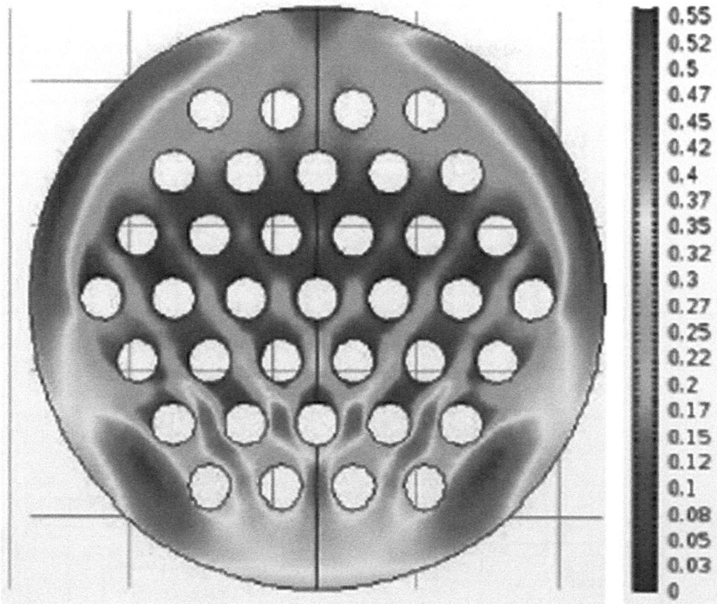

Fig. E.3.7.3 2D surface plots of mean velocity field at the mid-section of the shell-and-tube heat exchanger (Ahmed et al., 2017).

Quantitative estimation of the mean velocity is presented in Fig. E.3.7.3. The 2D plots prove a major portion of the shell-side fluid to pass through the outer peripheral region, where no significant heat transfer takes place due to low velocities.

The shell-side pressure drop and heat transfer coefficient (h_0) values were also plotted as a function of oil mass flow rate in the study of Ahmed et al. (2017). The corresponding pressure drop and h_0 values for 2 and 10 kg/s were read from the plots as 2–37 kPa and 220–560 W/m^2 K, respectively.

3.7.2.2 Algorithm-based simulation approach

Equations that are used to characterize flow (laminar vs turbulent, ΔP), heat transfer coefficients (h and U), heat transfer expressions (ΔT_{LM},Nu,$\dot{Q}$) all amount to a good number, like 13 equations as summarized by Maroulis and Saravacos (2003). One viable approach is to develop prepare an algorithm to reach easily the solution of interest. An example of such algorithm is developed in the study of Ribeiro Jr. and Andrade (2002) and presented in a modified form in Fig. 3.9. The algorithms are implemented by various computational programs such as FORTRAN, MATLAB, etc.

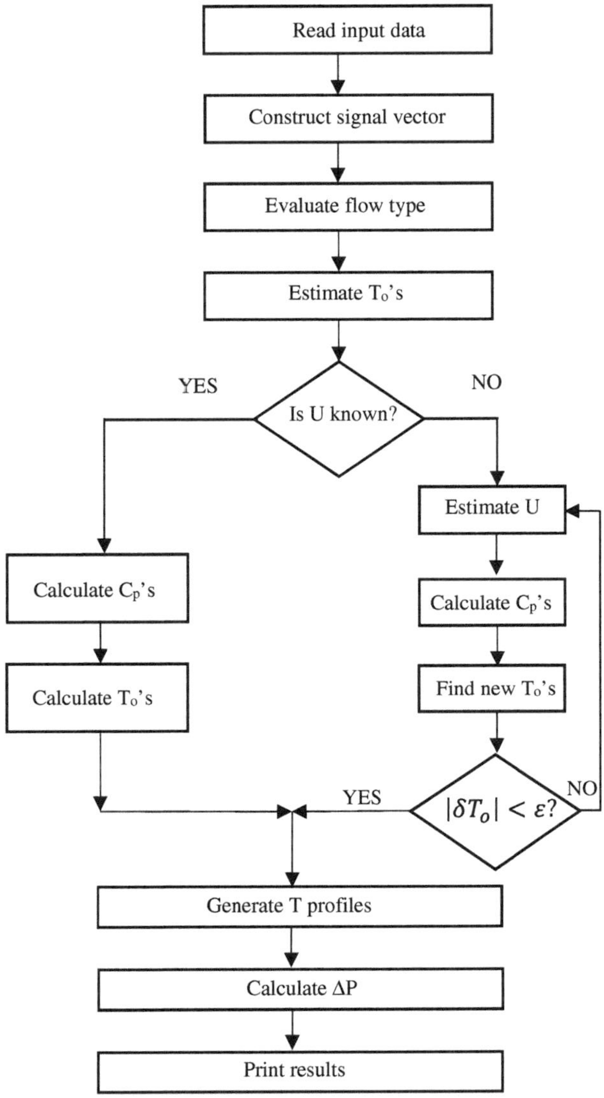

Fig. 3.9 Flow chart of an algorithm developed for simulation a plate heat exchanger. Adapted from Ribeiro Jr., C. P., & Andrade, M. H. C. (2002) An algorithm for steady state simulation of plate heat exchangers. *Journal of Food Engineering, 53*, 59–66.

Example 3.8

A pipe, which is 100 m long and 15 cm in diameter, is to provide hot water from an upstream process into a heat exchanger. The ambient temperature is 20°C and the hot water enters the pipe at 100°C. If the emissivity of the outside the pipe surface is 0.6:

(a) Calculate the wall and water temperatures and heat flux outside the pipe for flow rates of 0.4 and 1.2 kg/s.

(b) Comment on which mass flow rate is more appropriate to make use of the exit pipe stream in the heat exchanger.

Solution:

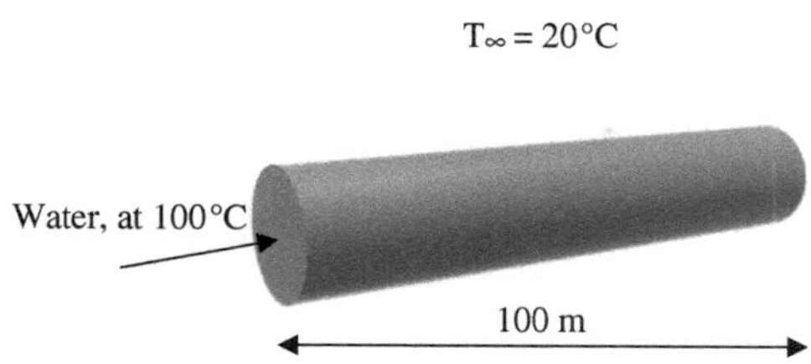

Assumptions: Steady-state conditions, negligible conduction resistance of the pipe wall, constant fluids properties.

Properties of water at 100°C:

$$\rho = 957.9\,\mathrm{kg/m^3};\ \mu = 0.284\times10^{-3}\,\mathrm{kg/m\,s};\ k = 0.682\,\mathrm{W/m\,K};\ c_p = 4219\,\mathrm{J/kg\,K};\ Pr = 1.75.$$

Applying energy balance to a control volume with length Δx and perimeter P:

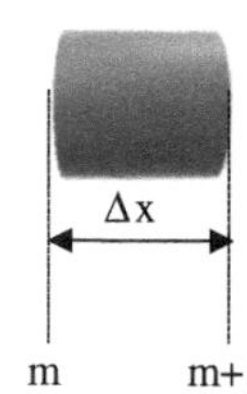

$$\dot{Q} = \dot{Q}_{in} = \dot{m}C_p\Delta T = hA\Delta T$$

$$\dot{m}_{water}C_p(T_{m,water} - T_{m+1,water}) = h_i P\Delta x(T_{m,water} - T_{m,wall}) \tag{3.64}$$

where h_i is the heat transfer coefficient inside the pipe. This type of flow is an example of forced convection in tubes. For this case, conditions of the Dittus-Boulter equation is valid when $0.7 \le Pr \le 160$ and $Re \ge 10{,}000$:

$$Re = \frac{D_h V\rho}{\mu} = \frac{4\dot{m}}{\pi D_h \mu}$$

where D_h is the hydraulic diameter and $D_h = D$ since the pipe is a circular tube.

$$Nu = \frac{h_i D}{k} = 0.023\,Re^{0.8}Pr^n$$

where $n = 0.3$ for cooling.

Thus, the heat transfer coefficients inside the pipe are calculated as 226.4 and 545.1 W/m^2 K for the mass flow rates of 0.5 and 2.0 kg/s, respectively.

Applying energy balance for the heat transfer through the duct wall:

$$\dot{q} = \dot{q}_{in} = \dot{q}_{\text{conv,out}} + \dot{q}_{\text{radiation,out}}$$

$$h_i(T_{m,\text{water}} - T_{m,\text{wall}}) = h_o(T_{m,\text{wall}} - T_{m,\infty}) + \sigma\varepsilon\left(T^4_{m,\text{wall}} - T^4_{m,\infty}\right) \tag{3.65}$$

where σ is the Stefan-Boltzmann constant, $\sigma = 5.67 \times 10^{-8}$ W/(m^2 K^4), and h_o is the heat transfer coefficient outside the pipe and can be calculated by the free convection equation:

$$h_o = 0.27\left(\frac{T_{m,\text{wall}} - T_{m,\infty}}{D}\right)^{1/4} \tag{3.66}$$

Inserting Eq. (3.66) into Eq. (3.65), we get:

$$h_i(T_{m,\text{water}} - T_{m,\text{wall}}) = \frac{0.27}{D^{1/4}}(T_{m,\text{wall}} - T_{m,\infty})^{5/4} + \sigma\varepsilon\left(T^4_{m,\text{wall}} - T^4_{m,\infty}\right) \tag{3.67}$$

And solving for $T_{m+1,\text{water}}$ from (3.64):

$$T_{m+1,\text{water}} = \left(1 - \frac{h_i P \Delta x}{\dot{m}_{\text{water}} C_p}\right) T_{m,\text{water}} + \left(\frac{h_i P \Delta x}{\dot{m}_{\text{water}} C_p}\right) T_{m,\text{wall}} \tag{3.68}$$

Now Eqs. (3.67) and (3.68) are solved by iteration. Once we calculate h_i and substitute 100°C as the initial temperature of water into Eq. (3.67), the temperature of the wall is calculated. This value is then substituted into Eq. (3.68), and the temperature of the water is calculated for one Δx increment.

Finally, the heat flux is calculated by:

$$\dot{q} = h_i(T_{m,\text{water}} - T_{m,\text{wall}})$$

The system equations of the problem was solved with MATLAB R2019b software (see codes as presented in Fig. E.3.8.3). For calculations, we chose $\Delta x = 1$ m. The temperature and heat flux changes along the pipe as plotted by MATLAB are shown in Figs. E.3.8.1 and E.3.8.2.

Comments:

When the final temperatures are compared, we see that water temperature decreases more sharply at 0.4 kg/s than that of 1.2 kg/s and reaches 360.7 K with low flow rate while the final temperature is 368.5 K when the flow is 1.2 kg/s. Therefore, we suggest that the stream at a higher flow (i.e., 1.2 kg/s) would be a better choice as an inlet to the heat exchanger. We also note that one should build a shorter pipe before the heat exchanger if the mass flow rate is restricted to 0.4 kg/s of water. A similar conclusion can be drawn from the heat flux plot that heat flux gets lower toward the end of the pipe for a mass flow rate of 0.4 kg/s.

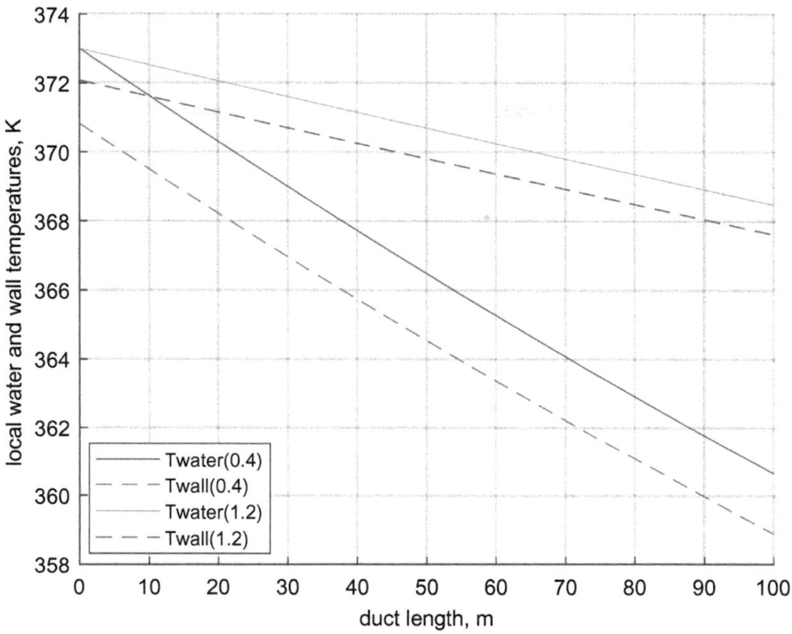

Fig. E.3.8.1 Temperature changes along the duct.

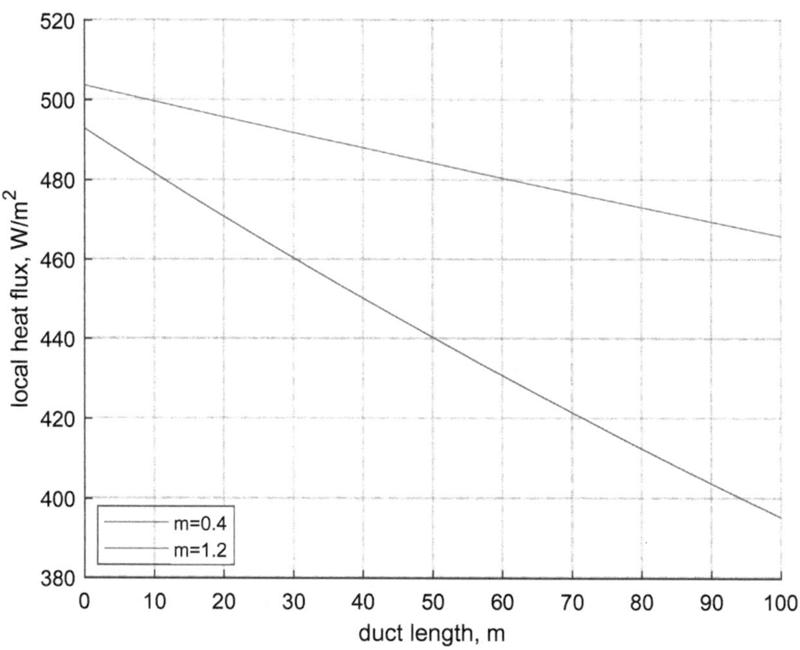

Fig. E.3.8.2 Heat flux changes along the duct.

```
clear all;
clc;
%%
%Constants and Variables
mass_flow_rates = [0.4, 1.2];          %mass flow rates in kg/s
length = 100;                          %duct length in m
dx = 1;                                %node spacing in m
d = 0.15;                              %diameter in m
Mu = 0.2838*10^-3;                     %viscosity in Pa*s
P = pi*d;                              %perimeter in m
k = 0.682;                            %thermal conductivity in W/m*K
Cp = 4219;                             %specific heat in J/kg*K


Tinf = 20 + 273;                      %environment temperature in K
e = 0.6;                               %emissivity
sigma = 5.67*10^-8;                  %Stefan-Boltzmann constant in
W/(m^2*K^4)

Re = 4*mass_flow_rates/(3.14*d*Mu);    %Reynolds number formula
Pr = 1.75 ;                            %Prandtl number

%Nusselt number and heat transfer coefficient calculations
Nu = 0.023*(Re.^0.8)*(Pr^0.3);         %Nusselt number
hi = Nu*k/d;                           %heat transfer coefficient inside
the pipe (W/m^2*K)

%%
%Temperature and Heat Flux Calculations
m = (length/dx)+1;                                %number of nodes
Twater_values = zeros(m,size(mass_flow_rates,2));
Twater_values(1,:) = 100+273;                    %water inlet temperature in
K
for k=1:size(mass_flow_rates,2)
    for i=1:m
        syms Twall
        eqn = hi(k)*(Twater_values(i,k)-Twall) == (0.27 *((Twall-
Tinf)^1.25))/(d^0.25) + sigma*e*(Twall^4-Tinf^4);
        Twall_values(i,k) = double(vpasolve(eqn, Twall));
        Twater_values(i+1,k) = (1-
(hi(k)*P*dx)/(mass_flow_rates(k)*Cp))*Twater_values(i,k)+(hi(k)*P*dx)/(ma
ss_flow_rates(k)*Cp)*Twall_values(i,k);
        heat_flux(i,k) = hi(k)*(Twater_values(i,k)-Twall_values(i,k));
    end
end
Twater_values(m+1,:) = [];
%%
%Plotting
x = 0:dx:length;
figure(1)
hold on
for k=1:size(mass_flow_rates,2)
    plot(x,heat_flux(:,k))
end
xlabel('duct length, m')
ylabel('local heat flux, W/m^2')
legend('m=0.4','m=1.2')
grid on
hold off

figure(2)
hold on
for k=1:size(mass_flow_rates,2)
    plot(x,Twater_values(:,k),x,Twall_values(:,k),'--')
end
xlabel('duct length, m')
ylabel('local water and wall temperatures, K')
legend('Twater(0.4)','Twall(0.4)','Twater(1.2)','Twall(1.2)')
grid on
hold off
```

Fig. E.3.8.3 MATLAB codes for Example 3.8.

Example 3.9

Cold water at 25°C flows through the smaller pipe in a double pipe heat exchanger to cool down hot water. This tube is 2 m long and made of stainless steel ($k=15$ W/m K) with internal and external diameters of 1.1 and 1.4 cm, respectively. If the heat transfer coefficient on the outer surface is 60 W/m^2 K, calculate the overall heat transfer coefficient based on the inner surface area of the tube U_i for cold water velocity values 1–20 m/s.

Solution:

Properties of water at 25°C are:

$$\nu = 0.894 \times 10^{-6}\ \text{m}^2/\text{s};\ k = 0.607\ \text{W/m K};\ Pr = 6.14.$$

Assumptions: (1) Water has constant properties, (2) water flow is fully developed, and (3) there is no fouling.

The inner and outer surface areas of the tube are:

$$A_i = \pi D_i L = \pi \times 0.011 \times 1 = 0.03454\ \text{m}^2$$

$$A_o = \pi D_o L = \pi \times 0.014 \times 1 = 0.04396\ \text{m}^2$$

The steps to solution of the problem:

1. Reynolds numbers are calculated for the given velocity values

2. h_i values are then calculated by:

$Nu = \frac{h_i D}{k} = 0.023\, Re^{0.8} Pr^n$, where $n=0.3$ for cooling.

3. The total thermal resistance of this heat exchanger per unit length, then, becomes:

$$R = \frac{1}{h_i A_i} + \frac{1}{h_o A_o} + \frac{\ln\left(\frac{D_o}{D_i}\right)}{2\pi k L}$$

4. Finally, the overall heat transfer coefficients based on the inner surface can be calculated with the relation below:

$$R = \frac{1}{U_i A_i}$$

The computational solution has been obtained with MATLAB R2019b software (Fig. E.3.9.1) The U_i changes with the velocity of the cold water as plotted in Fig. E.3.9.2. It is also seen from the plot that the rate of increase gets very low after an increase up to a certain point. This can be explained by the effect of thermal resistance. The total thermal resistance for this example stems from the different modes of heat transfer, which are convection inside the tube, conduction, and convection outside the tube. The velocity change of the water flowing inside the tube affects only the resistance inside the tube by changing the heat transfer coefficient inside the tube. Resistances arising from conduction and convection outside are constant, 0.0013 and 0.1895°C/W, respectively. As tabulated below, the resistance inside the tube becomes almost insignificant along with the conduction resistance. Convection taking place outside makes the most significant contribution to the thermal resistance (Table E.3.9.1).

```
clear all;
clc;

%properties of water at 25 C
kwater = 0.607;                                  %thermal conductivity in
W/m.C
Kv = 0.894*10^-6;                                %kinematic viscosity in m^2/s
Pr = 6.14 ;                                      %Prandtl number

%properties of the pipe
Din = 0.011;                                     %internal diameter in m
Dout = 0.014;                                    %external diameter in m
L = 2;                                           %length of th tube in m
kpipe = 15;                                      %thermal conductivity of the
pipe in W/m.C


V = 1:1:20;                                      %velocity in m/s
ho = 60;                                         %heat transfer coefficient
outside of the tube in W/m^2.K
Re = V*Din/Kv;                                   %Reynolds number

if Re > 10000
    disp ('flow is turbulent and Dittus-Boulter equation is valid')
    disp ('Nu = 0.023*(Re^0.8)*(Pr^0.4)')
    Nu = 0.023*(Re.^0.8)*(Pr^0.4);

else
end
 hi=kwater*Nu/Din;                                 %heat transfer coefficient
inside the tube in W/m^2.K

   Ai = pi*Din*L;                                 %the inner  surface area of
the tube in m^2
   Ao = pi*Dout*L;                                %the outer  surface area of
the tube in m^2

  for i = 1:size(V,2)
  Rconv_in(i) = 1/(hi(i)*Ai);                       %Thermal resistance due
to inside convection
  Rcond(i) = log(Dout/Din)/(2*pi*kpipe*L);          %Thermal resistance due
to conduction
  Rconv_out(i) = 1/(ho*Ao);                         %Thermal resistance due
to outside convection
  R(i) = Rconv_in(i)+Rcond(i)+Rconv_out(i);         %Total thermal resistance
in C/W
  Ui(i) = 1/(R(i)*Ai);                              %overall heat transfer
coefficient based on the inner surface in W/m^2.K
   end
   plot(V,Ui)
   xlabel('velocity, m/s')
   ylabel('overall heat transfer coefficient (inner surface), W/m^2.K')
   grid on
```

Fig. E.3.9.1 MATLAB codes for Example 3.9.

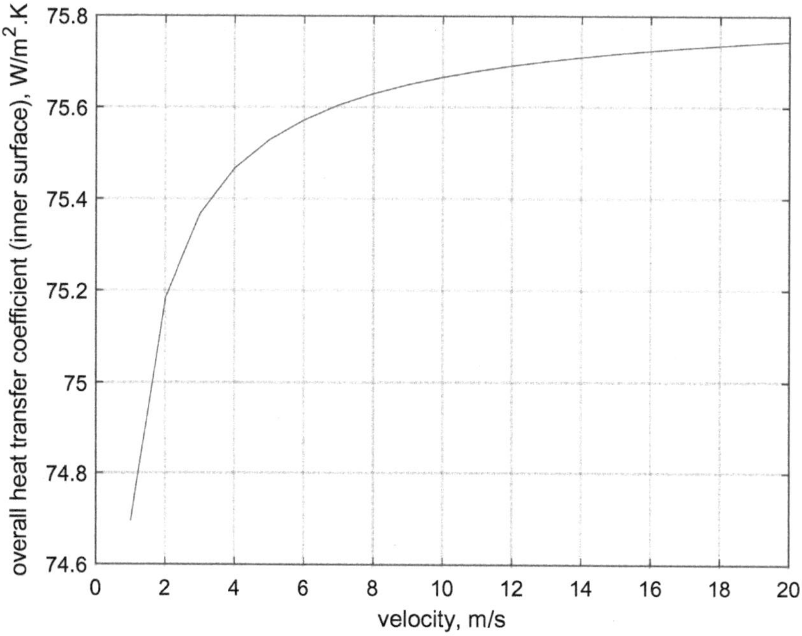

Fig. E.3.9.2 Change in *U* values with respect to velocity.

Table E.3.9.1 Thermal resistance change with velocity of cold water.

		Thermal resistances (°C/W)		
		Inside convection	**Conduction**	**Outside convection**
Velocity (m/s)	1 (m/s)	0.0029	0.0013	0.1895
	5 (m/s)	8.1361×10^{-4}	0.0013	0.1895
	10 (m/s)	4.6730×10^{-4}	0.0013	0.1895
	15 (m/s)	3.3785×10^{-4}	0.0013	0.1895
	20 (m/s)	2.6839×10^{-4}	0.0013	0.1895

References

Ahmed, A., Bhattacharjee, A., & Saha, S. (2017). Comparison of performance of shell-and-tube heat exchangers with plain and annular-finned tube bundle. *AIP Conference Proceedings*, *1919*, 020019. https://doi.org/10.1063/1.5018537.

Bastida, H., Ugalde-Loo, C. E., Abeysekera, M., & Qadrdan, M. (2017). Dynamic modeling and control of a plate heat exchanger. *IEEE conference on energy internet and energy system integration*. https://doi.org/10.1109/EI2.2017.8245736.

Biyanto, T. R., Roekmono, R., Rahmadiansyah, A., Aisyah, A. S., Darwito, P. A., Dhanardono, T., et al. (2015). Modelling and simulation of industrial heat exchanger networks under fouling condition using integrated neural network and HYSYS. *Jurnal Ilmiah Kursor*, *8* (1), 13–20.

Cengel, Y. A. (2003). *Heat transfer: A practical approach* (2nd ed.). New York, NY: McGraw-Hill Co, Inc.

Choi, W., Jun, S., Nguyen, L. T., Rungraeng, N., Yi, H., Balasubramanian, S., et al. (2013). 3-D milk fouling modeling of plate heat exchangers with different surface finishes using computational fluid dynamics codes. *Journal of Food Process Engineering*, *36*, 439–449.

Delplace, F., & Leuliet, J. C. (1995). Modelling fouling of a plate heat exchanger with different flow arrangement by whey protein solutions. *Trans. Am. Inst. Chem. Eng. C*, *73*, 112–119.

GBH Enterprises Ltd. *Process engineering guide: Selection of heat exchanger types*. www.GBHEnterprises.com. (Accessed 7 July 2020).

Geankoplis, C. J. (1993). *Transport processes and unit operations* (3rd ed.). Englewood Cliffs, NJ: Prentice-Hall International, Inc.

Georgiadis, M. C., & Macchietto, S. (2000). Dynamic modelling and simulation of plate heat exchangers under milk fouling. *Chemical Engineering Science*, *55*, 1605–1619.

Gut, J. A. W., & Pinto, J. M. (2003). Modeling of plate heat exchangers with generalized configurations. *International Journal of Heat and Mass Transfer*, *46*, 2571–2585.

Holman, J. P. (2010). *Heat transfer* (10th ed.). Jefferson City, MO: McGraw-Hill Co.

Kays, W. M., & London, A. L. (1964). *Compact heat exchangers* (2nd). New York: McGraw-Hill.

Larowski, A., & Taylor, M. A. (1983). Systematic procedure for selection of heat exchangers. *Proceedings of the Institution of Mechanical Engineers*, *197A*, 51–69.

Maroulis, Z. B., & Saravacos, G. D. (2003). *Food process design*. New York, USA: Marcel Dekker, Inc.

McKillop, A. A., & Dunkley, W. L. (1960). Heat transfer (plate heat exchangers). *Industrial and Engineering Chemistry*, 740–744.

Narataruksa, P., Triratana, P., Pana-Suppamassadu, K., Heggs, P. J., & Tia, S. (2008). *ScienceAsia*, *34*, 229–237.

Peng, B. T. (2005). *Experimental and numerical study of heat transfer and pressure drop for shell-and-tube heat exchangers with continuous helical baffles*. School of Energy and Power Engineering, Xi'an Jiaotong University.

Ribeiro, C. P., Jr., & Andrade, M. H. C. (2002). An algorithm for steady state simulation of plate heat exchangers. *Journal of Food Engineering*, *53*, 59–66.

Thirumarimurugan, M., & Kannadasan, T. (2009). Simulation studies on plate type heat exchanger using ANN. *International Journal of ChemTech Research*, *1*(2), 349–354.

Walker, P. (2005). *CFD modelling of heat exchanger fouling. A PhD Thesis in Department of Chemical Engineering and Industrial Chemistry* (pp. 71–74). Sydney, Australia: University of New South Wales.

Wan Mokhtar, W. M. F. (2012). Empirical modelling of pasteurization process using plate heat exchanger. *International Journal of Engineering Research and Technology*, *1*(3), 1–5.

Welty, J. R., Wicks, C. E., Wilson, R. E., & Rorrer, G. L. (2008). *Fundamentals of momentum, heat and mass transfer* (5th ed.). New Jersey: John Wiley and Sons, Inc.

Section B

Heat exchangers in the food processing industry

Plate heat exchangers in the food industry

4

Sara Hedayati[a,b], *Elham Ansarifar*[c], *and Seid Mahdi Jafari*[d]
[a]Nutrition Research Center, School of Nutrition and Food Sciences, Shiraz University of Medical Sciences, Shiraz, Iran, [b]Department of Food Hygiene and Quality Control, School of Nutrition and Food Sciences, Shiraz University of Medical Sciences, Shiraz, Iran, [c]Social Determinants of Health Research Center, Department of Public Health, School of Health, Birjand University of Medical Science, Birjand, Iran, [d]Department of Food Materials and Process Design Engineering, Gorgan University of Agricultural Sciences and Natural Resources, Gorgan, Iran

4.1 Introduction

Heat treatment is a common and important unit operation in food plants to increase the shelf life and assure the hygienic safety of food products (Jun & Puri, 2005; Sharma & Macchietto, 2021). Heat exchangers are extensively used in the food industry for heating, cooling, and heat recovery of food products (Gulenoglu, Akturk, Aradag, Uzol, & Kakac, 2014). Plate heat exchangers (PHEs) are appropriate for liquid–liquid heat-transfer operations that require rapid and uniform heating or cooling of heat-sensitive fluids such as liquid foods (Afonso, Hes, Maia, & Melo, 2003). PHEs are the most successful heat exchangers for liquids in the food industry because of their thermal performance, compactness, flexibility, low cost, ease of cleaning, assembly, and disassembly (Merheb, Nassar, Nongaillard, Delaplace, & Leuliet, 2007).

The first use of PHEs in the food industry dates back to the late 19th century (Magnusson, 1985). Nowadays, PHEs are extensively used for heat treatment of low-viscosity liquid foods without particles such as milk, fruit juice, beer, and liquid egg (Aguiar & Gut, 2014). Moreover, PHEs can be used in other food processing operations such as evaporation, heating, and cooling in pretreatment and fermentation (Wang, Sundén, & Richardson, 2004). Generally, the heat-transfer coefficient of a PHE is four to five times higher than that of a shell and tube heat exchanger. PHEs have a very compact heat-transfer area, and their space requirement is 10% of a typical shell and tube heat exchanger. When the fluid flows through a PHE, heat is transferred from hot to cold media. The countercurrent flow in PHEs maximizes the heat recovery, and a very close temperature approach can be obtained. Moreover, temperature cross is also possible in PHEs, meaning that the hot outlet can reach a lower temperature than the cold outlet that makes PHEs more thermally efficient than shell and tube heat exchangers (Alfa Laval, 2022).

The plates of PHEs should be manufactured from resistant materials because they are subjected to high pressure and temperature. Plates should be made from malleable

Thermal Processing of Food Products by Steam and Hot Water. https://doi.org/10.1016/B978-0-12-818616-9.00013-4

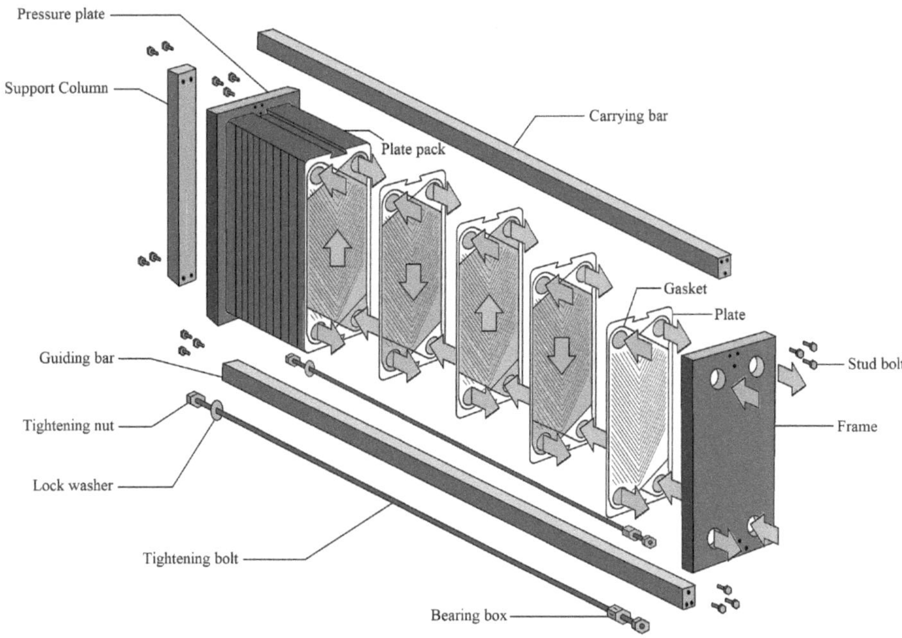

Fig. 4.1 Different components of a plate heat exchanger.

corrosion-resistant materials (Serth & Lestina, 2014). Nowadays, plates are made from different kinds of noble alloys such as AISI 316, AISI 301, nickel, Monel, titanium, Hastelloy, Incolloy, aluminum, silver, copper, tantalum, and so on (Thulukkanam, 2013). The basic form of a PHE is composed of wavy metal plates assembled in a frame (Gut, Fernandes, Pinto, & Tadini, 2004). Fig. 4.1 shows different components of a plate heat exchanger.

The plates form a series of channels for the fluid flow, and usually, the flow arrangement of hot and cold fluids in alternating chambers is countercurrent (Tamilselvan, Sivabalan, Prakash, Manojprasath, & Mahabubadsha, 2017). However, for very heat-sensitive food products, cocurrent flow is used in PHEs. In this flow design, the coldest fluid meets the hottest fluid when entering the PHE; therefore, the risk of overheating or freezing is minimized (Alfa Laval, 2022). To keep the fluids between the channels, plates are equipped with different sealing systems. The main drawback in the basic PHEs is the heat sensitivity of elastomer gaskets as their temperature limit is around 250°C. Moreover, some chemicals are not compatible with elastomer gasket materials. Therefore, alternative sealing systems have been used for sealing the plate channels.

PHEs used for liquid food and drink processing are classified into four groups based on their sealing systems: gasketed plate heat exchanger (GPHE), welded plate heat exchanger (WPHE), semiwelded plate heat exchanger (SWPHE), and brazed plate heat exchanger (BPHE). The type of sealing system is selected based on the application of PHEs (Wang et al., 2004). Additionally, new approaches such as double-wall plates, graphite plates, wide-gap designs, and plates designed for evaporating and condensing have been developed to increase the industrial utility of PHEs. However, the application of these new approaches is not common in the area of food processing. Therefore, they are not discussed in this chapter.

4.2 Different types of plate heat exchangers

4.2.1 Gasketed plate heat exchanger

GPHEs are the most common type of PHEs and were initially developed for the food processing where frequent cleaning is essential (Serth & Lestina, 2014). They are composed of a stack of rectangular corrugated metal plates which are fixed with gaskets that seal and direct the fluids into alternate channels (Fig. 4.2). The corrugation of plates provides turbulence and flow distribution. The thickness of plates is 0.5–1.2 mm, and the gap between the plates is 2–5 mm. The gaskets can be made from elastomers such as silicones, ethylene, propylene, natural rubber, and so on, but Teflon is not appropriate for the joints because of its viscoelasticity (Thulukkanam, 2013). The gaskets form channels between the plates for hot and cold fluid flow and heat exchange (Gut et al., 2004). Additionally, gaskets prevent the leakage of fluids to

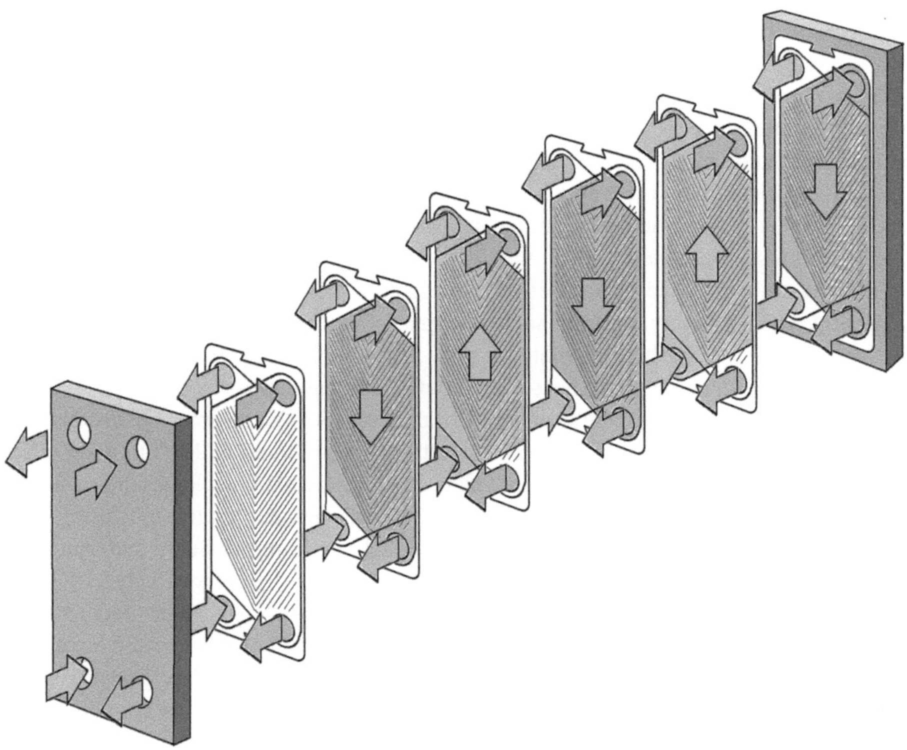

Fig. 4.2 Structure and fluid flow pattern in a gasketed plate heat exchanger.

the environment and seal the fluid ports on plate edges to prevent the mixing of hot and cold fluids.

The fluid flow pattern in hot and cold channels is usually countercurrent. GPHEs have a high thermal efficiency because of their large surface area, the small gap between the plates, and turbulence induced by the corrugations. However, the pressure and temperature in GPHEs should not exceed 25 bar and 250°C, respectively (Serth & Lestina, 2014). The GPHEs designed for different food products have different specifications, and factories should pay attention to the processing temperature and pressure when they want to order a GPHE. For instance, GPHEs with a high design pressure should be used for carbonated beverage processing. The fluids enter into the GPHEs through the portholes and connections of the plates and the sealing gaskets between the plates direct the fluids so that the hot and cold fluids pass countercurrently in alternating channels. As the fluid arrives between the plates, it passes through the distribution area which helps the fluids to quickly fill up the entire cross section of the plates, ensures an even distribution of fluids over the entire plate, maximizes the heat-transfer efficiency, and minimizes fouling (Alfa Laval, 2022).

4.2.2 Brazed plate heat exchangers

BPHEs were developed to increase the compactness, temperature, and pressure capabilities of PHEs. BPHEs are designed for high-efficiency applications, such as those applications with a high evaporation temperature and a low water/brine pressure drop (Alfa Laval, 2022). This results in reduced environmental impacts and lower costs. BPHEs can handle temperatures up to 400°C and pressures up to 30 bar. BPHEs are self-contained heat exchangers that do not require gaskets, tightening bolts, frames, and guide bars (Fig. 4.3). Therefore, the weight of BPHEs is much lower than that of the conventional plate and frame gasketed heat exchangers (Wang, & Manglik, 2007). However, in some cases, frames are used in BPHEs in order to improve their operating pressure range.

BPHEs are composed of a series of corrugated plates brazed together by vacuum soldering with a brazing material such as copper or nickel (Han, Lee, & Kim, 2003; Wang et al., 2004). The brazing material seals and holds the plates together at the contact points ensuring an optimal heat-transfer efficiency and pressure resistance. There are four to six apertures in the corners of brazed plates. The alternate plates are put together and work as a single plate because they are arranged at 180° to each other and form inlet and outlet channels. During brazing of the plate package, the capillary forces draw the brazing material to the contact points between plates and form braze junctions which improve the integrity of the BPHEs at high pressures and temperatures (Manigandan & Suresh, 2015). The performance of PHEs is affected by plate pattern design, printing depth, and corrugation pitch. Additionally, other factors such as port location, flow passage dimension around ports, and plate aspect ratio have substantial influences on the performance of BPHEs (Gürel et al., 2020). The advent of BPHEs dates back to 1990s, and these PHEs were initially used for condensation, evaporation, and air-conditioning systems. In the food industry, BPHEs are used for water heating and cooling, CO_2 heating and cooling, and refrigeration.

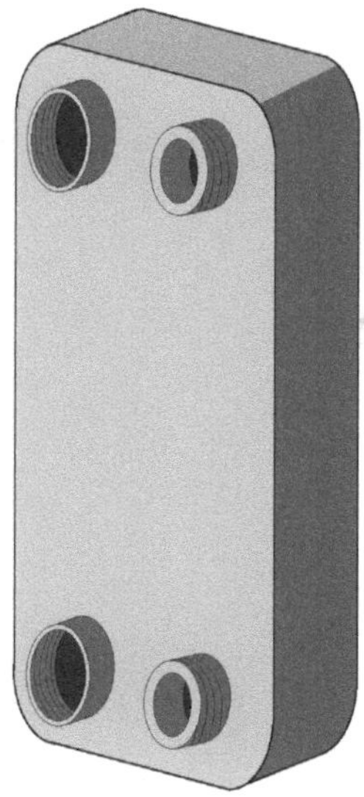

Fig. 4.3 Brazed plate heat exchanger.

4.2.3 Welded plate heat exchangers

WPHEs are suitable for high pressures, high temperatures, and corrosive fluids. They can be fully welded plate heat exchangers (FWPHEs) or SWPHEs.

4.2.3.1 Fully welded plate heat exchangers

In FWPHEs, the corrugated plates are welded by laser welding machines; therefore, elastomer gaskets are not required. The gasket-free pack of plates is compressed within a frame. Because of their specific design, all four sides of the FWPHEs can be disassembled, which facilitate the inspection and cleaning of the heat exchanger (Fig. 4.4). The exclusion of gaskets improves the reliability as well as pressure and temperature resistance of FWPHEs. The occupied space in FWPHEs is small, and they can have horizontal or vertical installation. The operating temperature range in FWPHEs is −50°C to 400°C, and the pressure is up to 40bar. Additionally, the welding of the plate pack in FWPHEs is in two dimensions only. Therefore, they expand and contract along the length of the plate pack as temperature and pressure changes take place. Subsequently, the plane of the plates can withstand rapid changes in temperature and pressure which make FWPHEs very flexible and fatigue-resistant. FWPHEs are suitable for corrosive fluids and severe duty processes. Heavily aggressive fluids can be used on either side of these heat exchangers. Typical applications of FWPHEs in the area of food processing include refrigeration, CO_2 heating or cooling, vegetable oil heaters, and so on.

Pillow-plate heat exchangers (PPHEs) are novel FWPHEs that are composed of a stack of wavy pillow-like parallel plates (Fig. 4.5). Pillow plates are made from two stainless-steel plates placed one each other and welded by either laser or resistance

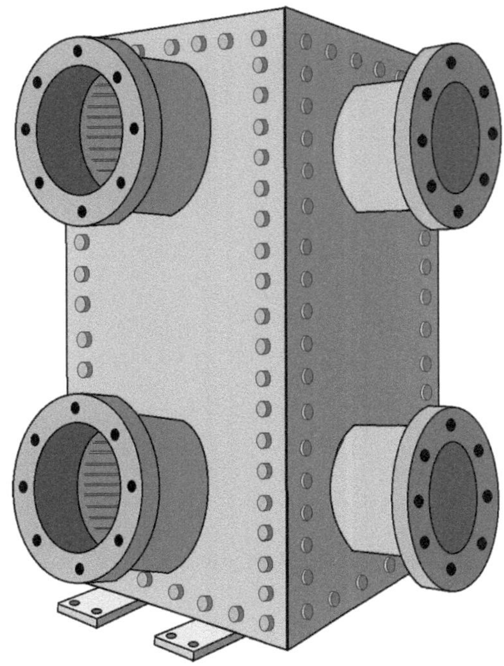

Fig. 4.4 Fully welded plate heat exchanger.

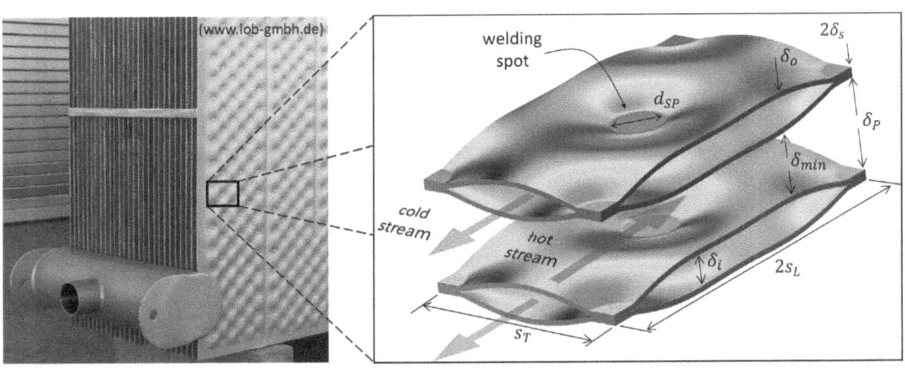

(a) (b)

Fig. 4.5 Pillow-plate fully welded heat exchanger (Piper et al., 2019).

welding to form the characteristic welding point grid. Subsequently, the inlet and outlet channels are attached and the edges are sealed with laser or resistance welding, and the final structure is obtained by hydroforming. In PPHEs, alternating inner and outer channels are arranged between the adjacent plates for heat transfer. The thermo-hydraulic performance of PPHEs depends on several factors such as the welding spot arrangement and the geometric parameters such as internal and external heat-transfer surface areas, cross-sectional areas, and characteristic lengths (Tran, Piper, Kenig, & Scholl, 2018).

4.2.3.2 Semiwelded plate heat exchangers

SWPHEs comprise pairs of corrugated plates welded on one side and traditional gasketed units on the other side. Similar to the conventional PHEs, the pair of plates are packed by a rigid frame and bolts in SWPHEs (Fig. 4.6). SWPHEs are disassembled easier than FWPHEs, and their capacity can be increased. The temperature range of SWPHEs is −47°C to 170°C, and the welded side can handle pressures up to 40 bar. The SWPHEs are resistant to most of the refrigerants on their weld surfaces and are specifically suggested for ammonia applications (TGT, 2022). SWPHEs are developed for handling aggressive materials. Generally, the aggressive material flows in the welded channel formed by the corrugations of two plates, whereas the nonaggressive secondary medium or the cooling medium flows in the channels of elastomer gaskets. The laser-welded channel allows use of fluids that are not compatible with normal gaskets and also allows for a higher design pressure than fully gasketed plate-and-frame heat exchangers (Alfa Laval, 2022). These heat exchangers are suitable for condensation, evaporation, and refrigeration cycles. The applications of SWPHEs in the food industry include water heating or cooling, glycol cooling, and refrigeration, and they are used in dairy, beverage, and sugar industries (danfoss, 2022).

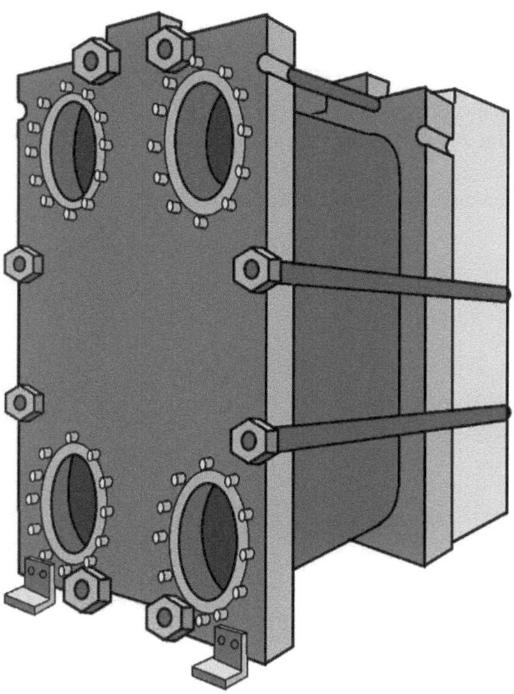

Fig. 4.6 Semiwelded plate heat exchangers.

4.3 Design of plates in heat exchangers

The design of plates is a critical issue as it has dramatic effects on heat transfer, cleaning, and catching of the system. The corrugated channels disturb the fluid flow by growing recirculation regions near the corrugated wall, which increases the turbulence and leads to a high heat-transfer coefficient with a minimum pressure drop (Tiwari, Ghosh, & Sarkar, 2014). PHEs can be classified based on their surface design, and >60 plate patterns have been developed for PHEs around the world. Because of the large number of corrugations and the diversity of plates, the design of PHEs is highly specialized and manufacturers develop exclusive design methods. As shown in Fig. 4.7, there are several categories of plate corrugations such as chevron or herringbone, washboard, zigzag, wavy, asterisk, protrusions and depressions, lung pattern, oblique washboard, and washboard with secondary corrugations (Durmuş, Benli, Kurtbaş, & Gül, 2009; Gürel et al., 2020; Mota, Carvalho, & Ravagnani, 2015).

Despite the diversity of plate designs, most of the PHEs in the food industry have the chevron pattern. The chevron corrugation, also known as the herringbone pattern, has shown better performance and is more commonly used by manufacturers. The strength of chevron plates is greater than that of corrugations such as washboard. Hence, chevron plates can withstand higher pressures with thinner plates (Thulukkanam, 2013).

Usually, there is a chocolate pattern distribution area at the two ends of plates. Fig. 4.8 illustrates a chevron plate with a chocolate pattern distribution area. This corrugation pattern has several advantages:

- Ensures the highest energy efficiency by uniform distribution of fluids across the entire plate.
- Eliminates the dead spots and stagnant zones that are the main causes of corrosion and fouling.
- Reduces the maintenance cost and increases the uptime (Alfa Laval, 2022).

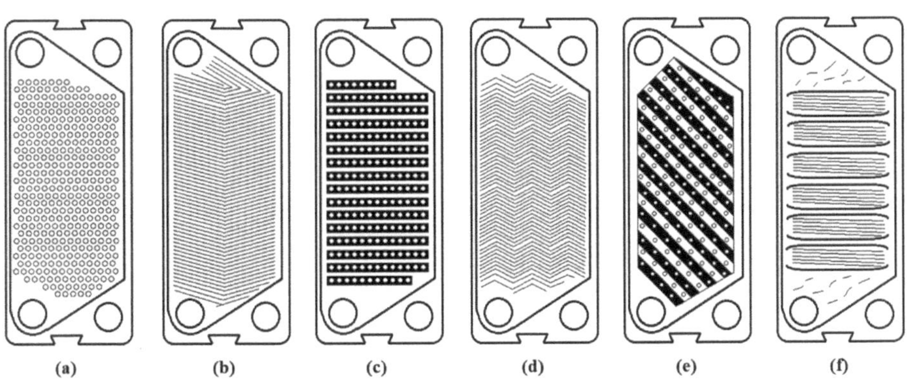

Fig. 4.7 Different plate surface designs: (A) protrusions and depressions; (B) chevron or herringbone; (C) washboard; (D) zigzag; (E) oblique washboard; (F) washboard with secondary corrugations.

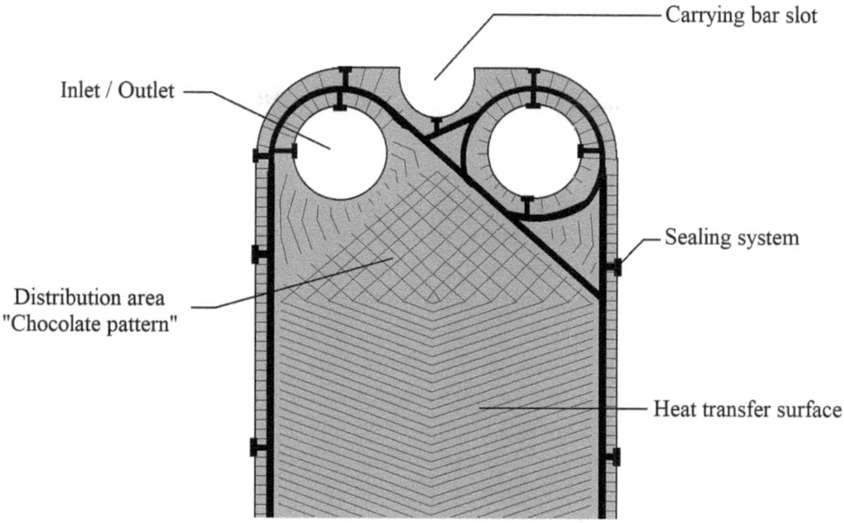

Fig. 4.8 Chevron plate with a chocolate pattern distribution area.

The geometrical parameters of chevron patterns have significant effects on the thermal–hydraulic performance of PHEs, which should be optimized by experimental and numerical studies (Zhang, Zhu, Mondejar, & Haglind, 2019). The depth of corrugations is between 3 and 5 mm, and the typical velocity of fluids is from 0.1 to 1 m/s (Thulukkanam, 2013). The chevron angle has substantial effects on the performance of PHEs; plates are divided into two categories of low-theta plates and high-theta plates based on their chevron angle (Fig. 4.9). Plates with obtuse chevron angles (60–65°) are considered as high-chevron angle or high-theta plates, whereas plates with an acute chevron angle (25–30°) are considered as low-chevron angle or

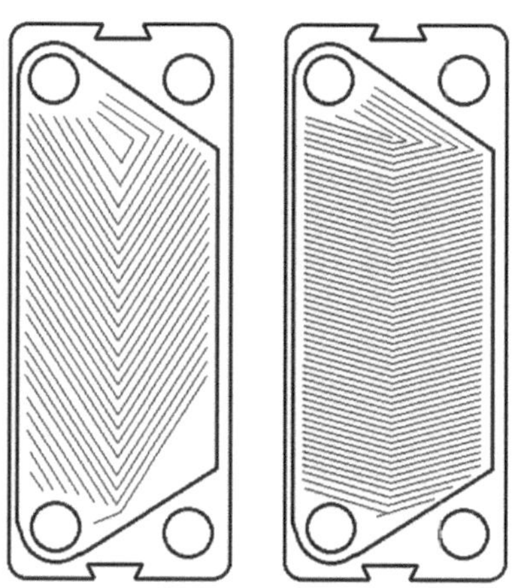

Fig. 4.9 Chevron plates with low and high theta angles.

low-theta plates. High-theta plates generate high turbulence, high heat transfer, and high pressure loss, whereas low-theta plates generate low turbulence, low heat transfer, and low pressure loss (Thulukkanam, 2013). The chevron angles of adjacent plates are reversed; therefore, when the plates are fixed in PHEs, the corrugations cross one another to form several contact points.

Kumar, Soni, and Singh (2018) determined the influence of geometrical parameters of chevron plates on thermal and hydraulic performance of PHEs within the Reynolds number range of 800–2300. They used two symmetrical plate sets with chevron angles of 60°/60° and 30°/30° and one unsymmetrical plate set with chevron angles of 60°/30°. The investigations were performed for a plate set having 21 channels (10 hot water and 11 cold water channels). The experimental results revealed that the chevron angle has significant influences on the fluid flow distribution in channels and consequently affects the power consumption and thermal performance of PHEs. The pressure drop in PHE channels increased in the following order: 60°/60° > 60°/30° > 30°/30°, indicating that increasing the chevron angle increases the pressure drop and has negative effects on hydraulic performance of PHEs. However, high chevron angles increase the uniformity of fluid flow distribution and improve the effectiveness of PHEs. The results revealed that increasing the chevron angle of plates from 60°/30° to 60°/60° caused an increase of 22.16% in effectiveness. It was also observed that the uniform flow distribution and effectiveness of the PHEs decreased by increasing the mass flow rate.

Similar results were reported by Kılıç and İpek (2017), who investigated the effectiveness, heat transfer, and pressure drop in corrugated PHEs with chevron angles of 30° and 60°, respectively. They stated that the effectiveness and heat-transfer rate of plates with a 60° chevron angle are higher than those with a chevron angle of 30°. However, the pressure drop in 60° chevron angle plates is higher than that in 30° chevron angle plates. In a study by Nilpueng, Keawkamrop, Ahn, and Wongwises (2018), contradictory results were reported. They investigated the performance of a commercial GPHE under different chevron angles (30° and 60°), plate surface roughnesses (0.95–2.75 μm), and Reynolds numbers (1200–3500) and observed that the pressure drop and heat-transfer coefficient increased by decreasing the chevron angle and increasing the surface roughness and Reynolds number. Under the experimental conditions, the best thermal performance of PHEs was obtained at a chevron angle of 30°, a plate surface roughness of 2.75 μm, and a Reynolds number of 1200. Based on these results, the thermal performance of a PHE is affected by several parameters and the chevron angle cannot be used to determine the thermal performance of PHEs alone.

4.4 Fouling in PHEs

Some of the food components are heat-sensitive and form fouling layers on the surface of heat processing equipment. Milk fouling is classified into two types:

Type A: A soft, voluminous, and curd-like material that is formed at a temperature above 75°C, maximizes at 95–110°C, and then reduces at higher temperatures. This type of fouling is composed of 50%–70% proteins and 30%–40% minerals.

The formation of fouling type A is mainly because of the denaturation of milk proteins. At temperatures around 75°C, β-lactoglobulin is the predominant component of the fouled layer, but at higher temperatures, it is dominated by casein.

Type B: A hard and granular form structure which is formed at higher temperatures. It contains 70%–80% minerals and 10%–20% protein. Calcium phosphate is the major component of type B fouling (Ansari, Chavan, & Bhatt, 2018).

Fouling depends on many chemical and physical variables such as milk composition, pH, milk age, temperature, air bubbles, plate coating, flow rate, microbial content, and so on. Some parameters such as the chemical composition of the liquid foods cannot be changed. However, parameters such as temperature and flow rate are determined by the operators and process designers. Fouling is an undesired phenomenon as it reduces the heat-transfer coefficient and results in increasing thermal and electrical energy consumption. Moreover, it can cause cross contamination if dislodged during subsequent heat treatments and increase the risk of microbiological problems in pasteurized food products (Verran, Boyd, Hall, & West, 2002). Therefore, PHEs require periodic cleanings to recover the hydraulic and thermal performance and prevent microbial contamination (Sharma & Macchietto, 2021).

Cleaning is invariably performed by Cleaning-in-Place (CIP) in PHEs in food plants to remove the foulants. CIP is an autonomous process that does not require dismantling for cleaning the internal surfaces of food processing equipment and is performed without any manual intervention (Escrig, Woolley, Rangappa, Simeone, & Watson, 2019). In the CIP technique, the time, temperature, chemicals, and mechanical forces are used for cleaning. Different chemicals or cleaning processes may be utilized in CIP depending on the physicochemical properties and the amount of fouling layer. However, generally, it is performed in five stages: 1—prerinse, 2—chemical cleaning, 3—postrinse, 4—sanitization, and 5—final rinse (Fryer, Christian, & Liu, 2006). CIP systems require large amounts of water, electrical energy, and acid and caustic solutions and produce large amounts of wastewater with a high biological oxygen demand level which must be treated to reduce the negative environmental impacts. The downtime for cleaning is about 4–6h a day; therefore, the useful operating time is reduced and the production costs are increased significantly (Jun & Puri, 2005; Rungraeng, Cho, Yoon, & Jun, 2012). Fouling is an unsolved problem in PHEs, but several strategies have been applied in order to reduce the fouling and its drawbacks in the food industry.

4.4.1 Strategies to reduce fouling and improve the heat transfer in PHEs

Long processing and a short cleaning time are favored in food processing operations, which can be achieved by optimization of PHE design and operation. Uniform fluid flow, turbulence, an appropriate plate surface, and production of plates from resistant materials can improve the product safety and reduce the need for frequent cleaning in PHEs (Thulukkanam, 2013). These strategies are effective in reducing fouling, production costs, and environmental pollution. In recent years, the energy costs have increased significantly and motivated the industries to use energy-saving approaches in their equipment. The heat-transfer enhancement in heat exchangers is an important issue that can be addressed to decrease the production costs in food plants. The heat-transfer enhancement techniques in PHEs can be active or passive. In active methods, external forces such as electric field, acoustic, or surface vibration are required, whereas in passive methods, the heat-transfer area is extended or the heat-transfer fluids in PHEs are modified (Tabari & Heris, 2015).

The approaches to reduce or eliminate fouling can be categorized into three groups:

- Modification to heat-transfer surfaces, such as changing the configuration of plates or utilization of surface coating.
- Modification to heat treatment by changing the temperature profiles or flow conditions or using emerging heating fluids such as nanofluids.
- Modification to liquid foods by addition of calcium phosphate inhibitors or oxidizing agents.

However, the third approach may be precluded in many cases as it affects the product specifications (Ansari et al., 2018).

4.5 Modification of heat-transfer surfaces

The high surface free energy and hydrophilic adhesion of the conventional metals such as SS316 have high hydrophilic adhesion and surface free energy that act as stimulants to milk fouling. Modification of stainless-steel plates in PHEs with hydrophobic coatings mitigates fouling. The heat-transfer performance in coated plates is kept for a longer time, and cleaning requirement is reduced. Moreover, harsh chemicals and water consumption and pumping are reduced (Bangavadi Munivenkatappa et al., 2022; Rungraeng et al., 2012).

Hydrophobic coating of plate surfaces is an effective approach to reducing the fouling on the PHEs (Balasubramanian & Puri, 2009; Ozden & Puri, 2010; Rosmaninho et al., 2007). The hydrophobic coatings can be classified into two groups of hydrophobic and superhydrophobic materials based on their water contact angle (WCA) values. The hydrophobic materials have WCAs greater than 90°, whereas the superhydrophobic materials have WCAs greater than 140° (Chiang & Lu, 2011; Klein, Biesheuvel, Ben, Meinhart, & Lange, 2003). The superhydrophobic surfaces have micro- and/or nanoscale roughness with very low surface energy. The roughness and surface energy of coatings have key roles on the cleaning efficiency of the fouling layer, and the foulants on hydrophobic surfaces are easily cleaned. Hydrophobic and superhydrophobic surface coatings in PHEs have been used for heat processing of liquid foods by several researchers.

Rungraeng et al. (2012) used a superhydrophobic nanoparticulate coating to inhibit the surface fouling of milk in PHEs. They dispersed carbon nanotubes (CNTs) in hydrophobic polytetrafluoroethylene (PTFE) by ultrasound treatment and high-temperature annealing and investigated the hydrophobicity of the CNT-PTFE-coated surface by water contact angle (WCA), foulant mass, correspondent surface energy, and energy efficiency and compared with conventional stainless-steel 316 and

PTFE-coated plates. The contact angle of the uncoated surface was 71.2, whereas it was 119.6 and 141.1 for PTFE and CNT-PTFE, respectively. The estimated surface energy of CNT-PTFE-coated PHEs was decreased by 97.0% of the uncoated plate value. The mass foulant in CNT-PTFE plates after 5h of milk processing was 70.3% lower than that in the control plate. Consequently, the total energy consumption of nanocomposite-coated plates decreased by 10.2%. Based on these results, CNT-PTFE coating can reduce the milk fouling on PHEs, the frequency of the CIP program, and the risk of milk underprocessing.

Zouaghi et al. (2019) investigate the influence of an antifouling amphiphilic silicone coating (Si-PEO) on the environmental impact of milk pasteurization using the life cycle assessment (LCA) approach and observed that the environmental impact was reduced more than 70%. Bangavadi Munivenkatappa et al., 2022 modified the surface of stainless-steel 316 (SS316) plates of PHEs by titanium dioxide (TiO_2) and PTFE nanoparticles and studied the surface properties and thermal performance of conventional SS316 and PTFE-TiO_2-coated plates during milk processing for 8h. They observed a significant reduction in contact angles of water and milk in PTFE-TiO_2-coated plates, indicating that the hydrophobicity was enhanced and the interaction between milk and the plate surface was weakened in coated films. Moreover, they reported that the fouling factor in uncoated plates was approximately 4 times higher than that in the coated plates. The overall heat-transfer coefficients of uncoated plates dropped by 62%, whereas this value was 19.4% for coated plates.

Balasubramanian and Puri (2009) used two commercially available coatings (Lectrofluor 641 and Ni-P-PTFE) for modification of stainless-steel plate surfaces in PHEs and investigated the influence of surface modification on milk fouling on PHEs after 6h heat treatment and compared the results with the amount of fouling on stainless-steel 316 as the control. The incidence of fouling in Lectrofluor 641-coated plate surfaces was 94% less than that of the uncoated plate, whereas it was 86% for Ni-P-PTFE-coated plates. The chemical analysis did not reveal incursion of the Lectrofluor 641 coating material into the deposited fouled layer, whereas some evidence of Ni-P-PTFE incursion was observed in the fouling. Huo, Xiao, Kirk, and Chen (2019) studied the effects of Fluorolink S10 coating on whey protein concentrate (WPC) fouling of stainless-steel plates at a very low fluid velocity and consequent cleaning. Jindal, Anand, Metzger, and Amamcharla (2018) modified the surface of stainless steel with Ni-P-polytetrafluoroethylene (Ni-P-PTFE) and compared the biofilm formation on plates of a PHE during 17h uninterrupted processing. The results revealed that the microbial count and biofilm formation on PTFE-coated plates were lower than those of uncoated plates. Also, the relative light units, an indication of the presence of organic matters, was lower on the surface of modified plates. They stated that the cleaning of uncoated plates required acid treatment, whereas the alkali treatment was sufficient for the cleaning of modified surfaces. Therefore, the use of Ni-P-PTFE-coated plates is cost- and time-effective for milk processing. Similarly, Huang and Goddard (2015) and Rosmaninho et al. (2007) confirmed that the Ni–P–PTFE surface improves the nonfouling properties of PHEs during heat treatment of milk-based fluids.

4.5.1 Modification to heat treatment by the incorporation of nanofluids

Conventional fluids such as water, ethanol, engine oil, liquid paraffin, ethylene glycol, refrigerants, lubricants, and so on have poor thermal conductivity. Application of fluids with higher thermophysical attributes and incorporation of fluid additives are new approaches to improving the heat transfer and reducing the processing time and energy consumption in PHEs (Pandya, Shah, Molana, & Tiwari, 2020). Nanofluids are colloidal suspensions of nanoscale particles with length scales of 1–100nm and preferably 10–50nm in conventional base fluids (Bhattad, Sarkar, & Ghosh, 2019). The dispersion of low concentrations of nanoparticles such as CuO, TiO_2, Al_2O_3, and so on in the base fluid extremely increases the heat-transfer characteristics such as thermal conductivity, heat capacity, density, viscosity, and so on in PHEs because of high stability, a high heat-transfer area, less pumping power, and agglomeration. The nanofluids are selected based on their shape, size, thermophysical properties, and stability of nanoparticles (Bhattad et al., 2019).

Tabari and Heris (2015) added multiwalled carbon nanotubes (MWCNTs) to water in corrugated PHEs in order to increase the heat-transfer capacity for milk processing. They measured the Nusselt number (*Nu*) and heat-transfer coefficient as the key parameters in convective heat transfer. The results revealed that *Nu* and the heat-transfer coefficient significantly increased by incorporation of MWCNTs even at concentrations $<1\%$ and increasing the concentration of nanoparticles increased these parameters. Additionally, they found that the nanoparticles were more effective at turbulent flow regimes compared to laminar flow. The increased heat-transfer characteristics caused more heat exchange of milk in a short time, which can reduce the occurrence of fouling in PHEs during pasteurization of milk. These findings are in agreement with the results reported by Tamilselvan et al. (2017), who added Al_2O_3 nanoparticles to demineralized water in order to produce nanofluids for milk processing in a PHE and stated that nanofluids have high thermal conductivity compared to the base fluid and the heat-transfer rate increased by increasing the concentration of the nanofluid.

Taghizadeh-Tabari, Heris, Moradi, and Kahani (2016) dispersed titanium dioxide nanoparticles (TiO_2) in distilled water at weight concentrations of 0.25%, 0.35%, and 0.8% in order to improve the heat-transfer capability in PHEs. The results revealed that increasing the concentration of TiO_2 enhanced the heat-transfer rate compared with distilled water. However, a continuous pressure drop was observed by increasing the concentration of TiO_2, which is a negative aspect of nanofluids. To evaluate the effectiveness of nanofluids for application in milk industry PHEs, the ratio between the enhancement of convective heat transfer and the pressure drop was determined under laminar and turbulent flow. The experimental results showed that the Nusselt number, heat-transfer coefficient, and pressure drop increased by increasing the concentration of TiO_2 nanoparticles and the changes were greater under turbulent flow. Bhattad et al. (2019) dispersed alumina–silver and magnesia–silver nanoparticles in water–ethylene glycol and water–propylene glycol mixtures and studied the effect of the nanoparticle concentration and flow rate of hybrid nanofluids on milk heating.

The results indicated that the heat-transfer coefficients, heat-transfer rate, pumping power, pressure drop, irreversibility, second law efficiency, entropy generation rate, and flow rate of milk increased, whereas the irreversibility distribution ratio, performance index, and nondimensional exergy destruction decreased by increasing the concentration and flow rate of the nanofluid. They reported that the performance improvement in alumina–silver was better than that of magnesia–silver and it can be recommended as an appropriate alternative for milk chilling.

4.6 Conclusion

Thanks to their properties, PHEs are widely used in food processing applications, from pasteurization to heat recovery systems. PHEs save floor space because of their compactness and are very energy-efficient. The maintenance of PHEs is easier than that of shell and tube heat exchangers because the heat-transfer surfaces are accessible for cleaning and inspection. PHEs are very flexible, and plates can be easily added or removed to adjust capacity. Food manufacturers select the suitable PHE based on their processing conditions such as the required temperature and pressure, type of food product, and production capacity. PHEs are constantly developed in order to increase their applications in the food industry. The developments can be in construction or performance of PHEs. New models of PHEs have been developed for the heat treatment of food products that contain particles, pulps, or fibers. The temperature resistance and pressure resistance of PHEs have increased, and new materials have been industrialized to reduce the corrosion of plates. Fouling is an important issue which imposes high costs to the food factories, and it has been distinguished that the traditional theories about fouling are not enough for design and operation of PHEs. Therefore, research about fouling formation and prevention should be addressed in the future. Moreover, methods should be developed for preventing maldistribution and mitigation of fouling. All these call for considerable theoretical work as well as experimental investigations.

References

Afonso, I. M., Hes, L., Maia, J. M., & Melo, L. F. (2003). Heat transfer and rheology of stirred yoghurt during cooling in plate heat exchangers. *Journal of Food Engineering*, *57*(2), 179–187.

Aguiar, H. F., & Gut, J. A. (2014). Continuous HTST pasteurization of liquid foods with plate heat exchangers: Mathematical modeling and experimental validation using a time–temperature integrator. *Journal of Food Engineering*, *123*, 78–86.

Alfa Laval. (2022). https://www.alfalaval.com/microsites/gphe/tools/how-gphes-work.

Ansari, M. I. A., Chavan, R. S., & Bhatt, S. B. (2018). Fouling of milk and cleaning-in-place in the dairy industry. In *Technological Interventions in Dairy Science* (pp. 179–206). Apple Academic Press.

Balasubramanian, S., & Puri, V. M. (2009). Reduction of milk fouling in a plate heat exchanger system using food-grade surface coating. *Transactions of the ASABE*, *52*(5), 1603–1610.

Bangavadi Munivenkatappa, M., Franklin, M. E. E., Dhotre, A. V., Pushpadass, H. A., Anthonysamy, A., Mandhyan, P. K., & Patil, P. G. (2022). Mitigation of fouling during milk processing in polytetrafluoroethylene-titanium dioxide coated plate heat exchanger. *Journal of Food Process Engineering*, *45*(6), e13836.

Bhattad, A., Sarkar, J., & Ghosh, P. (2019). Energetic and exergetic performances of plate heat exchanger using brine-based hybrid nanofluid for milk chilling application. *Heat Transfer Engineering*.

Chiang, C. K., & Lu, Y. W. (2011). Evaporation phase change processes of water/methanol mixtures on superhydrophobic nanostructured surfaces. *Journal of Micromechanics and Microengineering*, *21*(7), 075003.

danfoss. (2022). www.danfoss.com/en/products/dhs/heat-exchangers/gasketed-heat-exchangers/semi-welded-plate-heat-exchangers/#tab-overview.

Durmuş, A., Benli, H., Kurtbaş, İ., & Gül, H. (2009). Investigation of heat transfer and pressure drop in plate heat exchangers having different surface profiles. *International Journal of Heat and Mass Transfer*, *52*(5–6), 1451–1457.

Escrig, J., Woolley, E., Rangappa, S., Simeone, A., & Watson, N. J. (2019). Clean-in-place monitoring of different food fouling materials using ultrasonic measurements. *Food Control*, *104*, 358–366.

Fryer, P. J., Christian, G. K., & Liu, W. (2006). How hygiene happens: Physics and chemistry of cleaning. *International Journal of Dairy Technology*, *59*(2), 76–84.

Gulenoglu, C., Akturk, F., Aradag, S., Uzol, N. S., & Kakac, S. (2014). Experimental comparison of performances of three different plates for gasketed plate heat exchangers. *International Journal of Thermal Sciences*, *75*, 249–256.

Gürel, B., Akkaya, V. R., Göltaş, M., Şen, Ç. N., Güler, O. V., Koşar, M.İ., & Keçebaş, A. (2020). Investigation on flow and heat transfer of compact brazed plate heat exchanger with lung pattern. *Applied Thermal Engineering*, *175*, 115309.

Gut, J. A., Fernandes, R., Pinto, J. M., & Tadini, C. C. (2004). Thermal model validation of plate heat exchangers with generalized configurations. *Chemical Engineering Science*, *59*(21), 4591–4600.

Han, D. H., Lee, K. J., & Kim, Y. H. (2003). The characteristics of condensation in brazed plate heat exchangers with different chevron angles. *Journal-Korean Physical Society*, *43*(1), 66–73.

Huang, K., & Goddard, J. M. (2015). Influence of fluid milk product composition on fouling and cleaning of Ni–PTFE modified stainless steel heat exchanger surfaces. *Journal of Food Engineering*, *158*, 22–29.

Huo, J., Xiao, J., Kirk, T. V., & Chen, X. D. (2019). Effects of Fluorolink® S10 surface coating on WPC fouling of stainless steel surfaces and subsequent cleaning. *Food and Bioproducts Processing*, *118*, 130–138.

Jindal, S., Anand, S., Metzger, L., & Amamcharla, J. (2018). A comparison of biofilm development on stainless steel and modified-surface plate heat exchangers during a 17-h milk pasteurization run. *Journal of Dairy Science*, *101*(4), 2921–2926.

Jun, S., & Puri, V. M. (2005). 3D milk fouling model of plate heat exchangers using computational fluid dynamics. In *2005 ASAE Annual Meeting* American Society of Agricultural and Biological Engineers. p. 1.

Kılıç, B., & İpek, O. (2017). Experimental investigation of heat transfer and effectiveness in corrugated plate heat exchangers having different chevron angles. *Heat and Mass Transfer*, *53*(2), 725–731.

Klein, R. J., Biesheuvel, P. M., Ben, C. Y., Meinhart, C. D., & Lange, F. F. (2003). Producing super-hydrophobic surfaces with nano-silica spheres. *International Journal of Materials Research*, *94*(4), 377–380.

Kumar, B., Soni, A., & Singh, S. N. (2018). Effect of geometrical parameters on the performance of chevron type plate heat exchanger. *Experimental Thermal and Fluid Science*, *91*, 126–133.

Magnusson, B. (1985). *The origins and evolution of the alfa laval plate heat exchanger*. Stockholm, Sweden: Norstedts Tryckeri.

Manigandan, N., & Suresh, M. (2015). Numerical investigation of a Chevron type brazed plate heat exchanger. In *Vol. 787. Applied mechanics and materials* (pp. 222–226). Trans Tech Publications Ltd.

Merheb, B., Nassar, G., Nongaillard, B., Delaplace, G., & Leuliet, J. C. (2007). Design and performance of a low-frequency non-intrusive acoustic technique for monitoring fouling in plate heat exchangers. *Journal of Food Engineering*, *82*(4), 518–527.

Mota, F. A. S., Carvalho, E. P., & Ravagnani, M. A. S. S. (2015). Modeling and design of plate heat exchanger. In *Heat Transfer Studies and Applications*. Rijeka, Croatia: IntechOpen.

Nilpueng, K., Keawkamrop, T., Ahn, H. S., & Wongwises, S. (2018). Effect of chevron angle and surface roughness on thermal performance of single-phase water flow inside a plate heat exchanger. *International Communications in Heat and Mass Transfer*, *91*, 201–209.

Ozden, H. O., & Puri, V. M. (2010). Computational analysis of fouling by low energy surfaces. *Journal of Food Engineering*, *99*(3), 250–256.

Pandya, N. S., Shah, H., Molana, M., & Tiwari, A. K. (2020). Heat transfer enhancement with nanofluids in plate heat exchangers: A comprehensive review. *European Journal of Mechanics-B/Fluids*, *81*, 173–190.

Piper, M., Zibart, A., Djakow, E., Springer, R., Homberg, W., & Kenig, E. Y. (2019). Heat transfer enhancement in pillow-plate heat exchangers with dimpled surfaces: A numerical study. *Applied Thermal Engineering*, *153*, 142–146.

Rosmaninho, R., Santos, O., Nylander, T., Paulsson, M., Beuf, M., Benezech, T., … Melo, L. F. (2007). Modified stainless steel surfaces targeted to reduce fouling–evaluation of fouling by milk components. *Journal of Food Engineering*, *80*(4), 1176–1187.

Rungraeng, N., Cho, Y. C., Yoon, S. H., & Jun, S. (2012). Carbon nanotube-polytetrafluoroethylene nanocomposite coating for milk fouling reduction in plate heat exchanger. *Journal of Food Engineering*, *111*(2), 218–224.

Serth, R. W., & Lestina, T. (2014). *Process heat transfer: Principles, applications and rules of thumb*. Academic press.

Sharma, A., & Macchietto, S. (2021). Fouling and cleaning of plate heat exchangers: Dairy application. *Food and Bioproducts Processing*, *126*, 32–41.

Tabari, Z. T., & Heris, S. Z. (2015). Heat transfer performance of milk pasteurization plate heat exchangers using MWCNT/water nanofluid. *Journal of Dispersion Science and Technology*, *36*(2), 196–204.

Taghizadeh-Tabari, Z., Heris, S. Z., Moradi, M., & Kahani, M. (2016). The study on application of TiO2/water nanofluid in plate heat exchanger of milk pasteurization industries. *Renewable and Sustainable Energy Reviews*, *58*, 1318–1326.

Taha Ghaleb Toos (TGT). (2022). http://en.tgt-phe.com/product/list/TAHA-SW/9/1/view/.

Tamilselvan, K., Sivabalan, B., Prakash, R., Manojprasath, M., & Mahabubadsha, A. (2017). Experimental analysis of heat transfer rate in corrugated plate heat exchanger using nanofluid in milk pastuerization process. *International Journal of Engineering and Applied Sciences*, *4*(5), 257465.

Thulukkanam, K. (2013). *Heat exchanger design handbook* (2nd ed.). New York: CRC Press.

Tiwari, A. K., Ghosh, P., & Sarkar, J. (2014). Combined energy and exergy analysis of a corrugated plate heat exchanger and experimental investigation. *International Journal of Exergy*, *15*(4), 395–411.

Tran, J. M., Piper, M., Kenig, E. Y., & Scholl, S. (2018). Pillow-plate heat exchangers: Fundamental characteristics. In *Innovative heat exchangers* (pp. 233–245). Cham: Springer.

Wang, S., & B., & Manglik, R. M. (2007). *Plate heat exchangers: Design, applications and performance. Vol. 11*. Southampton, UK: WIT Press.

Verran, J., Boyd, R. D., Hall, K. E., & West, R. (2002). The detection of microorganisms and organic material on stainless steel food contact surfaces. *Biofouling*, *18*(3), 167–176.

Wang, L., Sundén, B., & Richardson, P. (2004). Optimising plate heat exchanger design and operation. In *Improving the thermal processing of foods* (p. 203).

Zhang, J., Zhu, X., Mondejar, M. E., & Haglind, F. (2019). A review of heat transfer enhancement techniques in plate heat exchangers. *Renewable and Sustainable Energy Reviews*, *101*, 305–328.

Zouaghi, S., Frémiot, J., André, C., Grunlan, M. A., Gruescu, C., Delaplace, G., … Jimenez, M. (2019). Investigating the effect of an antifouling surface modification on the environmental impact of a pasteurization process: An lca study. *ACS Sustainable Chemistry & Engineering*, *7*(10), 9133–9142.

Tubular heat exchangers in the food industry

5

Navneet Kumar[a] *and Navin C. Shahi*[b]
[a]Department of Processing and Food Engineering, Anand College of Agricultural Engineering and Technology, Agricultural University, Godhra, Gujarat, India, [b]Department of Post-Harvest Process and Food Engineering, College of Technology, Govind Ballabh Pant University of Agriculture and Technology, Pantnagar, Uttarakhand, India

5.1 Introduction

Heat transfer is required in almost every food-processing industry in several food-processing operations, such as pasteurization, baking, drying, and canning (Fryer & Robbins, 2005). Food ingredients, intermediate processed products, and finished products are heated for so many reasons, one of which is to inactivate the microbial growth and thus preserve and prolong longevity. Food processing is sensitive to sanitation requirements that include easy-to-clean heat exchangers to maintain product quality. An appropriate heat transfer mechanism is inevitable for restricting the potential destruction of organoleptic and nutritional attributes of food, which also saves time and reduces cost of energy required either electricity charges or fuel cost. In addition to the standard design of heat exchangers, other factors of utmost importance are to achieve efficient heat transfer, including avoiding any debris (organic/mineral) deposited on surface of heat exchangers, which is usually avoided using deionized water for heat transfer.

A tubular product-to-product heat exchange operation provides nutritional hygiene practices and thermal energy efficient for dairy and beverage industries. A variety of businesses have their expertise in specialized heat exchangers—offering reliable, cost-effective equipment that meet modern food-processing requirements. The design ensures food hygiene because during cleaning and sterilization phases, microorganisms cannot remain hidden. This function, combined with consistent turbulent flow pattern, generates an easy-to-clean tube heat exchanger. Appropriate design assures material sterility and removes food health hazards relative to conventional heat exchanger systems for product-to-product. Food fluids (e.g., meat, sauces, etc.) handled with machines must be completely eliminated at the end of each processing lot using appropriate cleaning practices. The compromised performance in terms of efficiency is expected in case of uncleaned heat exchangers. The heat exchanger is used to process milk and agro-food products, drinks, and alcohol for energy-efficiency purposes.

Heat exchangers play a major role in the process in the production of juice, sauce, edible oil, syrup, and soup. Evaporation with the support of heat exchangers is used to

Thermal Processing of Food Products by Steam and Hot Water. https://doi.org/10.1016/B978-0-12-818616-9.00006-7

increase the product's sugar concentration. Heat exchangers are used in cookies, peanut butter, fudges, sauces, mayonnaise, and other viscous foods. It has different sections that provide enough direct milk to the cream separator. A holding section exists for protein denaturation and microbial elimination. For minimal human intervention and flawless service, the exchangers have automated PLC control. Various software are also available to improve heat exchanger efficiency for transferring solid or fluid. The heat transfer equations are also utilized for energy-efficient system and control over the performance of system. In designing the equipment, heat transfer equations are utilized to calculate this energy transfer for efficient and controlled performance.

If one or both of the materials exchanging heat are air, moving constantly through the equipment and gaining/providing control, the system remains efficient and is known as continuous-flow heat exchanger. In processing industries, it is used often in pasteurization of milk/liquids, wherein water/steam is used as heating medium. Usually, two fluids pass through different tubes and transfer the energy through the intermediate surface between them. The heat transfer through the pipes remains dependent on the localized difference in temperatures and coefficient of heat transfers. The relative difference in the direction flow also determines the heat transfer rate because of temperature gradient. The parallel flow considers the flow of both the liquids in same direction, whereas opposite and cross-flow depict the flow in opposite and right-angle direction of flow, respectively. The flow direction may have distinct directions in various sections of heat exchangers and follows a mixed flow pattern.

The temperature difference between hot fluid and cold fluid at the entry point is highest in parallel heat exchanger, but temperature difference at exit become very little, which reduces the rate of heat transfer considerably; in comparison, counter-flow heat exchangers maintain the difference and effective heat transfer, therefore, counter-flow heat exchangers are often preferred. The corrugation on heat transfer surface also boosts the heat exchange process by creating turbulence in both the fluids and has a scope of increasing the efficiency up to 90%.

In the food industry, exchangers are used to remove microbials. Therefore, essentially, the device's primary purpose is to make food and beverages healthy to eat. It automatically increases food shelf-life. The exchanger is used to heat or cool different products before they are used for loading, processing, concentration, and other procedures. Models can vary based on particle viscosity and thickness. Based on these two considerations, together with the industry's requirement, the heat exchanger style is determined.

Those factors may also be discussed here. The cycle exchanger must have enough momentum across the plate to regulate the pressure drop during the heat transfer time. Recent years have seen a significant increase in demand for nutraceuticals industry, which is why heat exchangers are becoming increasingly necessary. More than ever, food safety regulations are being developed and enforced. Among the tools or machines that help reduce human intervention in processing nutraceuticals, mainly whey products and pharmaceutical products, various types of heat exchangers make their mark the most significant.

The tubular heat exchangers are used for heating oils, fuels, combusting air, and biogas heating systems. Engineers' works, designs, and provides tailor-made heat exchange systems and tubular heat transfer solutions for the field of application as double-tube, single-tube, shell-and-tube, shell-and-coil, and finned-tube heat exchangers. Preferably, the best quality material is used for the manufacturing of tubular heat exchangers. This is exhaustively used in food, oils, chemicals, biofuels, and agriculture products.

5.2 Scope and importance of tubular heat exchangers

Cylindrical tubes are the basic unit of tubular heat exchangers; however, some systems also use elliptical, round/flat twisted or triangular tubes. The design is quite versatile as the core configuration can easily be adjusted by changing various parameters like tube diameter, weight, and shape. Tubular heat exchangers can be designed for higher atmospheric pressure and high-pressure fluid variants. Tubular heat exchangers are most commonly used for liquid-to-liquid and liquid-to-phase (condensing otherwise evaporating) heat transfer applications. The main applications of such exchangers are gas-to-liquid and gas-to-gas heat transfer applications, when the functioning temperature or pressure is very high, or fouling on at least one fluid side is a thoughtful demerit and also no other forms of exchangers can operate. Such exchangers can be categorized into twin-pipe, shell-and-tube, and spiral-tube exchangers. They all are top surface bar exchangers with fins within/outside channels. Various types of tubular heat exchanger categorized are presented in Fig. 5.1.

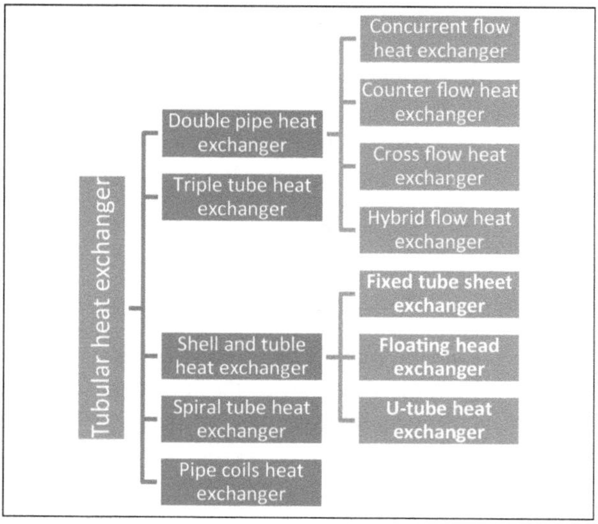

Fig. 5.1 Various types of heat exchangers.

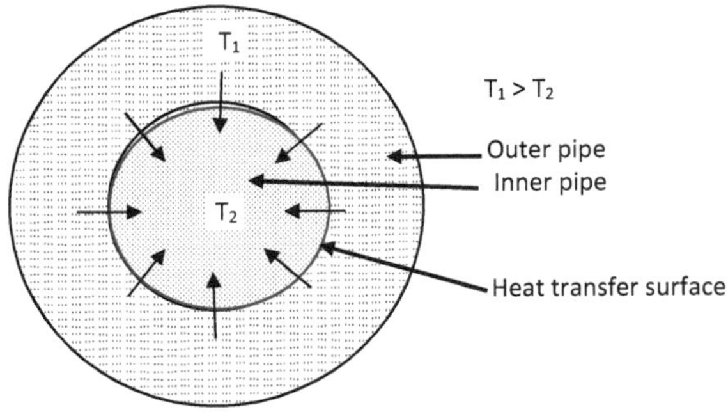

Fig. 5.2 Heat transfer mechanism in double-pipe heat exchanger.

5.3 Types of tubular heat exchangers

As above, all tubular heat exchangers work with transfer of heat using conduction, convection, or radiation and governed by the laws of thermodynamics. These are subcategorized on the basis of their features, construction methods, flow configurations, and heat transfer mechanism used for effective transfer of heat and product characteristics.

5.3.1 Double-pipe heat exchangers

The simplest kind of heat exchanger is the double-pipe heat exchanger, in which one pipe is concentrically contained inside a bigger diameter pipe, which relates with its name (Bhavsar, Matawala, & Dixit, 2013). Any one of the liquids for heat transfer travels in inner pipe, which transfers (gain/loose) energy through the conductive surface of the inner pipe to the liquid traveling in the annulus gap in between two pipes (Fig. 5.2). The outside fluid flows over the inner liquid, allowing heat exchange through the inner tube walls. The structure of double-pipe heat exchanger is performed in accordance to the flow arrangement, which relates to fluid motion direction within the heat exchanger relative with one another. Heat exchangers use four main flow configurations.

5.3.1.1 Concurrent/parallel-flow heat exchangers

Concurrent-flow heat exchangers, often referred to as parallel-flow heat exchangers, are heat exchangers in which the direction of the movement of flows is the same (Fig. 5.3). While this design is usually less powerful than counter-flow setup, it also permits for maximum thermal uniformity across the heat exchanger casing or walls. Nevertheless, there are areas where parallel flow has its benefits, such as required limited heat transfer. The temperature of the cooled fluid also reaches to a limiting temperature.

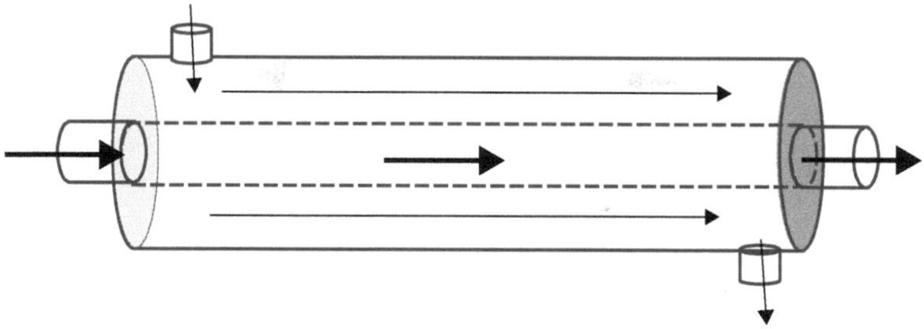

Fig. 5.3 Concurrent-flow heat exchanger.

5.3.1.2 Counter-current flow heat exchangers

Counter-current flow heat exchangers, often referred to as counter-flow heat exchangers, are structured to facilitate movement of antiparallel (i.e., parallel but in counter directions) flows inside the heat exchanger (Fig. 5.4). The most widely used of flow configurations, a counter-flow system usually exhibits the highest efficiencies as it enables the largest amount of heat transfer between fluids and thus the greater temperature change. It is advantageous to concurrent-flow heat exchanger due to providing uniform temperature difference between the fluids throughout the exchanger, approaching higher temperature of cold fluids at outlets near to hot fluid temperature, more uniform heat transfer and reduction in large thermal stresses due to lower temperature difference at the inlets.

5.3.1.3 Cross-flow heat exchangers

These heat exchangers provide the flow of one liquid perpendicularly to the flow of the other liquids (Fig. 5.5). These are preferred, where one fluid changes its state from liquid to vapor or vapor to liquid as in case of condensers, which condenses the steams and convert into water. It has the capability to handle more volume of vapors and

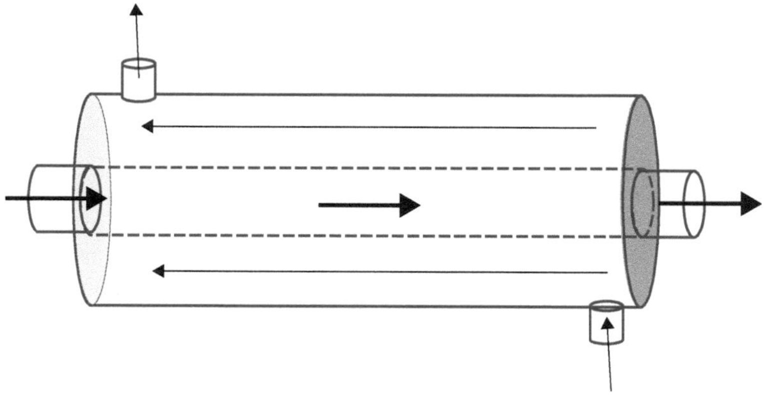

Fig. 5.4 Counter-current-flow heat exchanger.

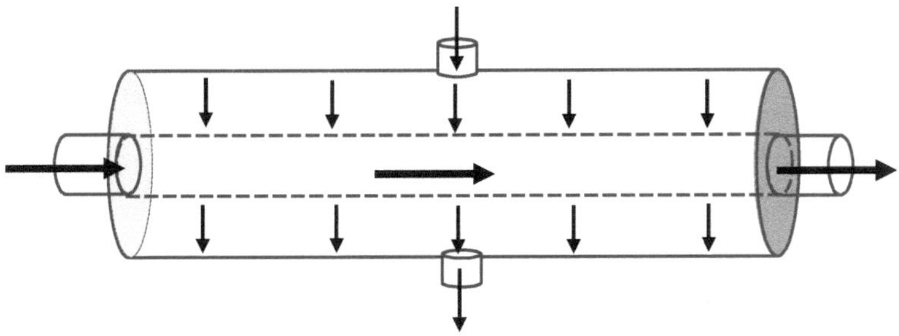

Fig. 5.5 Cross-flow heat exchangers.

heat exchanger efficiencies remained between concurrent and counter-current heat exchangers.

5.3.1.4 Hybrid-flow heat exchangers

These heat exchangers display a variation in exchanger characteristics for the concurrent, counter, and cross-flow heat exchanger configurations (Fig. 5.6). One liquid flows in a pipe, which has multiple placements across the flow of other liquid. It has concurrent/counter-current with cross-flow arrangement in a single unit. These are designed to meet constraints in application, such as availability of space for accommodating the size, expense, or temperature and pressure specifications require for the product.

5.3.2 Triple-tube heat exchangers

These are fabricated by placing three concentric tubes (Fig. 5.7). The heating or cooling mediums are allowed to flow from the innermost tube and the outermost annulus gap, while the food material is kept in the middle annulus space provided (Dwivedee, Arya, Diwakar, & Khare, 2019). The direction of flow of heating and cooling medium is kept opposite to the flow of food material. In several applications,

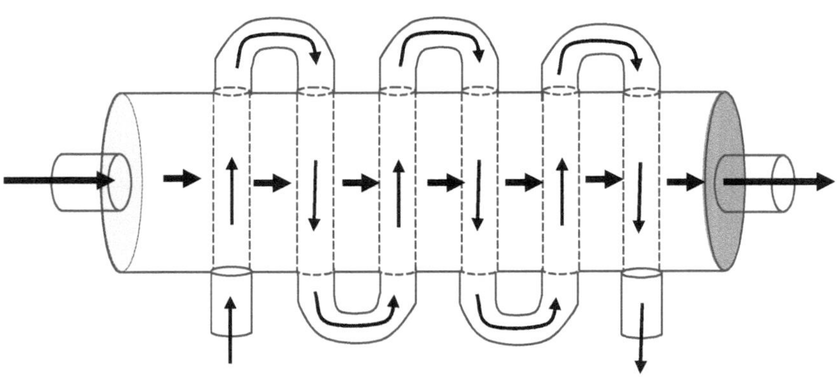

Fig. 5.6 Hybrid (cross and concurrent)-flow heat exchangers.

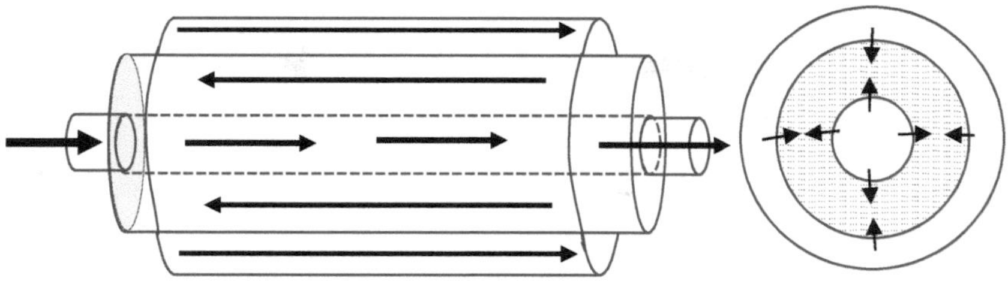

Fig. 5.7 Triple-tube heat exchanger.

the warm liquid travels in one direction across the inside and outside tubing, while the cooling fluid moves through the middle channel in the opposite direction. It requires smaller floor area to attain similar heat transfer rates, higher heating or cooling rates, and reduction in heat losses due to shorter processing time in comparison to double-tube heat exchangers. These were preferred for the pasteurization of milk earlier, looking into the higher heat transfer rates and safe and validated processing. Triple-tube heat exchangers are also preferred in rapid regeneration systems by providing good surface area since heat transfer occurs on all three tubes. Other added advantage of handling high pulp or viscous product shows its versatility in processing industries.

5.3.3 Shell-and-tube heat exchangers

These exchangers function by bringing two working fluids in thermal contact of each other using tubes enclosed in exterior cylindrical shell (Fig. 5.8). These are manufactured using highly conducive materials like steel and aluminum. One fluid is allowed to flow in tubes through the inlet provide on one side and discharge through the out end (tube side), while another fluid is passed through the container (shell side). A collection/bundle of tubes consisting of numerous tubes according to the surface area required for desirable heat transfer rate is provided in a shell. These are compact and easily manufactured, easier in operation, and provide excellent heat transfer (Hewitt, Shires, & Bott, 1994). These are effectively used in preheating sections, manufacturing units, evaporators, condensers, coolers, and so on.

The basic concept or principle of these exchangers is to transfer heat from a hot fluid to a cold fluid. It can be seen from Fig. 5.8 that the feed is filled in inlet header section, which is separated by tube plates and flow through the tubes and exits from the outlet

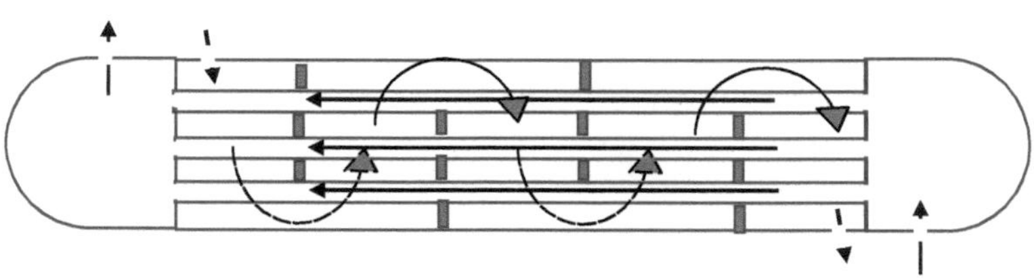

Fig. 5.8 Shell-and-tube heat exchangers.

header sections. However, other fluid is passed through shell input and exits the shell. The tubes may be coupled with turbulators to induce the turbulent action and reduction in fouling, which enhances heat transfer rates. Baffles are also provided to create turbulence in the fluid passing in shell section for effective heat transfer.

5.3.3.1 Fixed-tube sheet exchangers

The tube sheets are made stationary and welded to the plate directly, which makes it the most cost-effective option due to its simplicity in various shell-and-tube exchangers (Fig. 5.9). However, these do not work well with higher temperature variation between the temperature of fluids flowing in shell and tube, which may cause expansion of tubes. Also, the shell side cannot be cleaned externally; therefore, it should not be used for liquids creating the problem of foul and decreasing the efficiency of the exchanger.

5.3.3.2 Floating-head exchangers

These exchangers incorporate fine aspects for other exchangers by incorporating floating tube sheet at one end, although other end is kept stationary similar to earlier exchangers (Fig. 5.10). The floating parts allow tubes to extend freely while heat transfer took place at higher temperatures and minimizing the risk of damaging the structure. The floating-head exchanger is the optimal heat exchanger for performance and operation but has a higher cost in making the head moveable.

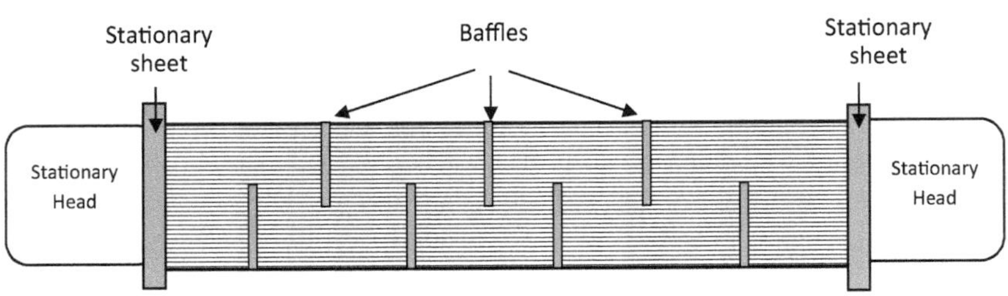

Fig. 5.9 Fixed-tube sheet heat exchanger.

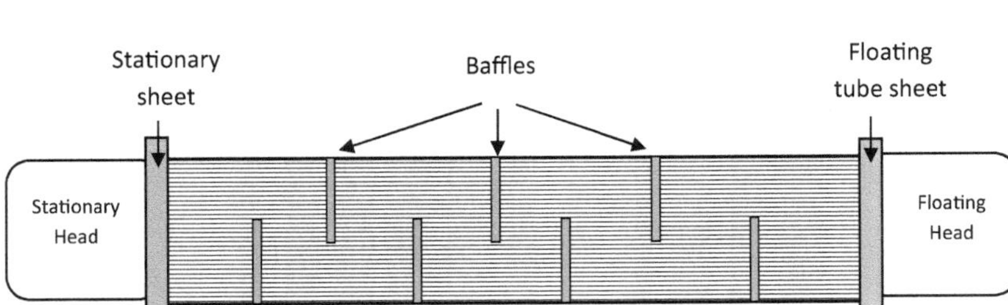

Fig. 5.10 Floating-head heat exchanger.

5.3.3.3 U-tube heat exchangers

These consist of tubes in the form of the alphabetic character "U," which remain static on one tube plate (Fig. 5.11). The header portion is divided into two halves for providing multiple pass implicitly. The tube bend allows thermal expansion without any sort of damage in expansion joints, as the bend side in the shell is free-floating and has space for extension/contract. They are efficient in working and excellent in performance at higher temperatures, where expansion is expected, however cleaning of bands remains a limitation.

5.3.4 Spiral-tube heat exchanger

These heat exchangers have a coil system, which is mounted on a small case for maximizing heat transfer efficiency in smaller space. These coils are made of stainless steel and welded at joints. The coil assembly remains compact in a shell. The tubes are joined using various configurations and the annulus space left between tubes work as shell side. The fluids also exert centrifugal force while flowing to the heat transfer surface, which enhances the rate of heat transfer. Heat-transfer performance is also optimized without the use of any baffle. A spiral pattern is very compact, working in a shell-and-tube configuration. Tube bundle removal space requirements are virtually eliminated as the tube bundle coiled. It also minimizes the material used in various parts, namely, manifolds, heads, tube sheets, and so on. The shell side remains smaller than other designs and supports for the tubes are not required. The basic heat transfer mechanism is represented in Fig. 5.12. The fluids (hot and cold) are separated by tubes and high heat transfer rates can be obtained. The fluids are allowed to flow in a counter-current direction.

5.3.5 Pipe/shell coils heat exchangers

These are very simple and easily manufactured. A cope coil, which is helically rolled, is kept inside a circular shell or pipe and is generally placed at the centerline (Fig. 5.13). The cold fluid flows through the pipe and hot fluid passes through the tube. The temperature of both fluids can be measured. Generally, pipe/shell and coil

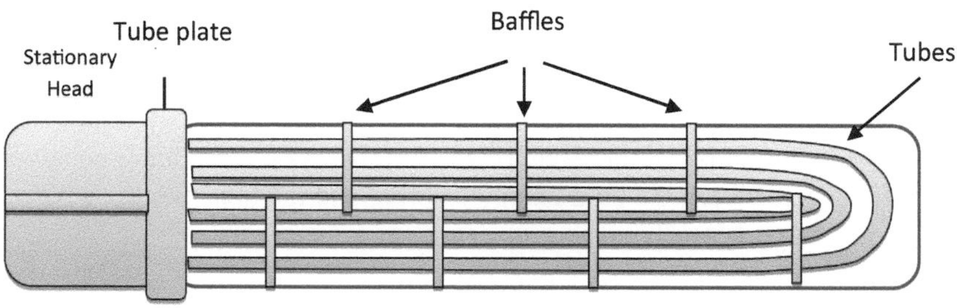

Fig. 5.11 U-tube heat exchanger.

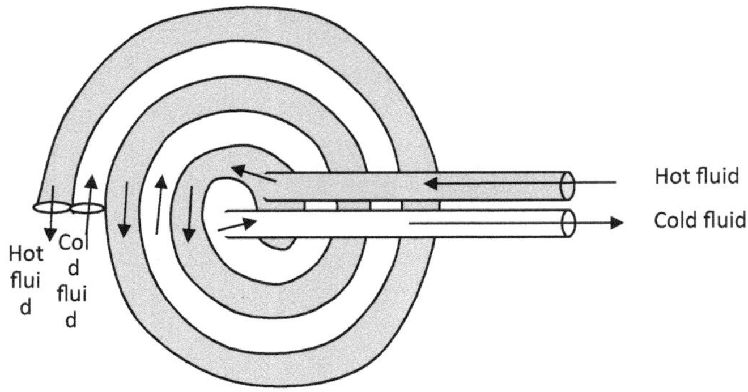

Fig. 5.12 Spiral-tube heat exchanger.

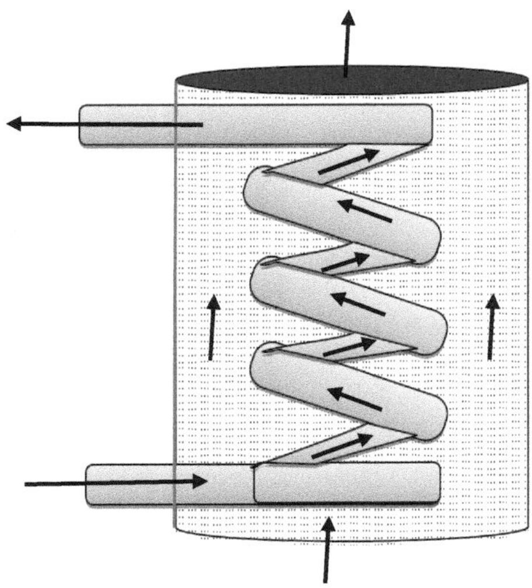

Fig. 5.13 Pipe/shell coil heat exchanger.

heat exchanger is usually employed to heat smaller systems. These can be placed in pilot plants, diary plants, and food-processing plants. These are easier in construction and operation.

5.4 Energy balance

According to principle of conservation of energy, "the net change either increase or decrease in the total energy of the system during a process or a reaction is equal to the energy difference between the total energy entering and the total energy leaving the system during that process."

$$\text{Total energy of the system} - \text{Total energy leaving that system} = \text{Change in the total energy of the system}$$

$$E_{in} - E_{out} = \Delta E_{system}$$

Energy change in a system, ΔE_{system}

$$\text{Energy Change} = \text{Energy at final state} - \text{Energy at initial state}$$

$$\Delta E_{system} = E_{final} - E_{initial} = E_2 - E_1$$

where $\Delta E = \Delta U + \Delta KE + \Delta PE$; ΔE, change in energy; E_{in} amount of input energy; E_{out} Amount of output energy.

Modes of Energy

Energy is versatile. It can be transferred to or from a system in mainly three forms as work and mass flow, and heat transfer.

(i) Heat transfer, Q

Heat transfer to a system increases molecular energy and thus system internal energy, and the heat transfer from the system (heat loss) decreases. The energy transferred out as heat comes from the molecules of the system. The transfer of heat may be performed using (a) conduction, (b) convection, and (c) radiation.

(ii) Work, W

It is an energy interaction that is not caused by a temperature difference between a system and its surroundings. It is the mechanical interaction with in the system.

(iii) Mass flow, m

The energy transfer also occurs due to flow of mass entering and exiting the system. The increase in the energy of a system is obvious due to entering some mass due to the amount of energy held within the mass. Similarly, energy of the system decreases with the exiting of mass from the system. Energy can be transferred in the forms of work, mass, and heat (Fig. 5.14), and the net transfer of a quantity is equal to the difference between the amounts transferred in and out, the energy balance can be written as

$$E_{in} - E_{out} = (Q_{in} - Q_{out}) + (W_{in} - W_{out}) + (E_{mass,in} - E_{mass,out}) = \Delta E_{system}$$

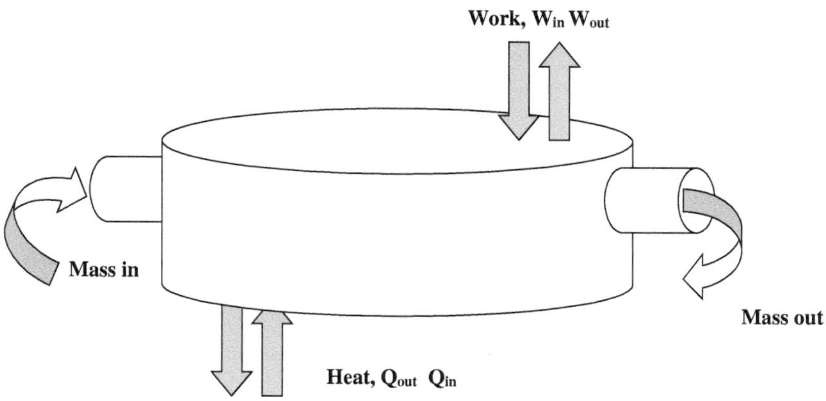

Fig. 5.14 Different modes of energy transfer of a system.

Example 5.1

In a dairy plant having the capacity to process the milk at a rate of 1000 kg/h for heating from 40 to 72°C in a heat exchanger, the heating medium (water) enters the heat exchanger at 95°C and leaves at 80°C. If heat lost to the ambiance is 2 kW, estimate mass flow rate of the heating is medium. The heat capacities of water and milk are 4.2 and 3.9 kJ/kg°C, respectively.

Solution:

The mass and energy flow diagram (Fig. 5.15) represent:

$$\text{Rate of energy input} = \dot{m}_{m,\text{in}} \times H_{m,\text{in}} + \dot{m}_{w,\text{in}} \times H_{w,\text{in}}$$

$$\text{Rate of energy output} = \dot{m}_{m,\text{out}} \times H_{m,\text{out}} + \dot{m}_{w,\text{out}} \times H_{w,\text{out}} + q$$

where $\dot{m}_{m,\text{in}}$, $\dot{m}_{m,\text{out}}$ is input and output mass flow rate of milk, respectively; $H_{m,\text{ in}}$, $H_{m,\text{out}}$ enthalpy of input and output milk, respectively; $\dot{m}_{w,\text{in}}$, $\dot{m}_{w,\text{out}}$ input and output mass flow rate of water, respectively; $H_{w,\text{in}}$, $H_{w,\text{out}}$ enthalpy of input and output water, respectively.

In steady-state condition.

Rate of energy input should be equal to rate of energy output, therefore,

$$\dot{m}_{m,\text{in}} \times H_{m,\text{in}} + \dot{m}_{w,\text{in}} \times H_{w,\text{in}} = \dot{m}_{m,\text{out}} \times H_{m,\text{out}} + \dot{m}_{w,\text{out}} \times H_{w,\text{out}} + q.$$

Given

$$\dot{m}_{m,\text{in}} = \dot{m}_{m,\text{out}} = 1000\,\text{kg/h}$$
$$H_{m,\text{in}} = C_pT = 3.9 \times 40 = 156\,\text{kJ/kg}$$
$$H_{m,\text{out}} = C_pT = 3.9 \times 72 = 288.8\,\text{kJ/kg}$$
$$\dot{m}_{w,\text{in}} = \dot{m}_{w,\text{out}} = \text{to be calculated } (\dot{m}_w)$$
$$H_{w,\text{in}} = C_pT = 4.2 \times 95 = 399\,\text{kJ/kg}$$
$$H_{w,\text{out}} = C_pT = 4.2 \times 80 = 336\,\text{kJ/kg}$$
$$q = 2\,\text{kW} = 2 \times 60 \times 60 = 7200\,\text{kJ/h}$$
$$1000 \times 156 + \dot{m}_w \times 399 = 1000 \times 288.8 + \dot{m}_w \times 336 + 7200$$
$$\dot{m}_w \times 63 = 1000 \times 132.8 + 7200$$
$$\dot{m}_w = 2222.22\,\text{kg/h}$$

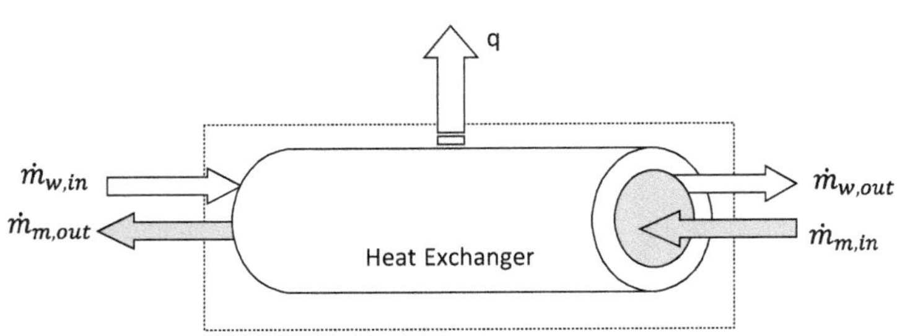

Fig. 5.15 Flow chart of energy balance in heat exchanger.

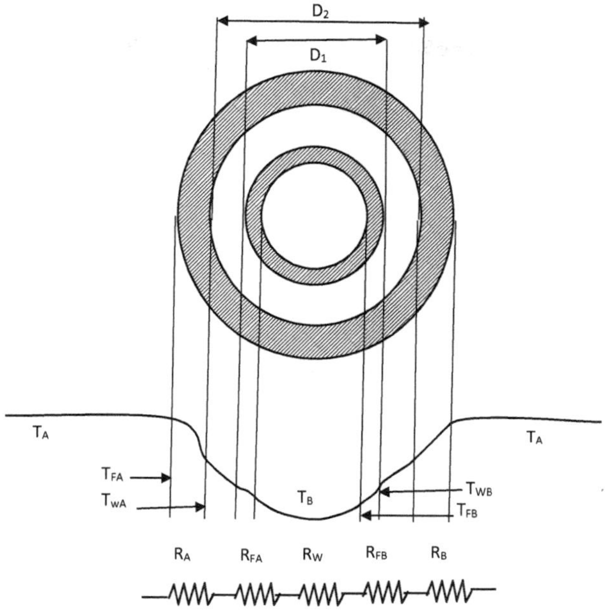

Fig. 5.16 Temperature variation in tubular heat exchanger.

5.5 Overall heat transfer coefficient

Heat is being transported from fluid A, which has bulk temperature, T_A, through a planer wall of thickness x into fluid B, whose bulk temperature is T_B. The fouling layers exist on both the sides of the wall (Fig. 5.16). Heat flux, q, is due to the temperature driving force $(T_A - T_B)$ by introducing a total resistance to heat transfer R_{Total}. Let us assume that D_1 and D_2 are the inner and outer diameter and L is the length.

$$\text{Heat flux, } q = \frac{T_A - T_B}{R_{Tolal}} = U(T_A - T_B) \tag{5.1}$$

$$R_{Total} = R_A + R_B + R_W + R_{FA} + R_{FB}$$

R_A and R_B are resistances due to convective heat transfer in two fluids; R_{FA} and R_{FB} Fouling resistance in the respective fouling layers.

$$R_w = \frac{\log\left(D_1/D_2\right)}{2\pi k_w L}$$

where k_w is the thermal conductivity of the tube wall material.

Since, heat transfers from outside surface to fluid B

$$Q = \pi D_2 L \cdot U(T_A - T_B) \tag{5.2}$$

where U is overall heat transfer coefficient.

From Eqs. (5.1) and (5.2)

$$q = \frac{T_A - T_B}{R_{Total}} = U(T_A - T_B) = \frac{Q}{\pi D_2 L} \tag{5.3}$$

$$\frac{1}{U} = R_A + R_B + R_{FA} + R_{FB} + R_W \tag{5.4}$$

$$\frac{1}{U} = R_{FA} + R_{FB} + \frac{1}{h_A} + \frac{1}{h_B} + \frac{x}{k_w} \tag{5.5}$$

where h_A and h_B are convective filer heat transfer coefficient.

It is assumed that the fouling layer thickness is small:

$$Q = \pi D_2 LU(T_B - T_A) = \pi D_2\, L \propto_A (T_{FA} - T_A) = \pi D_2 L \frac{(T_{WA} - T_{FA})}{R_{FA}}$$
$$= \frac{T_{WB} - T_{WA}}{R_W}$$

$$= \pi D_1 L \frac{(T_{FB} - T_{WB})}{R_{FB}} = \pi D_1\, L \propto_B (T_B - T_{FB}) \tag{5.6}$$

$$T_{FA} - T_A = \frac{Q}{\pi D_2 L \alpha_A} \tag{5.7}$$

$$T_{WA} - T_{FA} = \frac{Q \cdot R_{FA}}{\pi D_2 L} \tag{5.8}$$

$$T_{WB} - T_{WA} = Q \cdot \frac{\log \left(D_1/D_2 \right)}{2\pi k_w L} \tag{5.9}$$

$$T_{FB} - T_{WB} = \frac{Q \cdot R_{FB}}{\pi D_1 L} \tag{5.10}$$

$$T_B - T_{FA} = \frac{Q}{\pi D_1 L\, \alpha_B} \tag{5.11}$$

Sum of equations from (7) to (11)

$$(T_B - T_A) = \frac{Q}{\pi D_2 L} \left[\frac{1}{\propto_A} + R_{fA} + \frac{D_2 \log \left(D_1/D_2 \right)}{2k_w} + \frac{D_2}{D_1} R_{fB} + \frac{D_2}{D_1} \frac{1}{\propto_B} \right] \tag{5.12}$$

From Eqs. (5.3) and (5.12)

$$\frac{1}{U}=\frac{1}{\propto_A}+R_{fA}+\frac{D_2\log\left(^{D_1}/_{D_2}\right)}{2k_w}+\frac{D_2}{D_1}R_{fB}+\frac{D_2}{D_1}\frac{1}{\propto_B} \tag{5.13}$$

Example 5.2

A double-pipe heat exchanger, consisting of 2 cm internal diameter and 2.5 cm external diameter tube mounted in a concentric outer tube, is used to cool a hot oil. The oil flows in the inside tube and has a heat transfer coefficient of 200 W/m^2 K and a fouling resistance of 0.008 m^{2}K/W. The cooling water flows in the annulus between the inner and outer pipe. Its heat transfer coefficient is 5000 W/m^2 K and *h* fouling resistance is 0.0002 m^2 K/W. The tube wall thermal conductivity is 100 W/m K. Calculate the overall heat transfer coefficient for this heat exchange.

Solution:

$$\frac{1}{U}=R_{fA}+R_{fB}\frac{D_2}{D_1}+\frac{D_2}{2kw}\ln\frac{D_2}{D_1}+\frac{1}{\alpha A}+\frac{D_2}{D_1\alpha_B}$$

$$=0.0002+0.0008\times\frac{0.025}{0.020}+\frac{0.025}{2\times 100}\ln\frac{0.025}{0.020}+\frac{1}{5000}+\frac{0.025}{200\times 0.020}=0.007678$$

$U=130.2\,\text{W/m}^2\,\text{K}$.

5.6 Log mean temperature difference

5.6.1 Counter-flow heat exchangers

Inlet temperature of hot fluid $T_{h,\text{in}}$.

Outlet temperature of hot fluid $T_{h,\text{out}}$.

Inlet temperature of cold fluid $T_{c,\text{in}}$.

Outlet temperature of cold fluid $T_{c,\text{out}}$.

ΔT_{LM}—log mean temperature deference

$$\Delta T_{LM}=\frac{\Delta T_1-\Delta T_2}{\ln\left(\Delta T_1/\Delta T_2\right)}$$

$$\Delta T_{LM}=\frac{\left(T_{h,\text{in}}-T_{c,\text{out}}\right)-\left(T_{h,\text{out}}-T_{c,\text{in}}\right)}{\ln\left(\left(T_{h,\text{in}}-T_{c,\text{out}}\right)/\left(T_{h,\text{out}}-T_{c,\text{in}}\right)\right)} \tag{5.14}$$

Heat balance

$$Q=UA\Delta T_{LM}=M_cC_{pc}\left(T_{c,\text{out}}-T_{c,\text{in}}\right)=M_hC_{ph}\left(T_{h,\text{in}}-T_{h,\text{out}}\right) \tag{5.15}$$

where M_c is the mass flow rate of the cold fluid (kg/s), M_h is the mass flow rate of the hot fluid (kg/s), C_{pc} is the specific heat of cold fluid (J/kg K), and C_{ph} is the specific heat of hot fluid (J/kg K).

ΔT_{LM} is expressed in the term of two temperature ratio R and P.

1. Heat capacity rate ratio,

$$R = \frac{T_{h\ \text{in}} - T_{h\ \text{out}}}{T_{c\ \text{out}} - T_{c\ \text{in}}} = \frac{M_c\, C_{pc}}{M_h\, C_{ph}} \tag{5.16}$$

2. Temperature effectiveness,

$$P = \frac{T_{c\ \text{out}} - T_{c\ \text{in}}}{T_{h\ \text{in}} - T_{c\ \text{in}}} \tag{5.17}$$

$$PR = \frac{\cancel{T_{c\ \text{out}} - T_{c\ \text{in}}}}{T_{h\ \text{in}} - T_{c\ \text{in}}} \times \frac{T_{h\ \text{in}} - T_{h\ \text{out}}}{\cancel{T_{c\ \text{out}} - T_{c\ \text{in}}}}$$

$$PR = \frac{T_{h\ \text{in}} - T_{h\ \text{out}}}{T_{h\ \text{in}} - T_{c\ \text{in}}}, \tag{5.18}$$

From Eqs. (5.14), (5.16)–(5.18)

$$P(R - 1) = (T_{h,\text{in}} - T_{c,\text{out}}) - (T_{h,\text{out}} - T_{c,\text{in}}) \tag{5.19}$$

$$\frac{(1 - P)}{(1 - PR)} = \frac{T_{h\ \text{in}} - T_{c\ \text{out}}}{T_{h\ \text{out}} - T_{c\ \text{in}}} \tag{5.20}$$

$$\Delta T_{LM} = \frac{P(R-1)}{\ln\left[(1-P)/1-PR\right]}(T_{h\ \text{in}} - T_{c\ \text{in}}) = \frac{Q}{UA} = \frac{M_c\, C_{pc}(T_{c\ \text{out}} - T_{c\ \text{in}})}{UA}$$

$$\frac{P(R-1)}{\ln\left[(1-P)/1-PR\right]} \times \frac{(T_{h\ \text{in}} - T_{c\ \text{in}})}{(T_{c\ \text{out}} - T_{c\ \text{in}})} = \frac{M_c\, C_{pc}}{UA}$$

$$\frac{M_c\, C_{cp}}{UA} = \frac{(R-1)}{\ln\left[(1-P)/1-PR\right]} \tag{5.21}$$

Similarly

$$\frac{M_h\, C_{ph}}{UA} = \frac{(R-1)/R}{\ln\left[(1-P)/(1-PR)\right]} \tag{5.22}$$

1. When U, A, R, and either M_c or M_h are known, then calculate P directly and obtain $T_{c\ \text{out}}$ for the $T_{h\ \text{out}}$.
2. When U, R, P, and either M_c or M_h are known, then calculate A and the length of the pipe.

5.6.2 Parallel/series arrangement of double pipe heat exchangers

The parallel arrangement of double-pipe exchangers (Fig. 5.17) represents assumption (Çengel & Ghajar, 2011; Holman, 2010):

(1) The overall heat transfer coefficient U is constant.

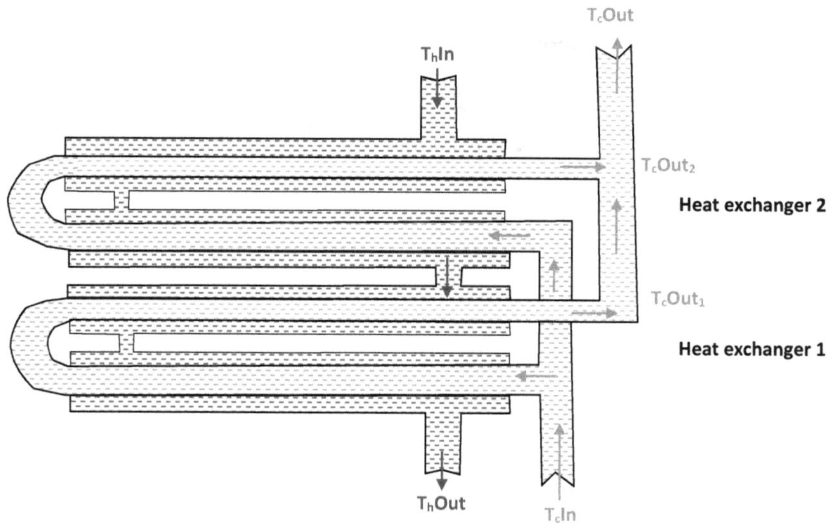

Fig. 5.17 Parallel/Series arrangement of double-pipe heat exchangers.

(2) The specific heat of the two fluids is constant.
(3) The cold stream mass flow is split equally between the two units.
(4) The heat exchanger areas of the two units are equal.

Rate of heat transfer, $Q = UA\,\Delta T_m$

Consider two tubular heat exchangers are used and following temperature and fluid flow temperatures are listed below:

Cold fluid in parallel and hot fluid in series		
	HE1	**HE2**
Hot fluid inlet temperature	$T_{h,\text{int}}$	$T_{n\ \text{in}}$
Hot fluid outlet temperature	$T_{h,\text{out}}$	$T_{h,\text{out}}$
Cold fluid intel temperature	$T_{c\ \text{in}}$	$T_{c\ \text{in}}$
Cold fluid outlet temperature	$T_{c,\text{out1}}$	$T_{c,\text{out2}}$
Hot fluid flow rate	M_h	M_h
Cold fluid flow rate	$M_c/2$	$M_c/2$
Surface area	$A/2$	$A/2$

Ratio of the heat capacities (R') of two fluid streams is same in each HE, that is,

$$R' = \frac{(M_c/2)C_{pc}}{M_h C_{ph}} = \frac{1}{2}\frac{M_c C_{pc}}{M_h C_{ph}} = \frac{1}{2}R \tag{5.23}$$

As per assumption, U is same for both HE.
From Eq. (5.21)

$$\frac{M_c C_{pc}}{UA} = \frac{(R-1)}{\ln((1-P)/(1-PR))} \text{ and } U\&R \text{ are constant}$$

Hence, the value of temperature ratio P' also must be same for both HE.

$$P' = \frac{T_{c,\text{out1}} - T_{c,\text{in}}}{T_{h,\text{int}} - T_{c,\text{in}}} = \frac{T_{c,\text{out2}} - T_{c,\text{in}}}{T_{h,\text{in}} - T_{c,\text{in}}} \tag{5.24}$$

For HE 1, heat transfer rate, Q_1

$$Q_1 = U\left(\frac{A}{2}\right).\Delta T_{LM,H.E.1} = U\left(\frac{A}{2}\right).\left[\frac{(T_{h,\text{int}} - T_{c,\text{out1}}) - (T_{h,\text{out}} - T_{c,\text{in}})}{In\left[\frac{T_{h,\text{int}} - T_{c,\text{out1}}}{T_{h,\text{out}} - T_{c,\text{in}}}\right]}\right]$$
$$= M_h.C_{ph}(T_{h,\text{int}} - T_{h,\text{out}}) \tag{5.25}$$

From Eqs. (5.22) and (5.25)

$$\frac{M_h\, C_{ph}}{U\left(\frac{A}{2}\right)} = \frac{(R'-1)/R'}{In\left[\frac{T_{h,\text{int}} - T_{c,\text{out1}}}{T_{h,\text{out}} - T_{c,\text{in}}}\right]} \tag{5.26}$$

Overall heat transfer rate, Q

$$Q = U.A.\Delta T_{LM} = M_h.C_{ph}(T_{h,\text{in}} - T_{h,\text{out}}) \tag{5.27}$$

From Eqs. (5.26) and (5.27)

$$\Delta T_{LM} = \frac{M_h.C_{ph}(T_{h,\text{in}} - T_{h,\text{out}})}{U.A} = \frac{[(R'-1)/2R'](T_{h,\text{in}} - T_{h,\text{out}})}{In\left[\frac{T_{h,\text{int}} - T_{c,\text{out1}}}{T_{h,\text{out}} - T_{c,\text{in}}}\right]} \tag{5.28}$$

P' and R' also have the same value in both HE1 and HE2
From Eq. (5.18)

$$P'R' = \frac{T_{h,\text{in}} - T_{h,\text{int}}}{T_{h,\text{in}} - T_{c,\text{in}}} \text{ (for H.E.2)}$$

$$P'R' = \frac{T_{h,\text{int}} - T_{h,\text{out}}}{T_{h,\text{int}} - T_{c,\text{in}}} \text{ (For H.E.1)}$$

$P'R'$ have the same value.

$$\frac{T_{h,\text{in}} - T_{h,\text{int}}}{T_{h,\text{in}} - T_{c,\text{in}}} = \frac{T_{h,\text{int}} - T_{h,\text{out}}}{T_{h,\text{int}} - T_{c,\text{in}}}$$

$$\cancel{T_{h,\text{in}}.T_{h,\text{int}}} - T_{h,\text{int}}^2 - T_{h,\text{in}}.T_{c,\text{in}} + T_{h,\text{int}}.T_{c,\text{in}}$$
$$= \cancel{T_{h,\text{in}}.T_{h,\text{int}}} - T_{h,\text{in}}.T_{h,\text{out}} - T_{c,\text{in}}.T_{h,\text{int}} + T_{c,\text{in}}.T_{h\,\text{out}}$$

$$T_{h,\text{int}}^2 - 2T_{c,\text{in}}.T_{h,\text{int}} + T_{c,\text{in}}(T_{h,\text{in}} + T_{h\,\text{out}}) - T_{h,\text{in}}.T_{h,\text{out}} = 0$$

Since $x = \frac{-b \pm \sqrt{b^2 - 4ac}}{2a}$ for $ax^2 + bx + c = 0$

$$T_{h,\text{int}} = T_{c,\text{in}} \pm \sqrt{T_{c,\text{in}}^2 - T_{c,\text{in}}T_{h,\text{in}} - T_{c,\text{in}}T_{h,\text{out}} + T_{h,\text{in}}.T_{h,\text{out}}}$$

$$T_{h,\text{int}} = T_{c,\text{in}} \pm \sqrt{T_{c,\text{in}}(T_{c,\text{in}} - T_{h,\text{in}}) - T_{h,\text{out}}(T_{c,\text{in}} - T_{h,\text{in}})}$$

$$T_{h,\text{int}} = T_{c,\text{in}} \pm \sqrt{(T_{c,\text{in}} - T_{h,\text{in}})(T_{c,\text{in}} - T_{h,\text{out}})}$$

$$T_{h,\text{int}} = T_{c,\text{in}} \pm \sqrt{(T_{h,\text{in}} - T_{c,\text{in}})(T_{h,\text{out}} - T_{c,\text{in}})}$$

Since $T_{h,\text{int}} > T_{c\text{in}}$

$$T_{h,\text{int}} = T_{c,\text{in}} + \sqrt{(T_{h,\text{in}} - T_{c,\text{in}})(T_{h,\text{out}} - T_{c,\text{in}})}$$

$$\frac{T_{h,\text{int}} - T_{c,\text{in}}}{T_{h,\text{out}} - T_{c,\text{in}}} = \sqrt{\frac{T_{h,\text{in}} - T_{c,\text{in}}}{T_{h,\text{out}} - T_{c,\text{in}}}} \tag{5.29}$$

For HE1 (from Eq. 5.26),

$$\frac{M_h\, C_{ph}}{U\left(\frac{A}{2}\right)} = \frac{(R' - 1)/R'}{\ln\left[\frac{T_{h,\text{int}} - T_{c,\text{out1}}}{T_{h,\text{out}} - T_{c,\text{in}}}\right]} \tag{5.30}$$

where $R' = \frac{T_{h,\text{int}} - T_{h,\text{out}}}{T_{c,\text{out1}} - T_{c,\text{in}}}$

$$\frac{T_{h,\text{int}} - T_{c,\text{out1}}}{T_{h,\text{out}} - T_{c,\text{in}}} = \frac{(T_{h,\text{int}} - T_{c,\text{in}}) - (T_{c,\text{out1}} - T_{c,\text{in}})}{(T_{h,\text{out}} - T_{c,\text{in}})}$$
$$= \frac{R'[(T_{h,\text{int}} - T_{c,\text{in}}) - (T_{c,\text{out1}} - T_{c,\text{in}})]}{R'(T_{h,\text{out}} - T_{c,\text{in}})}$$

$$= \frac{R'.(T_{h,\text{int}} - T_{c,\text{in}}) - (T_{h,\text{int}} - T_{h,\text{out}})}{R'.(T_{h,\text{out}} - T_{c,\text{in}})}$$

$$\frac{T_{h,\text{int}} - T_{c,\text{out1}}}{T_{h,\text{out}} - T_{c,\text{in}}} = \frac{R'.(T_{h,\text{int}} - T_{c,\text{in}}) - (T_{h,\text{int}} - T_{c\ \text{in}} - T_{h\ \text{out}} + T_{c\ \text{in}})}{R^1(T_{h,\text{out}} - T_{c,\text{in}})}$$

$$= \frac{(T_{h,\text{int}} - T_{c,\text{in}})(R' - 1) + (T_{h\ \text{out}} - T_{c\ \text{in}})}{R^1(T_{h,\text{out}} - T_{c,\text{in}})}$$

$$\frac{T_{h,\mathrm{int}} - T_{c,\mathrm{out1}}}{T_{h,\mathrm{out}} - T_{c,\mathrm{in}}} = \frac{(R' - 1)}{R'} \left(\frac{T_{h,\mathrm{int}} - T_{c,\mathrm{in}}}{T_{h,\mathrm{out}} - T_{c,\mathrm{in}}} \right) + \frac{1}{R'} \tag{5.31}$$

From Eqs. (5.29) and (5.31)

$$\frac{T_{h,\mathrm{int}} - T_{c,\mathrm{out1}}}{T_{h,\mathrm{out}} - T_{c,\mathrm{in}}} = \frac{(R' - 1)}{R'} \sqrt{\frac{T_{h,\mathrm{in}} - T_{c,\mathrm{in}}}{T_{h,\mathrm{out}} - T_{c,\mathrm{in}}}} + \frac{1}{R'}$$

Put the value in Eq. (5.28)

$$\Delta T_{\mathrm{LM}} = \frac{[(R' - 1)/2R'](T_{h,\mathrm{in}} - T_{h,\mathrm{out}})}{\ln\left[\frac{T_{h,\mathrm{int}} - T_{c,\mathrm{out1}}}{T_{h,\mathrm{out}} - T_{c,\mathrm{in}}}\right]} = \frac{[(R' - 1)/2R'](T_{h,\mathrm{in}} - T_{h,\mathrm{out}})}{\ln\left[\frac{(R'-1)}{R'}\sqrt{\frac{T_{h,\mathrm{in}} - T_{c,\mathrm{in}}}{T_{h,\mathrm{out}} - T_{c,\mathrm{in}}}} + \frac{1}{R'}\right]}$$

$$\Delta T_{\mathrm{LM}} = \frac{[(R' - 1)/2R'](T_{h,\mathrm{in}} - T_{h,\mathrm{out}})}{\ln\left[\frac{(R'-1)}{R'}\sqrt{\frac{T_{h,\mathrm{in}} - T_{c,\mathrm{in}}}{T_{h,\mathrm{out}} - T_{c,\mathrm{in}}}} + \frac{1}{R'}\right]} \tag{5.32}$$

Leaving temperature of mixed out cone of two cold streams:

$$\overline{T}_{c,\mathrm{out}} = (T_{c,\mathrm{out1}} + T_{c,\mathrm{out2}})/2$$

$$R^1 = \frac{R}{2} = \frac{1}{2}\left(\frac{T_{h,\mathrm{in}} - T_{h,\mathrm{out}}}{T_{h,\mathrm{out}} - T_{c,\mathrm{in}}}\right)$$

$$P = \frac{\overline{T}_{c,\mathrm{out}} - T_{c,\mathrm{in}}}{T_{h,\mathrm{in}} - T_{c,\mathrm{in}}}$$

$$\sqrt{\frac{T_{h\ \mathrm{in}} - T_{c\ \mathrm{in}}}{T_{h\ \mathrm{out}} - T_{c\ \mathrm{in}}}} = \frac{1}{(1 - PR)^{1/2}}$$

Put the above value in Eq. (5.32),

$$\Delta T_{\mathrm{LM}} = \frac{[(R' - 1)/2R'](T_{h,\mathrm{in}} - T_{h,\mathrm{out}})}{\ln\left[\frac{(R'-1)}{R'}\frac{1}{(1-PR)^{1/2}} + \frac{1}{R'}\right]} \tag{5.33}$$

Put $R' = R/2$

$$\Delta T_{\mathrm{LM}} = \frac{\left[\left(\frac{R}{2} - 1\right)/R\right]}{\ln\left[\frac{\left(\frac{R}{2}-1\right)}{\frac{R}{2}}\frac{1}{(1-PR)^{1/2}} + \frac{1}{\frac{R}{2}}\right]}(T_{h,\mathrm{in}} - T_{h,\mathrm{out}})$$

Let us assume that heat transfer rate is expressed as $Q = U.\ A.\ \theta.\ (T_{h\ \mathrm{in}} - T_{c\ \mathrm{in}})$ since $\frac{T_{h\ \mathrm{in}} - T_{h\ \mathrm{out}}}{T_{h\ \mathrm{in}} - T_{c\ \mathrm{in}}} = PR$

$$\theta = \frac{P\left(\frac{R}{2} - 1\right)}{\ln\left[\frac{\left(\frac{R}{2}-1\right)}{\frac{R}{2}}\left(\frac{1}{(1-PR)^{1/2}}\right) + \frac{1}{\frac{R}{2}}\right]}$$

Case 1:
Hence, the heat exchanger comprising n number of heat exchanger units with cold fluid in parallel flow arrangement and the hot fluid in series flow arrangement can be represented as:

$$\theta = \frac{P\left(\frac{R}{n} - 1\right)}{\ln\left[\left(\frac{\left(\frac{R}{n}-1\right)}{\frac{R}{n}}\right)\left(\frac{1}{(1-PR)^{1/2}}\right) + \left(\frac{1}{\frac{R}{n}}\right)\right]}$$

Case 2:
Similarly, the heat exchanger comprising n number of heat exchanger units with hot fluid in parallel flow arrangement and the cold fluid in series flow arrangement can be represented as:

$$\theta = \frac{P(1 - n.R)/n}{\ln\left[(1 - n.R)\left(1/(1 - P)^{(1/n)}\right) + n.R\right]}$$

5.7 Design steps of a tubular heat exchanger

The heat exchanger should be designed according to specific requirements of temperature and residence time in food-processing industries. Therefore, the following step-by-step procedure is generally adopted to design a tubular heat exchanger:

Step 1: Analysis of food-processing application
The application of heat exchange in food-processing industries involves thermal processing (blanching/pasteurization/sterilization), drying, dehydration, and so on. The specifications of requirement are necessary and will meet the requirements of the application, namely, drying temperature and holding time.

Step 2: Identification of fluid properties
The fluids should be identified for determining the flow properties, which decides the flow characteristics and heat transfer properties of fluid, namely, viscosity, density, thermal conductivity, and specific heat. For more efficient heat exchanger design, information of the physical properties of fluids is essential.

Step 3: Analysis of energy balance
The material and energy balance of the fluid provides information of the amount of energy, which is required for deciding the energy source and mode of heat transfer. Generally, the data of produce flow rate and temperature at the exit of produces are defined, which provide information of inlet and outlet temperature and flow rate of liquids.

Step 4: Identification of dimensions and numbers
The diameter of the tubes, number of tubes, diameter of shell, wall thickness, and length of tubes are prime geometrical observations, which is required to be defined for the design of heat exchangers. This also takes care about the size and materials available for the fabrication of heat exchangers.

Step 5: Thermal calculation
Thermal calculations are needed to find out heat transfer coefficients on both the sides, which is also affected by fluid properties and velocity of fluid passing through the tubes according to the designed geometry. Consideration of tubular surface is also needed. The surface area is calculated by dividing the amount of heat transfer (duty) (kcal/h), with product of overall heat transfer coefficient (kcal/h m^2°C) and log mean temperature difference (°C). The pressure drop may also be calculated for more specific calculations.

Step 6: Thermal calculation
The estimated area and the projected predefined area using the heat exchanger configurations should meet the requirement. In case of exceeding the estimated area, the design of heat exchangers must be reconfigured for more surface area. On the contrary, design for lower surface area is also inevitable in case of lower area estimated. These steps of designing are repeated till appropriate designed geometry is obtained.

Step 7: Mechanical design calculations
Detailed calculation for thickness of wall, nozzles, tubes, joints, and tube is performed to withstand the fluid pressure at temperatures during heat-transfer process. Any alteration, in material and dimension used, will affect dimensions and thermal calculations; suitable correction should be reassured or recalculated.

Step 8: Manufacturing drawings
The drawing is prepared using the information obtained from thermal and mechanical calculation and consists of main and specific parts, namely, tubes, shells, connections, and joints.

References

Bhavsar, J. J., Matawala, V. K., & Dixit, S. (2013). Design and experimental analysis of spiral tube heat exchanger. *International Journal of Mechanical and Production Engineering*, *1* (1), 37–42.

Çengel, Y. A., & Ghajar, A. J. (2011). *Heat and mass transfer: Fundamentals & applications*. New York: McGraw-Hill.

Dwivedee, K. C., Arya, R. A., Diwakar, N., & Khare, A. (2019). A review of triple concentric heat exchanger. *International Journal of Current Engineering and Scientific Research*, *6* (9), 12–16.

Fryer, P. J., & Robbins, P. T. (2005). Heat transfer in food processing: Ensuring product quality and safety. *Applied Thermal Engineering*, *25*, 2499–2510.

Hewitt, G. F., Shires, G. L., & Bott, T. R. (1994). *Process heat transfer*. New York: CRC Press.

Holman, J. P. (2010). *Heat transfer* (10th ed.). New York: McGraw-Hill.

Web references

http://www.genemco.com/aloe/tripletube.html (accessed 27 August 2020).

http://www.rometec.it/uploads/File/sentry/heat/Spiral.pdf (accessed 27 August 2020).

https://sentry-equip.com/sentry-spiral-tube-heat-exchanger.html (accessed 27 August 2020).
https://www.comsol.com/blogs/modeling-coil-heat-exchanger/ (accessed 27 August 2020).
https://www.enggcyclopedia.com/2011/05/heat-exchanger-types/ (accessed 27 August 2020).
https://www.enggcyclopedia.com/2019/05/shell-tube-heat-exchanger-types/ (accessed 27 August 2020).
https://www.geurtsheatexchangers.com/products/heat-exchangers/coil-shell-heat-exchangers/ (accessed 27 August 2020).
https://www.hrs-heatexchangers.com/resource/design-tubular-heat-exchanger/ (accessed 27 August 2020).
https://www.thomasnet.com/articles/process-equipment/shell-and-tube-heat-exchangers/ (accessed 27 August 2020).
https://www.thomasnet.com/articles/process-equipment/understanding-heat-exchangers/ (accessed 27 August 2020).

Shell and tube heat exchangers in the food industry

6

Pooja Nandakumar[a], *Dharani Loganathan*[b], *Deepa Priya Natarajan*[c], *and Periyasamy Manikandan*[d]
[a]Department of Plastics Technology, Central Institute of Petrochemicals Engineering & Technology, Chennai, Tamil Nadu, India, [b]Department of Chemical Engineering, KPR Institute of Engineering and Technology, Coimbatore, Tamil Nadu, India, [c]Department of Chemical Engineering, Sri Venkateswara College of Engineering, Sriperumbudur, Tamil Nadu, India, [d]Department of Chemical Engineering, Kongu Engineering College, Perundurai, Tamil Nadu, India

6.1 Introduction

Thermal treatment is one of the significant applications in food industry to change food properties and broaden timeframe of realistic usability without the addition of chemicals. Thermal treatments such as drying and evaporation lower food moisture content, whereas chilling, freezing, pasteurization, and sterilizing limit microbial composition. Thermal treatments are determined by the structure, physicochemical, biological, and microbiological features of the raw materials and the intended end product (Fesanghary, Damangir, & Soleimani, 2009). Heat exchangers are devices that allow thermal energy to flow between two or more fluids at differing temperatures. Whatever the condition of the fluids involved (gas or liquid), and regardless of whether there is a phase shift in the course of treatment, the principles of heat transfer are conduction, convection, evaporation, condensation, and radiation (Chauduri, Diwekar, & Logsdon, 1997). Heat exchangers are utilized in a variety of applications including power generation, industrial, chemical, and food processing, etc., as shown in Fig. 6.1 (Kern, 1950; Rohsenow & Hartnett, 1973; Selbas, Kizilkan, & Reppich, 2006). Sterilization, pasteurization, blanching, frying, drying, evaporation, and irradiation are some of the heat applications in the food industry that use a variety of heat transfer equipment (Hewitt, 1998). Heat exchangers can be used for the food sector to cool down a variety of different products. Many food substances, such as hazelnut paste and other forms of food pastes, must be refrigerated or heated before they may be handled further (Shah & Bell, 2000).

The general classification of heat exchangers based on their application is given in Fig. 6.2. Despite the wide variety of heat exchangers availability, Shell-and-tube Heat Exchangers (STHX) are often used in food industries. It is a form of heat exchanger that includes a series of tubes containing the fluid that needs to be heated or cooled, with the second fluid flowing over the tubes that have to be heated or cooled. Because of their toughness, STHX are commonly utilized for high pressure applications

Thermal Processing of Food Products by Steam and Hot Water. https://doi.org/10.1016/B978-0-12-818616-9.00004-3

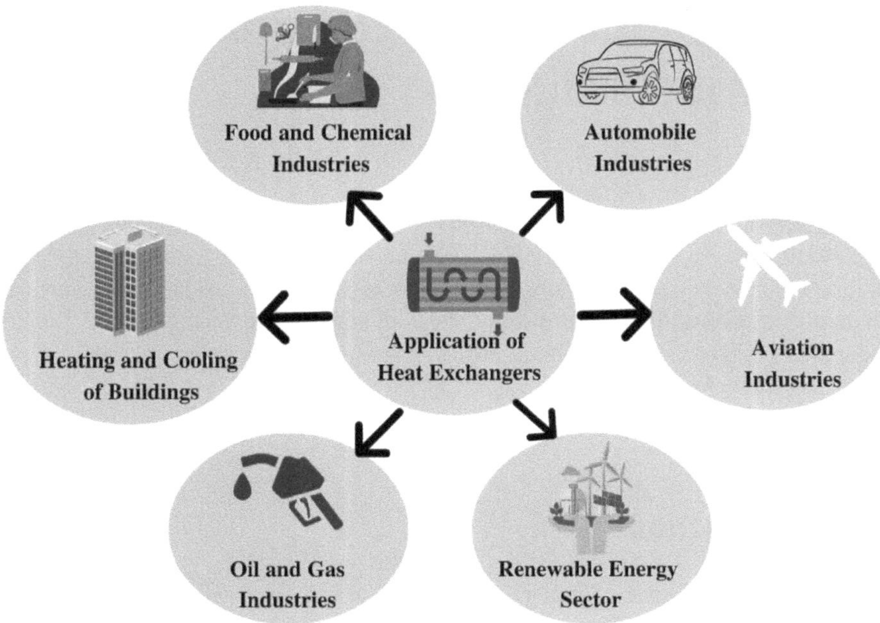

Fig. 6.1 Application of heat exchangers.

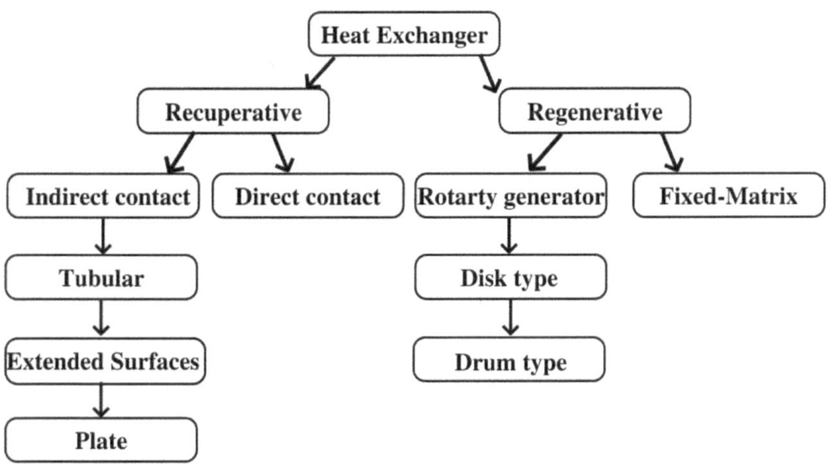

Fig. 6.2 Classification of heat exchangers based on application.

(pressures >30 bar and temperatures >260 °C) (Keblinski, Phillpot, Choi, & Eastman, 2002). Because of their broad assortment of configuration and working conditions, STHX are the most commonly used type of heat exchanger in food industry. They are utilized in power plants, chemical industries, and process industries, among other places (Jafari, Saremnejad, & Dehnad, 2017). The fundamental design of STHX originally created in the early 1900s to fulfill the demand of power plants such as condensers and feed water heaters that operate at relatively high pressures. During the 1920s, STHX production technology advanced rather rapidly, owing primarily to the efforts of a small number of significant firms (Jafari, Saremnejad, Dehnad, & Rashidi, 2017). The constantly expanding use of STHX led in a surge of research activity in the 1940s and 1950s. These comprised not just shell side flow calculations, but also calculations of tube sheet strength and structure, as well as true mean temperature difference. The designer must understand the mechanical components of STHXs and how they influence thermal design. Shell, channel cover, tube sheet, shell cover, baffles, tubes, channels, and nozzles are the most important components of a STHX (Kern, 1983).

The important parameters that determine STHX performance are: (i) tube width; (ii) temperature difference between two fluids: (iii) thermal conductivity of the fabrication material; (iv) physical features of the exchanger and surface area of the tubes; (v) flow direction, such as counter-current, co-current, or mixed flow; (vi) liquid properties such as viscosity and heat capacity, among others (Kahani, Heris, & Mousavi, 2013).

In the food industry, heat treatment of fluid foods in a STHX is a common application. Pasteurization is one such applications. It is a food preservation technique that uses a high-temperature treatment. Depending on the necessity, fluid foods are heated to a temperature of 62.8°C for 30 min or 71.7°C for 15 s. For the thermal treatment of fluid foods such as whole milk, hazy orange juice, and apple and apricot puree, heat exchangers are commonly utilized (Kang, Kim, & Oh, 2006). It's best to employ a STHX with helically corrugated walls. Helically curved heat exchangers are chosen over smooth walls because they improve heat transmission by convection, boosting efficiency (Mare et al., 2011). Jafari et al. investigated the design of a STHX using alumina-water nanofluids to improve heat transfer in food products, as well as changes in temperature-time profiles and energy consumption rates during the production of two food products namely tomato and water melon using the designed heat exchanger (Nasiri, Etemad, & Bagheri, 2011). Rozzi et al. investigated both Newtonian and non-Newtonian fluid's convective heat transfer and friction losses in helically enhanced tubes. The results show that helical corrugated tubes are especially useful in boosting convective heat transfer for generalized Reynolds numbers range from around 800 to the transitional flow regime's limit (Pandey & Nema, 2012).

In a STHX with turbulent flow, Farajollahi et al. evaluated the heat transfer characteristics of γ-Al_2O_3/water and TiO_2/water nanofluids. The impacts of Peclet number, suspended nanoparticle volume concentration, and particle type on thermal properties were studied. According to the findings, adding nanoparticles to the base fluid improves the heat transmission characteristic significantly (Peyghambarzadeh, Hashemabadi, Hoseini, & Seifi Jamnani, 2011). Elias et al. analyzed the effect of various particle forms, such as cylinders, bricks, blades, and platelets, on the total heat transfer coefficient, heat transfer rate, and entropy generation of STHX with various baffle angles and segmental baffles. The results demonstrate that cylindrical shape nano particles outperformed all other shapes in terms of overall heat transfer coefficient and heat transfer rate for all baffle angles, including segmental baffle

(Taghizadeh-Tabari, Zeinali Heris, Moradi, & Kahani, 2016). The current book chapter discusses the various classifications and applications of STHX in the food industry, as well as the different heat transfer enhancement techniques used in the processing of food products, along with fouling of STHX and the prevention methods.

6.2 Components and classifications

6.2.1 Classification of shell and tube heat exchangers

Heat exchangers are typically categorized based on the type of transfer mechanism that occurs within them. The following is a general classification of heat exchangers.

6.2.1.1 Classification according to transfer processes

Heat exchangers are classified according to transfer processes into indirect- and direct contact types.

6.2.1.1.1 In-direct-transfer type exchangers

The fluid streams stay separate in an indirect-contact heat exchanger, and heat is transferred continuously via an impervious separating wall or transiently into and out of a wall. As a result, there should be no direct contact between thermally interacting objects fluids (Zeinali Heris, Noie, Talaii, & Sargolzaei, 2011). This form of heat exchanger, also known as a surface heat exchanger, can be used in a variety of applications and classified as direct-transfer, storage, and fluidized-bed exchangers are the three types of exchangers.

6.2.1.1.2 Direct-transfer type exchangers

Through a separating wall, heat is continually transferred from the hot to the cold fluid in this kind. Although the exchanger requires the simultaneous flow of two (or more) fluids, the two (or more) fluids do not mix since they flow through separate fluid tubes. The majority of such heat exchangers have no moving components. A recuperative heat exchanger, or simply a recuperator, is one of several types of exchangers. Tubular, plate-type, and extended surface exchangers are examples of direct transfer heat exchangers (Gulyani, 2000).

6.2.1.1.3 Direct-contact heat exchangers

Through a separating wall, heat is continually transferred from the hot to the cold fluid in this kind. Although the exchanger requires the simultaneous flow of two (or more) fluids, the two (or more) fluids do not mix since they flow through separate fluid tubes. The majority of such heat exchangers have no moving components. A recuperative heat exchanger, or simply a recuperator, is one of several types of exchangers. Tubular, plate-type, and extended surface exchangers are examples of direct transfer heat exchangers (Master, Chunangad, Boxma, Kral, & Stehlík, 2006).

6.2.1.2 Classification according to number of fluids

The majority of heating, cooling, heat recovery, and heat rejection processes involve heat transfer between two fluids (Bell, 2004). As a result, two-fluid heat exchangers are the most often used. Cryogenics and various chemical processes employ three fluid heat exchangers (for example, air separation systems, a helium-air separation unit, hydrogen purification and liquefaction, and ammonia gas synthesis). In some chemical process applications, heat exchangers with up to 12 fluid streams have been employed.

6.2.1.3 Classification according to construction features

Construction aspects are widely used to describe heat exchangers. Tubular, plate-type, extended surface, and regenerative exchangers are the four basic construction types. Other types of heat exchangers, such as scraped surface exchangers, tank heaters, cooler cartridge exchangers, and others, are also available. Although some of these are tube exchangers, they have several distinguishing characteristics when compared to traditional tubular exchangers (Lutcha & Nemcansky, 1990).

6.2.1.3.1 Tubular heat exchangers

For these exchangers circular tubes are most commonly employed, however elliptical, rectangular, and round/flat twisted tubes have also been utilized in specific applications. Because the core shape may be readily changed by altering the tube diameter, length, and arrangement, the design has a lot of versatility. Tubular exchangers can be built for large pressure differences between the fluids and high-pressure differences relative to the environment. Tubular exchangers are typically utilized for heat transfer from liquid to liquid and liquid to phase transition. They are employed in gas-to-liquid and gas-to-gas heat transfer applications when the operating temperature and/or pressure are extremely high, or fouling is a serious issue on at least one fluid side, and no other form of exchanger will operate (Master, 2004).

6.2.1.3.2 Shell-and-tube heat exchangers

STHX as shown in Fig. 6.3 are the most widely used heat exchange devices in the chemical industry, out of all types of exchangers (Lei, He, Li, & Gao, 2008). They are mostly used as heat transfer equipment, but they can also be used as reactors, falling film absorbers, and other applications (Gaddis & Gnielinski, 1997; Lei et al., 2008). While there are several different designs and materials of heat exchangers, certain elements are universal. The tubes are mechanically connected to tube sheets, which are enclosed within a shell with ports for inlet and outlet fluid or steam, in all STHX. They are made to keep the liquid flowing within the tubes from mixing with the fluid flowing outside. Tube sheets may be glued to the shell or allowed to expand and contract in response to temperature changes. An expansion bellows is used in the above design, or one tube sheet is allowed to float within the shell. The nonfixed tube sheet method allows for the removal of the entire tube package assembly from the shell for cleaning the shell circuit (Wang, Chen, Xu, & Ji, 2010).

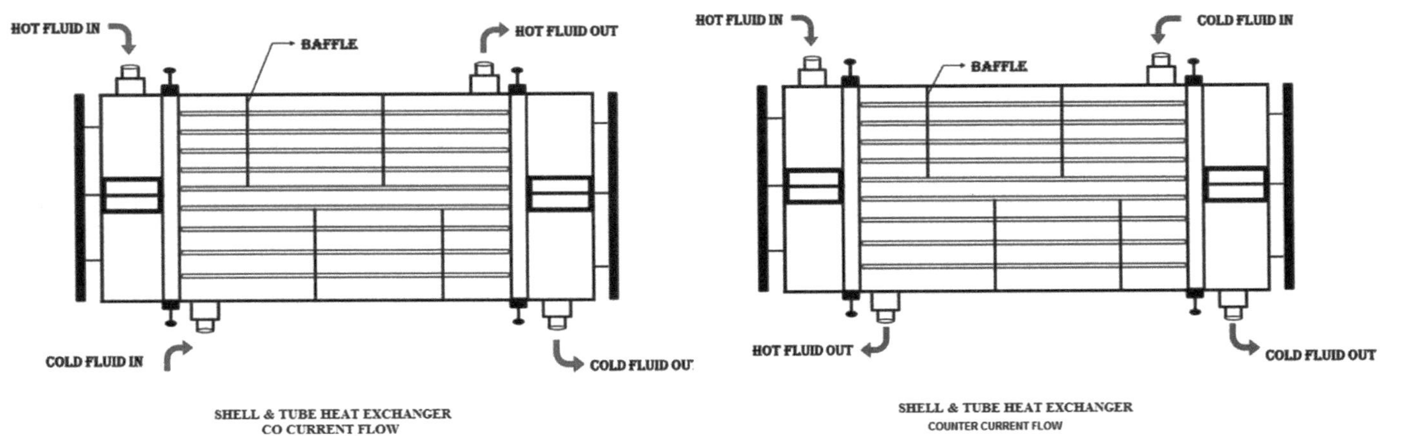

Fig. 6.3 Cocurrent flow (*left*), and counter-current flow (*right*) shell and tube heat exchangers.

Fixed tubesheet design, U-tube design, and floating-head design are the three most frequent forms of shell-and-tube exchangers (El Maakoul et al., 2006). The front-end head is stationary in all three forms, whereas the rear-end head can be either stationary or floating, depending on thermal stresses in the shell, tube, or tubesheet caused by temperature variations due to heat transmission.

6.2.2 Components of shell and tube heat exchangers

6.2.2.1 Shells

The shell is a container for the shell fluid that is usually cylindrical in shape with a circular cross section, while alternative shell forms are employed in certain applications and nuclear heat exchangers to adapt to the tube bundle shape. If the shell diameter is less than roughly 0.6 m (2 ft.), the shell is manufactured from a circular pipe; if the shell diameter is higher than 0.6 m (2 ft.), the shell is constructed from a metal plate rolled and welded lengthwise (2 ft.) (Li & Kottke, 1998).

6.2.2.2 Tubes

Shell-and-tube exchangers employ round tubes of various shapes. Tube bundles of straight and U-tubes are the most popular in process and power sector exchangers. In modern nuclear exchangers, however, sine-wave bends, J-shape, L-shape, or hockey sticks, and inverted hockey sticks are employed to handle considerable thermal expansion of the tubes (Li & Kottke, 1998).

6.2.2.3 Nozzles

The nozzles, or entrance and exit apertures for the shell-and-tube fluids, are pipes with a constant cross section that are welded to the shell and channels. They are utilized to evenly distribute or collect fluid on the shell-and-tube sides. They vary from a nozzle used for fluid metering or in jet engines, which has a changeable flow area along the length of the flow.

6.2.2.4 Front- and rear-end heads

These are used for tube fluid entry and exit, and many rear-end heads have a provision to account for tube thermal expansion. Depending on the thermal tensions between the tubes and shell, the front-end head is stationary, while the rear-end head might be either fixed (allowing for no tube thermal expansion) or floating.

6.2.2.5 Baffles

There are two types of baffles: transverse and longitudinal. The goal of longitudinal baffles is to manage the overall flow direction of the shell fluid so that the two fluid streams flow in the same direction. Plate baffles and grid (rod, strip, and other axial-flow) baffles are two types of transverse baffles. Plate baffles are used to support the tubes during assembly and operation, as well as to route the fluid in the tube bundle at a roughly right angle to the tubes, resulting in greater heat transfer coefficients.

6.2.2.6 Tube-sheets

These are used to keep the ends of tubes together. A tube-sheet is a circular metal plate having holes drilled through it for the desired tube design, tie rod holes (for spacing and holding plate baffles), gasket grooves, and bolt holes for flanging to the shell and channel. Tube-to-tube-sheet joints are made in a variety of ways to prevent leakage of the shell fluid at the tube-sheet through a clearance between the tube hole and tube, such as expanding the tubes, rolling the tubes, hydraulic expansion of tubes, explosive welding of tubes, stuffing of the joints, or welding or brazing of tubes to the tube sheet.

6.2.2.7 Tube pitch

The distance between the center of one tube and the center of another tube is known as tube pitch. The tubes are arranged in a triangle or square arrangement, with the square shape being the easiest to clean and causing the least turbulence. Unlike triangular and rotational square pitch tubing, the square pitch design enables vapors to rise between the tubes.

6.2.2.8 Tie rods and spacers

Tie rods and spacers are structural and support components that keep the baffles in place and maintain the distance between them. The number of baffles and the diameter of the shell dictate the number of rods and spacers. Tie rods are fastened to the tube sheet and stretch the length of the bundle all the way to the final baffle.

6.2.3 Construction and working principle

A STHX has a basic idea and operation that is based on the flow and thermal contact of two liquids. The method of exchanging temperature between two fluids is described by the name of a STHX. A heated or hot fluid flows around a cold fluid in a heat exchanger, transferring heat in the direction of the cold fluid's movement (Saffar-Avval & Damangir, 1995). There will be a heat exchange or transfer across a conductive surface in every case when two pieces of material make touch. A STHX is a device that allows two fluids to exchange or transfer heat through the use of conductive metals. One fluid passes through the tubes while the other fluid flows through the shell in STHX operation. The shell inlet for the shell fluid to enter is at the top of the schematic of a straight tube STHX, with the tube inlet at the bottom right (Lei et al., 2011).

A STHX is divided into two portions or compartments: the shell side and the tube side. When dealing with a STHX, it is critical to choose which side the hot fluid will enter and which side the cool fluid will enter; this choice is known as fluid allocation (Jafari, Saremnejad, & Dehnad, 2017). Also because tubes are intended to bear high pressure, when there is a pressure differential between the fluids, the lower pressure fluid enters via the shell inlet (Lemos, Costa, & Bgajewicz, 2017).

6.2.3.1 Determination of fluid allocation

When calculating fluid flow for the shell side, note that the shell is more expensive to manufacture than the tubes and is more difficult to clean. Baffles on the shell side direct the flow of fluid through the tube bundles. Viscous fluids and fluids with a high flow rate are processed through the shell side, which has more turbulence and a higher transfer coefficient, resulting in better heat transfer. Normally, large temperature changes are accomplished on the shell side. On the tube side, turbulent flow is required, which is accomplished by placing turbulators within the tubes through the perforations in the tube sheet. Turbulence in the tubes, like turbulence in the shell, enhances heat transmission capacity. The turbulators' secondary duty is to maintain the inside of the tubes clean and free of debris. Tubes have less turbulence and pressure drop, resulting in a more streamlined flow (Rodriguez & Smith, 2007; Srinivasan, 2005).

6.3 Heat transfer aspects and coefficients

STHX operates on the idea that heat is transmitted from the tube side fluid to the shell side fluid through the tube walls, without the two liquids being in physical contact with each other. To put it another way, it transfers heat without moving the heat-carrying fluid. The evaluation of the heat transfer coefficient is an important aspect of STHX examination (Fig. 6.4). It is desirable to have a high heat transfer coefficient. The following correlation is used to compute the convective heat transfer coefficient of STHX,

$$Q = U/(A \times LMTD)$$

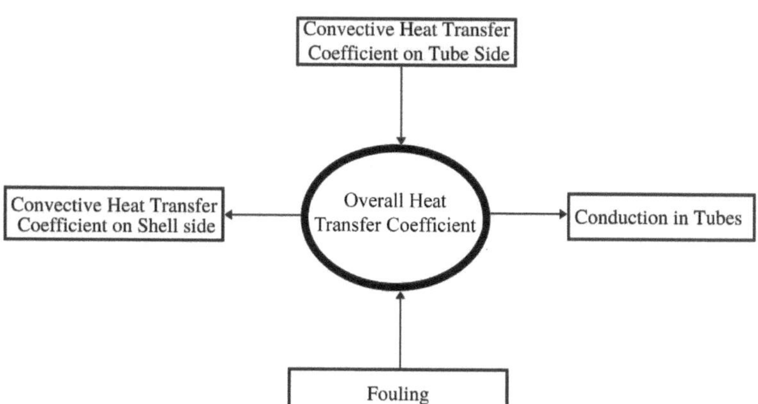

Fig. 6.4 Factors influencing the overall heat transfer coefficient.

where Q is the amount of heat transferred from hot stream to cold stream (W), A is total heat transfer area (m^2), LMTD is the logarithmic mean temperature difference (K).

The heat duty Q can be written as follows assuming only sensible heat is transmitted,

$$Q = m_{\text{hot}} C_{p,\text{hot}} (T_{h1} - T_{h2}) = m_{\text{cold}} C_{p,\text{cold}} (T_{c1} - T_{c2}).$$

where m_{hot}, m_{cold}, $C_{p,\text{hot}}$, and $C_{p,\text{cold}}$ are the mass flow rate and specific heat capacity of hot and cold fluid, respectively.

$$\text{LMTD} = (T_{h1} - T_{c2})/(T_{h2} - T_{c1})/(\ln((T_{h1} - T_{c2})/(T_{h2} - T_{c1}).$$

where T_{h1}, T_{h2}, T_{c1}, and T_{c2} are the inlet and outlet temperature of hot and cold stream, respectively (Fesanghary et al., 2009).

The following generic result can be written for the overall heat transfer coefficient U_o based on the outside surface area of the tubes, which is the heat transfer surface,

$$(1/U_o) = (1/h_o) + (\Delta r/k) \times (A_o/A_{lm}Alm) + (1/h_i) \times (A_o/A_i) + R_{f,o} + R_{f,i}$$

where h_o the heat transfer coefficient for liquid passing in the shell is, h_i is the heat transfer coefficient for liquid passing through the tubes, A_i and A_o are the inner and outside surface areas of tubes, respectively, and A_{lm} is their log mean in the above equation. $R_{f,0}$ for the shell side and $R_{f,i}$ for the tube side are the fouling resistances on a unit area basis. The standards produced by TEMA contain a wealth of information on fouling resistances.

The inner heat transfer coefficient h_i can be calculated using the standard method for forecasting heat transfer in tubes, which includes the use of a viscosity correction when applicable. Because of the significant increase in heat transfer afforded by eddy transport, turbulent flow is usually expected, and a good design would attempt to create turbulent flow. Because the flow passage is not simple, even in the absence of baffles, predicting the shell-side heat transfer coefficient o h is more difficult. When calculating the fluid velocity across the tube bank, the presence of baffles must be taken into account. Heat transfer correlations for flow via tube banks, such as those provided in Holman's book, are used (Aminian & Shahhosseini, 2008). These correlations presuppose flow normal to the long axis of a geometrical array of tubes. Holman's book provides the following correlation,

$$Nu = \left(\frac{h_o D_o}{k}\right) = C\,Re^n Pr^{1/3}.$$

The Reynolds number $Re = (D_o V_{\max} \rho)/\mu$, where D_o is the outside diameter of a tube. $V_{\max}$ is the "maximum" velocity of the fluid through the tube side. As a rough rule of thumb, the optimum heat transfer coefficient h_o for flow across the tube bank determined using a proper correlation is multiplied by a conservative correction factor of

0.6 in the end because of these numerous corrections (Ishiyama, Heins, Paterson, Spinelli, & Wilson, 2010). The provided equations make it simple to calculate heat transfer coefficient values for process design. In order to perform valuable process sensitivity analysis, the range of variation of this unknown coefficient can also be acquired. When experimental data is absent, estimation can be done using the proposed equations.

Heat transfer coefficient during frying is important in the modeling and calculation of food fryer systems, according to Alvis et al. The methods for measuring the heat transfer coefficient during deep fat frying available in the literature had some drawbacks, such as the absence of standard approaches. So, the authors looked at the primary effects of deep-fat frying on fried food and went over the most common methods for calculating the convective heat transfer coefficient (Mirzaei, Hajabdollahi, & Fadakar, 2017). Heat transfer data for the cooling and/or freezing of foods was evaluated by Becker et al. An industrial study yielded 777 cooling curves for 295 food items, and a unique iterative approach based on the notion of "equivalent heat transfer dimensionality" was created to extract heat transfer coefficients from the cooling curves. A selection of the 777 heat transfer coefficients generated by this technique, as well as 144 heat transfer coefficients for 13 food items taken from the literature, were used to create nine Nusselt-Reynolds-Prandtl correlations (Liu et al., 2016). Debnath et al. investigated the effect of binary blending of five frying oils on kinematic viscosity and convective heat transfer coefficient at elevated temperatures (170, 180, and 190°C) and discovered a good correlation between these frying oils' kinematic viscosity ($R^2 > 0.996$) or convective heat transfer coefficient ($R^2 > 0.987$) and their chosen binary blends with optimal temperature range (Abd & Naji, 2017).

Vidyarthi et al. investigated the development and application of inverse methods for calculating the fluid-to-particulate heat transfer coefficient of a liquid-particulate mix (sweet potato puree and chickpea) during aseptic processing. The research could help with rheological specifications for carrier fluids, processing equipment design, and process settings for commercial aseptic systems (Shinde & Chavan, 2017). Afonso et al. conducted an experiment to develop a correlation for calculating the convective heat transfer coefficients of stirred yoghurt in a plate heat exchanger. A rheological-research was conducted to characterize the flow behavior of stirred yoghurt, examining its dependence on both shear rate and temperature (Eryener, 2006). Under turbulent flow conditions, Coronel et al. investigated the determination of convective heat transfer coefficient in both helical and straight tubular heat exchangers. The tests were carried out in helical heat exchangers using coils with two different curvature ratios, and the heat transfer coefficient results were compared (Yu, Ren, & Zeng, 2018). For the equivalent thermo-physical characteristics and surface heat transfer coefficient of bulk layers of fruits such as (strawberries, raspberries, apricots, peaches, grapes, and apples) in trays, Fikiin et al. created experimental engineering data and prediction equations (Dizaji, Jafarmadar, & Asaadi, 2017). Marra and colleagues created a mathematical model of a solid food material being heated in a cylindrical batch Ohmic heating cell. The Ohmic heating process' temperature profiles and distribution were simulated and analyzed (Alimoradi & Veysi, 2017).

6.4 Pressure drop—Tube side and shell side

Correlations were created based on experimental data for common heat exchangers in the early attempts to give techniques for determining shell-side pressure drop and heat transfer coefficient. The well-known Kern technique was an effort to correlate data for conventional exchangers using a simple equation akin to equations for flow in tubes. This technique, however, is limited to a set baffle cut (25%) and cannot account for baffle-to-shell and tube-to-baffle leakages. Although the Kern equation is not especially exact, it does allow for a very simple and quick computation of shell-side coefficients and pressure drop and has been used effectively since its conception (Gao, Bi, Nie, & Wu, 2015).

6.4.1 Calculation of shell-side and tube side pressure drop

The total nozzle-to-nozzle pressure drop of a shell-and-tube heat exchanger with bypass and leakage streams is computed as the sum of the three components:

1. By considering the pressure drop in the interior cross-flow section (baffle tip to baffle tip), the combined pressure drop of all the interior cross-flow section is (Sinnott, 1993)

$$\Delta p_c = \Delta p_{bi}(N_b - 1)R_l R_{b.}$$

where N_b is the number of baffles, R_l is the leakage correction factor, R_b is the bypass correction factor. Typically, R_b=0.5–0.8, depending on the construction type and number of sealing strips, and R_l = 0.4–0.5.Δp_{bi} is the pressure drop in an equivalent ideal tube bank in one baffle compartment of central baffle spacing and it is calculated from (Sinnott, 1993)

$$\Delta p_{bi} = 4f_i \frac{G_s^2}{2\rho_s}\left(\frac{\mu_{s,w}}{\mu_s}\right)^{0.14} N_{tcc}$$

where f_i is the friction coefficient, G_s is the mass velocity of the shell-side fluid, ρ_s is the shell-side fluid density, μ is the shell-side fluid viscosity, $\mu_{s,w}$ is the viscosity of shell-side fluid evaluated at wall surface temperature, and N_{tcc} is the number of tube rows crossed in one cross-flow section, that is, between the baffle tips.

2. Leakage has an effect on the pressure decrease in the window, while bypass has no effect. The Delaware technique provides two distinct correlations for, one for turbulent flow and one for laminar flow, both based on the shell-side cross-flow for simplicity. Both correlations are used for mass velocity.

$$m_w = \frac{m_s}{\sqrt{S_m\, S_w}}$$

where m_s is the mass flow rate of the shell-side fluid, and m_w is the shell-side flow mass velocity through segmental baffle window.

3. Bypass affects pressure drop in the entry and exit portions but not leakage. The end zone flow area varies from the center cross-flow zone in the following ways: (a) the number of tube

rows crossed includes those in the entry or departure window, (b) in the end zones, the leakage streams have not yet evolved (entry) or have only just joined the main stream (exit), and hence the leakage correction factor is inapplicable. (c) The baffle spacing in the end zones may differ from the center spacing, particularly with U-tube bundles. As a result, an end zone adjustment factor is used. Then the pressure drop is given as

$$\Delta p_e = (\Delta p_{bi})\left\langle 1 + \frac{N_{tcw}}{N_{tcc}} \right\rangle R_b R_s$$

where R_b is the bypass correction factor and R_s is the end zone correction factor. For all baffle spacings equal, $R_s = 2$.

Finally, the total shell-side pressure drop Δp_s, excluding nozzles, is (Sinnott, 1993; Tkachuk et al., 2019):

$$\Delta p_s = \Delta p_c + \Delta p_w + \Delta p_e.$$

The pressure drop in the nozzles must be calculated separately and added to the total pressure drop.

6.5 Fouling

The accumulation of undesirable material deposits on heat transfer surfaces during process heating and cooling is referred to as fouling. It happens in all sectors and in the vast majority of heat exchanger designs, with consequences ranging from heat transfer deterioration to flow resistance and pressure decreases. Fouling reduces heat exchange efficiency by generating a sort of insulation on heat-transfer surfaces and lowering heat transmission. This section discusses the types, and causes of fouling, as well as fouling detection, heat exchanger cleaning, and fouling mitigation.

6.5.1 Types of fouling

The four most prevalent kinds of fouling and aging in sanitary processing are as follows:

The deposition of a crust or coating of processed fluids, minerals, or cleaning chemicals on the surface of heat exchanger components is referred to as corrosion. Incrustation generated by calcium carbonate, calcium sulphate, and silicates is known as scaling. *Sediment*: a byproduct of corrosion, metal oxides, silt, alumina, and diatomic creatures (microalgae) and their feces. *Biological growth*: Bacteria, nematodes, and protozoa are all sources of bio-fouling.

6.5.2 Factors influencing fouling

Fouling is caused by a variety of factors, including water pH, product viscosity, and the roughness of component surfaces, among many others. The factors may be combined to form a fouling factor, which numerically depicts heat transfer resistance—or thermal resistance—in your system. Food industry applications contain fats, carbohydrates, and proteins that contribute to fouling. Particulate matter in food and beverage applications can clump and clog, while cosmetic and pharmaceutical particles created during processing can adhere and collect in pharmaceutical applications. Operators focus on four critical processing parameters to assist reduces fouling.

6.5.2.1 The velocity of a fluid

In most circumstances, fouling reduces with increased flow velocity because increasing flow velocity increases fluid shear stress, which induces greater deposit removal. In the case of larger deposits, raising the flow velocity beyond a certain limit may not considerably reduce fouling. In the event of extremely dense deposits, increasing flow velocity may have no impact.

6.5.2.2 The temperature of the fluid

Scaling can be caused by minerals such as calcium carbonate in water ($CaCO_3$). The amount of salt deposited on the heat exchanger surface increases as the temperature rises. Similarly, biological growth can occur when the temperature rises during food preparation. Finally, if the temperature differential between the hot product output and the hot medial inlet is too great, proteins and fats in food goods might burn.

6.5.2.3 Fluid chemistry

Calcium phosphate and whey protein can accumulate on heat exchanger surfaces during applications such as milk processing. Fouling raises pressure drop by limiting the region through which the product may flow. Proteins, lipids, carbohydrates, and minerals may all come out of solution in dairy products and deposit on heat exchanger surfaces and foul pipes. If a fluid undergoes chemical changes, the difference can cause a fouling layer to form on the tube or plate surfaces. Suspended salts, for example, may solidify when the temperature rises.

6.5.2.4 Fabrication materials

Utilize units made of corrosion-resistant stainless steels and alloys to avoid corrosion fouling from layers of thermal-resistant material. AISI 304 or AISI 316L stainless steel can be used to make plates, tubes, and full heat exchanger units. Parts of units may be made of titanium, stainless steel alloys, or other corrosive, high-salt materials for more corrosive, high-salt products.

6.5.3 Detection of fouling in shell and tube heat exchangers

Physical inspection or monitoring system performance is often used to detect fouling. Physical inspection entails measuring the thickness of fouling on plates, pipes, or tubes. Heat transfer: Because of the insulating qualities of fouling, heat transfer may exceed operational requirements. Increased pressure drop between the intake and output of a heat exchanger may indicate frictional resistance or restricted flow routes produced by fouling deposits. The best technique to troubleshoot and identify fouling in heat exchangers is to monitor temperature and pressures. Critical variables that might indicate heat exchanger fouling are measured and reported by instruments.

6.5.3.1 Temperature transmitters

Temperatures of fluids moving through heat exchangers should be monitored. Temperature reductions may indicate heat transfer loss due to fouling.

6.5.3.2 Pressure transmitters and flow meters

Pressure transmitters detect pressure decreases between the intake and output of the heat exchanger. Flow meters indicate declines that might be caused by fouling material buildup in pipes and plates.

6.5.4 Heat exchanger cleaning and cleaning agents

Cleaning is essential to keep heat exchangers running efficiently. Cleaning-in-place (CIP) technology circulates cleaning chemicals and rinses to flush heat exchanger internal surfaces without dismantling them. The optimum flow rate ensures that fluids have an efficient mechanical action during cleaning. Some manufacturers propose a minimum velocity of 1 ft./s between heat exchanger plates, however specifications vary by company.

6.5.4.1 Incrustation, scaling

Cleaning incrustation, also known as scaling, is the removal of calcium carbonate, calcium sulfate, or silicates from plate surfaces. Cleaning products must be compatible with both plate metal and gasket composition. Hydrochloric acid should never be used on titanium or stainless steel plates. In addition, never use water with more than 300 ppm of chlorine when preparing cleaning solutions. Consult your equipment provider if chlorination of nontitanium equipment cannot be avoided.

6.5.4.2 Removing sediment

Metal oxides, silt, alumina, and diatomic creatures and their feces are the most prevalent types of sediment. Heat releases minerals and other particles from fluids throughout processing cycles, which settle and deposit on heat transmission surfaces, forming sediment. Since hydrochloric acid produces general corrosion, pitting, and stress corrosion cracking, it should not be used with stainless steel or titanium plates.

6.5.4.3 Bacterial growth removal

Biological growth such as bacteria, nematodes, and protozoa can develop when heat exchangers are used to raise the temperature of processed meals. Incrustation removal necessitates the same level of attention to plate and gasket composition as incrustation removal. For individual equipment, manufacturers can give more thorough information on cleaning and sterilizing. Shell-and-tube heat exchangers are more prone to fouling than plate heat exchangers because they operate at lower internal fluid velocities. Keep the shell-and-tube heat transfer surfaces clean to ensure effective operation and greater transfer coefficients. Cleaning chemicals for a plate-and-frame heat exchanger are dependent on the same factors, and cleaning compounds must be compatible with the metallurgy of the heat exchanger. Operators must use suitable protective equipment, such as safety boots, gloves, and eye protection, during all cleaning activities to avoid injury.

6.6 Heat transfer enhancement techniques

Heat transfer enhancement strategies are several approaches used to increase the rate of heat transfer without significantly affecting the overall performance of STHX system. Heat transfer enhancement techniques minimize thermal resistance by enhancing heat transfer surface area or creating turbulence in the fluid flowing through STHX. Rough surfaces or extended surfaces are employed to increase the effective surface area, meanwhile inserts, winglets, turbulators, and other turbulence generating devices are used to create the turbulence. The thermal performance factor (TPF), which indicates the ratio of the relative effect of change in heat transfer rate to change in friction factor, can be used to assess the effectiveness of a heat transfer improvement technology. It is defined by the equation below.

$$\eta = \frac{\frac{Nu}{Nu_0}}{\left(\frac{f}{f0}\right)^{\frac{1}{3}}}$$

where η is thermal performance factor, the Nusselt numbers and friction factors for a duct layout with and without inserts are Nu, f, Nu_0, and f_0, respectively. The head loss or pumping power is measured by the friction factor (Heat Exchangers, n.d.). There are three types of heat transfer enhancement techniques: passive methods, active methods, and compound methods (Fig. 6.5). By increasing surface area, roughness, and modifying the boundary conditions, the rate of heat transfer can be passively increased. Whereas, the active method involves adding nanosized, high thermal conductivity, and metallic powder to the base fluid to accelerate the heat transfer rate. In comparison to active approaches, passive procedures, which use inserts in the flow path to increase the heat transfer rate, are the best. Since the insert manufacturing process is straightforward and these strategies can be handily applied in a current

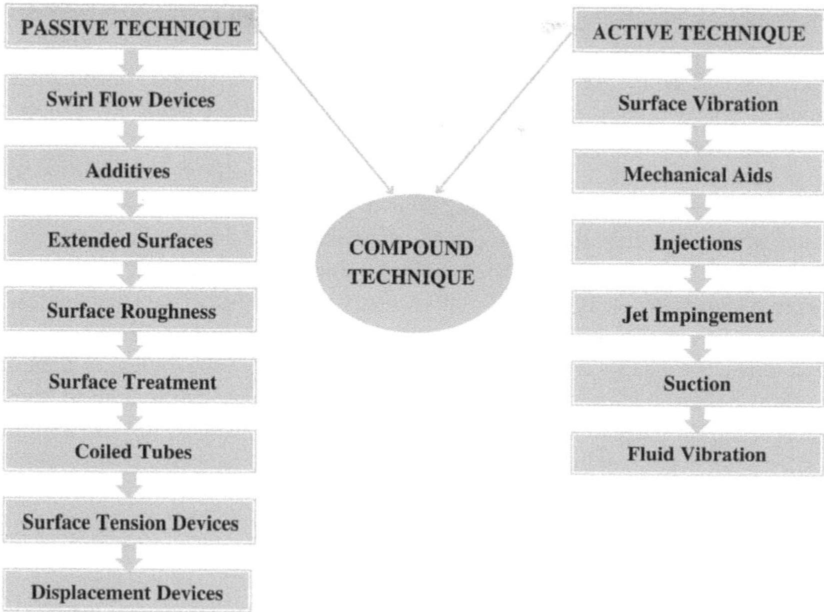

Fig. 6.5 Classification of heat transfer enhancement techniques used in shell and tube heat exchangers.

application. When two or more approaches, passive and active, are used concurrently to increase the heat transfer of any device to a level greater than that produced by either of those processes independently, the phrase compound enhancement technique is used (Kays & London, 1998).

6.6.1 Heat transfer enhancement by passive method

In general, heat transfer enhancement is concerned with improving the thermos-hydraulic performance of heat exchangers. The passive heat transfer enhancement method requires no external power. Increasing the effective surface area and residence time of the heat transfer fluid is one strategy to improve passive heat transfer efficacy. The employment of this technology induces swirling in the bulk of the fluids and disturbs the actual boundary layer, resulting in a rise in surface area, given time, and similarly heat transfer coefficient in the present system (Kays & London, 1998). The passive heat transfer enhancement technology for STHX used in the food industry is detailed below.

6.6.1.1 Swirl flow devices

They create swirl flow or secondary recirculation in a channel and superimpose it on the axial flow. Helical strip or cored screw type tube inserts, as well as twisted tapes, are examples of these devices. They can be utilized in heat exchangers with single-phase or two-phase flows. The best examples of swirl flow devices are twisted tape inserts. Twisted tapes are metallic strips that have been twisted using effective methods to achieve the necessary shape and dimension before being introduced into the flow. Besides promoting heat transfer rates with less friction factor penalty on pumping power, twisted tape inserts are popular and frequently employed in STHX for heat transfer enhancement. By generating swirl into the bulk flow and interrupting the boundary layer at the tube surface due to the constant changes in the surface shape, the insertion of twisted tapes in a tube provides a simple passive strategy for increasing convective heat transfer. Because of the changes in geometry of twisted tape inserts (Table 6.1), the heat transfer coefficient and Nusselt number are improved. When twisted-tape is used as an insert, the pressure drop inside the tube increases at the same time. Many experiments and numerical simulations have been carried out to study the required design in order to attain improved thermal performance with lower frictional losses. The pitch and twist ratio affects the heat transfer enhancement of twisted tape inserts (Shah, 1991).

6.6.1.2 Additives for liquids and gases

In the case of liquids, these involve the incorporation of solid particles, soluble trace additives, and gas bubbles into single phase flows, as well as trace additives that lower the liquid's surface tension in boiling systems. In the case of gases, these can be liquid droplets or solid particles that are injected into single-phase gas flows as a dilute phase (gas-solid suspensions) or as a dense phase (fluidized beds) (Shah, 1991).

Table 6.1 Classification of most commonly used twisted tapes in shell and tube heat exchanger (Kern, 1950).

Types of twisted tapes	Description
• Typical twisted tape	• The length of these tapes corresponds to the exchanger tube's length
• Varying length, alternate-axes and pitches twisted tape	• These are short length tapes connected by alternate axes with varied pitches spaced or twist types
• Multiple twisted tapes	• In a single heat exchanger tube, many twist tapes are employed
• Twisted tape with rod and varying spacer	• Twisted tape with a rod and a spacer to improve heat transfer
• Twisted tape with attached fins and baffles	• In order to achieve more enhancement, baffles are affixed to the twisted tape at various intervals
• Twisted tapes with slots, holes, cuts	• To increase turbulence, slots and holes of appropriate dimensions are cut into the twisted tape
• Helical left-right twisted tape with screw	• Twisted tapes are left-right helical in shape

6.6.1.3 Extended surfaces

The goal of implementing these techniques in STHX is to produce swirl turbulence and secondary flow. To keep the system in steady state conduction, the heat transmitted through solids, walls, or boundaries must be continually dissipated to the surrounds or environment. Large amounts of heat must be dispersed from small regions in many engineering applications especially in food industry. Heat transmission through convection between a surface and the fluid environment can be improved by attaching thin metal strips called fins to the surface. The fins enhance the surface's effective area, allowing for more heat transmission via convection. Fins can also be referred to as "extended surfaces." Extended surfaces (fins) are a common heat-transfer device. The fin's surface area influences the rate of heat transfer. For example, a heat sink with fins increases the contact surface area. The difficulty of determining the heat flow through a fin necessitates information of the temperature distribution through it. This is achieved by considering of the fin as a metallic plate attached to a heated wall at its base and transmitting heat to a fluid via convection. Conduction transports heat through the fin. As a result, the temperature distribution in a fin is influenced by the qualities of the fin material as well as the surrounding fluid (Evans, 1974).

6.6.1.4 Surface roughness

These are surface changes that create turbulence in the flow field, especially in single-phase flows, but do not enhance the heat transfer surface area. Random sand-grain roughness to discrete three-dimensional surface protuberances is among their geometric properties (TEMA, 1999).

6.6.1.5 Surface treatment

They are heat transfer surfaces with a fine-scale finish or coating modification. The change might be continuous or discontinuous, and the roughness is significantly lower than what influences single-phase heat transfer, hence they are mostly employed for boiling and condensing (TEMA, 1999).

6.6.1.6 Coiled tubes

These approaches are best suited to heat exchangers that are quite small. In single phase flow and most boiling zones, coiled tubes produce secondary flows and vortices that encourage a higher heat transfer coefficient (Walker, 1990).

6.6.1.7 Surface tension devices

These are wicking or grooved surfaces that improve the boiling and condensing surfaces directly. The most common application for these devices is heat exchanger phase transformation (TEMA, 1999).

6.6.1.8 Displaced enhancement devices

These are the primary insert strategies utilized in confined force convection. By displacing fluid from the heated or cooled surface of the duct/pipe with bulk fluid to the core flow, these devices promote energy transfer indirectly at the heat exchange surface (TEMA, 1999).

6.6.1.9 Treated surfaces

The treated surfaces are made up of different structured surfaces like continuous or discontinuous integral surface roughness and coatings. The roughness produced by this treatment has no discernible influence on single-phase heat transfer. These are only applicable in the situation of two-phase heat transfer. This method makes use of pits, cavities, or scratches. They are mostly used for boiling and condensing duties. Boiling is a convective heat transfer process that is utilized in boilers to produce steam, heat absorption in evaporators in refrigeration systems, food dehydration and drying, and liquid distillation. When very wetting fluids such as refrigerants, organic liquids, cryogens, and alkali liquid metals are used, the usual cavities in heated surfaces are flooded with subcooled liquid. Coatings of nonwetting material like Teflon, on either the heated surface or its pits and cavities were shown to be effective in nucleate boiling for high surface tension fluids. The heated surface can be covered with a stainless-steel surface and Teflon to generate no-wetting material patches, resulting in three to four times better heat transfer coefficients. When vapors condensate on treated surfaces, drop wise condensation occurs, which is good for reducing surface wetting and breaking up the condensate layer into droplets? At the cold heat transfer interface, this method enables greater drainage and more effective vapor removal. When compared to Bergle's proposed film wise condensation, this approach enhances heat transfer by a ratio of 10–100 in drop wise condensation. Nonwetting inorganic compounds, noble metals, or organic polymers can be utilized to coat heat transfer surfaces effectively. Organic coatings, for example, have been widely employed in steam systems (TEMA, 1999).

6.6.1.10 Inserts

Inserts necessitate additional fluid flow configurations that increase and augment heat transfer. Twisted tape, wire coils, ribs, baffles, plates, helical screw inserts, mesh inserts, convergent—divergent conical rings, conical rings, and so on are examples of inserts. Twisted tape and wire coil inserts (Fig. 6.6) are widely used than other type of inserts. Twisted tapes are metallic strips that have been twisted using various processes to get the desired shape and dimension, and they are put into the flow to improve heat transfer. Twisted tape inserts improve heat transfer rates by reducing friction. The use of twisted tapes in a tube provides a simple passive strategy for improving convective heat transfer by swirling heavy flow, breaking the boundary layer at the tube surface due to rapid changes in surface geometry. Because of the changes in geometry of twisted tape inserts, the heat transfer coefficient and Nusselt number are improved. When twisted-tape is used as an insert, the pressure drop inside

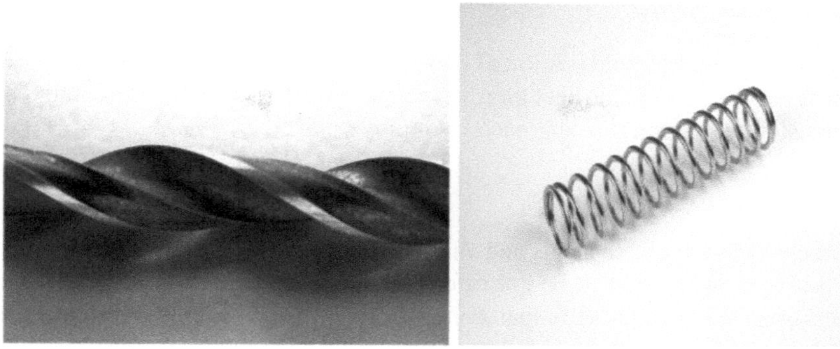

Fig. 6.6 Twisted tape (*left*) and wire coil (*right*).

the tube increases at the same time. Many experiments and numerical simulations have been carried out to study the required design in order to attain improved thermal performance with lower frictional losses. The Pitch and Twist ratio affects the heat transfer enhancement of twisted tape inserts. Wire coil inserts are commonly used in heat transfer equipment and have been used as a passive improvement technique. They have a number of advantages over other enhancing techniques including:

1) Simple installation and removal.
2) Low-cost, simple manufacturing technique.
3) Mechanical strength of the original simple tube is preserved.
4) Installation in a smooth tube heat exchanger that already exists (retrofit).
5) Containment of fouling (in refineries, chemical plants, and marine applications).

6.6.2 Heat transfer enhancement by active method

Active methods entail the use of external power to improve heat transmission. Because of the equipment requirements, this strategy can only be used in a narrow variety of aspects. These techniques have not demonstrated significant potential in comparison to passive techniques because it is challenging to provide external power input in many circumstances.

6.6.2.1 Surface vibration

They are mostly being used in single-phase flows thus far. The surface vibrations are aided by applying a low or high frequency, leading to higher convective heat transfer coefficients.

6.6.2.2 Mechanical aids

Mechanically or by rotating the surface, these devices agitate the fluid. Rotating tube exchangers and scraped surface heat and mass exchangers are examples of mechanical aids.

6.6.2.3 Injections

The same or a different fluid is injected into the main bulk fluid via a porous heat transfer barrier or upstream of the heat transfer section in this procedure. For a single-phase heat transfer procedure, this approach is employed (Walker, 1990).

6.6.2.4 Jet impingements

Heat exchange in micro channels and various microprocessors used in electronic circuits has been discovered to be extremely important, since excessive heat creation might shorten the material's life span owing to overheating. The research on the usage of synthetic jets in heat dissipation inside micro channels and microstructures, which is a fluidic device working as a Helmholtz resonator, has been discovered to be quite amazing. An actuated cavity is defined by a constantly moving diaphragm/diaphragm affected by voltage fluctuations and a small opening (Orifice) membrane. It has been discovered to be a very effective mechanism of heat transmission since it penetrates thermal barrier layers and can be simply incorporated into structures, unlike other systems that require complicated plumbing. In a study by Chandratilleke et al., heat transmission was boosted by 4.3 times for low Reynold number values ranging from 0 to 70 when compared to when no synthetic jet was used (Keblinski et al., 2002).

6.6.2.5 Suction

This method involves injecting the same or a different fluid into the main bulk fluid via a porous thermal energy interface or upstream of the heat transfer segment. This method is employed in the single-phase heat transfer procedure (Rohsenow & Hartnett, 1973).

6.6.2.6 Fluid vibration

Rather than vibrating the surface, pulsations are formed within the fluid itself. For single phase flows, this type of vibration augmentation approach is used (Rohsenow & Hartnett, 1973).

6.6.3 Experimental works on the enhancement of heat transfer

Salam et al. explored the measurement of tube-side heat transfer coefficient, friction factor, and heat transfer enhancement efficiency of water in a circular tube with rectangular-cut twisted tape insert for turbulent flow in a circular tube. As a smooth tube, a copper tube with stainless steel rectangular-cut twisted tape inserts was used. The efficiencies of heat transfer enhancement were found to be in the range of 1.9–2.3, and they increased as the Reynolds number increased (Hewitt, 1998). Using water as the working fluid, Eiamsa-ard et al. investigated the heat transfer, flow friction, and thermal performance factor characteristics in a tube fitted with delta winglet twisted tape. The obtained results demonstrate that as the twisted ratio decreases and the depth of the wing cut ratio increases, the mean Nusselt number and mean friction factor in the tube with the delta-winglet twisted tape increase (Shah & Bell, 2000). Sundar et al. calculated the turbulent convective heat transfer and friction factor of magnetic Fe_3O_4 nanofluid flowing through an evenly heated horizontal circular tube with and without twisted tape inserts in an experiment. Under turbulent flow conditions, the authors presented a generalized regression equation for the estimate of Nusselt number and friction factor including both water and Fe_3O_4 nanofluid in a plain tube and with twisted tape inserts (Chauduri et al., 1997).

Krishnan et al. compiled a list of the most relevant papers on the effect of heat transfer properties in STHX using Nanofluids. Nano fluids, according to the researchers, have a lot of potential for improving heat transmission in heat exchangers. Nano fluid is a next-generation enhanced heat transfer fluid (Jafari, Saremnejad, & Dehnad, 2017). Fares et al. evaluated the impact of graphene nanofluids on convective heat transfer in a vertical STHX in an experimental setting. Graphene flakes were made from graphite foam, which is made from sugar as a starting ingredient. Using 0.2% graphene/water nanofluids, a maximum increase in the heat transfer coefficient of 29% was attained. Furthermore, utilizing graphene/water nanofluid increased the heat exchanger's mean thermal efficiency by 13.7% (Jafari, Saremnejad, Dehnad, Rashidi, 2017). Rao et al. investigated the heat transfer rate of nano fluids utilizing a STHX in single and multitubes during turbulent flow employing a forced convection mode utilizing alumina nanoparticles. The authors found that on comparison to traditional fluids, the heat transmission rate is excellent with the nano fluids (Kern, 1983). Chen et al. investigated the heat transfer performance of a flat aluminum tube micro channel heat exchanger with spray cooling using experimental analysis. The effect of evaporative cooling was reduced, whereas flow friction rose. The water spray can greatly improve the air-side heat transfer performance, according to the test findings just presented (Kahani et al., 2013).

Heat exchangers are common heating and cooling equipment in a variety of industries, including the food industry. Heat exchanger inefficiency in terms of heat transfer has been a major issue, driving up manufacturing costs. Recent technological advancements have introduced new ways to enhance these systems. Nano fluids have recently become increasingly popular as an alternative to standard passive and active heat transfer improvement approaches. The following results are drawn from an exhaustive analysis of numerous active heat transfer augmentation methods employed by various researchers in heat exchangers and other comparable applications. For enhancement, the active method requires an external power source, and different parameters involved in these methods, such as flow regulation, electrical supply, ultrasonic wave intensity, and so on, must be employed and/or altered according to the enhancement required.

- Active methods also raise the friction factor, which is undesirable, and the ability to alter the parameters involved in these methods can be utilized to optimize heat transfer with the least amount of friction factor increase, which is a restriction of passive techniques.
- In mechanical aids such as scrapers, an external power source is not required, but the pumping power is raised as the friction factor rises.
- With improvements in efficiency, the design of heat exchangers in passive approaches stays innovative and simple.

- Because very little work has been reported on compound approaches, there is a great scope for future research.
- Magnetic and electroactive materials are found to be particularly promising for future research because they can be employed with passive techniques as well (Keblinski et al., 2002).

References

Abd, A. A., & Naji, S. Z. (2017). Analysis study of shell and tube heat exchanger for clough company with reselect different parameters to improve the design. *Case Studies in Thermal Engineering*, *10*, 455–467.

Alimoradi, A., & Veysi, F. (2017). Optimal and critical values of geometrical parameters of shell and helically coiled tube heat exchangers. *Case Studies in Thermal Engineering*, 73–78.

Aminian, J., & Shahhosseini, S. (2008). Evaluation of ANN modeling for prediction of crude oil fouling behavior. *Applied Thermal Engineering*, *28*(7), 668–674.

Bell, K. J. (2004). Heat exchanger design for the process industries, transactions of the ASME. *Transactions of the ASME, Journal of Heat Transfer*, *126*(6), 877–885.

Chauduri, P. D., Diwekar, U., & Logsdon, J. (1997). An automated approach for the optimal design of heat exchangers. *Industrial & Engineering Chemistry Research*, *36*, 3685e3693.

Dizaji, H. S., Jafarmadar, S., & Asaadi, S. (2017). Experimental exergy analysis for shell and tube heat exchanger made of corrugated shell and corrugated tube. *Experimental Thermal and Fluid Science*, 475–481.

El Maakoul, A., et al. (2006). Numerical comparison of shell side performance for shell and tube heat exchangers with trefoil-hole, helical and segmental baffles. *Applied Thermal Engineering*, *109*, 175–185.

Eryener, D. (2006). Thermoeconomic optimization of baffle spacing for shell and tube heat exchangers. *Energy Conversion and Management*, *47*(11 – 12), 1478–1489.

Evans, F. L. (1974). *Equipment design handbook for refineries and chemical plants* (2nd ed.). Houston, TX: Gulf Publishing Company.

Fesanghary, M., Damangir, E., & Soleimani, I. (2009). Design optimization of shell-and tube heat exchangers using global sensitivity analysis and harmony search algorithm. *Applied Thermal Engineering*, *29*, 1026e1031.

Gaddis, E. S., & Gnielinski, V. (1997). Pressure drop on the shell side of shell-and-tube heat exchangers with segmental baffles. *Chemical Engineering and Processing*, *36*(2), 149–159.

Gao, B., Bi, Q., Nie, Z., & Wu, J. (2015). Experimental study of effects of baffle helix angle on shell-side performance of shell-and-tube heat exchangers with discontinuous helical baffles. *Experimental Thermal and Fluid Science*, 48–57.

Gulyani, B. B. (2000). Estimating number of shells in shell and tube heat exchangers: A new approach based on temperature cross. *Transactions of the ASME, Journal of Heat Transfer*, *122*(3), 566–571.

Hewitt, G. F. (1998). *Heat exchanger design handbook*. New York: Begell House.

Ishiyama, E. M., Heins, A. V., Paterson, W. R., Spinelli, L., & Wilson, D. I. (2010). Scheduling cleaning in a crude oil preheat train subject to fouling: Incorporating desalter control. *Applied Thermal Engineering*, *30*(13), 1852–1862.

Jafari, S. M., Saremnejad, F., & Dehnad, D. (2017). Nano-fluid thermal processing of watermelon juice in a shell and tube heat exchanger and evaluating its qualitative properties. *Innovative Food Science & Emerging Technologies*, *42*(Suppl. C), 173–179.

Jafari, S. M., Saremnejad, F., Dehnad, D., & Rashidi, A. M. C. (2017). Evaluation of performance and thermophysical properties of alumina nanofluid as a new heating medium for processing of food products. *Journal of Food Process Engineering*, *40*(5), 1–9. https://doi.org/10.1111/jfpe.12544.

Kahani, M., Heris, S. Z., & Mousavi, S. M. (2013). Effects of curvature ratio and coil pitch spacing on heat transfer performance of Al2O3/water nanofluid laminar flow through helical coils. *Journal of Dispersion Science and Technology*, *34*(12), 1704–1712.

Kang, H. U., Kim, S. H., & Oh, J. M. (2006). Estimation of thermal conductivity of nanofluid using experimental effective particle volume. *Experimental Heat Transfer*, *19*(3), 181–191.

Kays, W. M., & London, A. L. (1998). *Compact heat exchangers, reprint* (3rd ed.). Malabar, FL: Krieger Publishing.

Keblinski, P., Phillpot, S. R., Choi, S. U. S., & Eastman, J. A. (2002). Mechanisms of heat flow in suspensions of nano-sized particles (nanofluids). *International Journal of Heat and Mass Transfer*, *45*, 855–863.

Kern, D. Q. (1950). *Process heat transfer*. New York: McGraw-Hill.

Kern, D. Q. (1983). *Process heat transfer* (international student ed.). Japan: McGraw-Hill Book Company.

Lei, Y. G., He, Y. L., Li, R., & Gao, Y. F. (2008). Effects of baffle inclination angle on flow and heat transfer of a heat exchanger with helical baffles. *Chemical Engineering and Processing*, *47*(12), 2336–2345.

Lei, C., Peng, Z., Day, T., Yan, X., Bai, X., & Yuan, C. (2011). Experimental observation of surface morphology effect on crystallization fouling in plate heat exchangers. *International Communications in Heat and Mass Transfer*, *38*(1), 25–30.

Lemos, J. C., Costa, L. H., & Bgajewicz, J. (2017). Linear method for the design of shell and tube heat exchangers including fouling modeling. *Applied Thermal Engineering*, *125*, 1345–1353.

Li, H., & Kottke, V. (1998). Effect of baffle spacing on pressure drop and local heat transfer in shell-and-tube heat exchangers for staggered tube arrangement. *International Journal of Heat and Mass Transfer*, *41*(10), 1303–1311.

Liu, L., Ding, N., Shi, J., Xu, N., Guo, W., & Wu, C. (2016). Failure analysis of tubeto-tubesheet welded joints in a shell-tube heat exchanger. *Case Studies in Engineering Failure Analysis*, 32–40.

Lutcha, J., & Nemcansky, J. (1990). Performance improvement of tubular heat exchangers by helical baffles. *Chemical Engineering Research and Design*, *68*(3), 263–270.

Mare, T., Halelfadl, S., Sow, O., Estelle, P., Duret, S., & Bazantay, F. (2011). Comparison of the thermal performances of two nanofluids at low temperature in a plate heat exchanger. *Experimental Thermal and Fluid Science*, *35*(8), 1535–1543.

Master, B. I. (2004). *Heat exchanger*. United States, US 6827138 B1.

Master, B. I., Chunangad, K. S., Boxma, A. J., Kral, D., & Stehlík, P. (2006). Most frequently used heat exchangers from pioneering research to worldwide applications. *Heat Transfer Engineering*, *27*(6), 4–11.

Mirzaei, M., Hajabdollahi, H., & Fadakar, H. (2017). Multi-objective optimization of shell-and-tube heat exchanger by constructal theory. *Applied Thermal Engineering*, *125*, 9–19.

Nasiri, M., Etemad, S. G., & Bagheri, R. (2011). Experimental heat transfer of nanofluid through an annular duct. *International Communications in Heat and Mass Transfer*, *38*(7), 958–963.

Heat Exchangers Prepared by Bob Heaslip KESCO For Queens University CHEE 470—Fall 2008, 1–11.

Pandey, S. D., & Nema, V. K. (2012). Experimental analysis of heat transfer and friction factor of nanofluid as a coolant in a corrugated plate heat exchanger. *Experimental Thermal and Fluid Science*, *38*, 248–256.

Peyghambarzadeh, S. M., Hashemabadi, S. H., Hoseini, S. M., & Seifi Jamnani, M. (2011). Experimental study of heat transfer enhancement using water/ethylene glycol based nanofluids as a new coolant for car radiators. *International Communications in Heat and Mass Transfer*, *38*(9), 1283–1290.

Rodriguez, C., & Smith, R. (2007). Optimization of operating conditions for mitigating fouling in heat exchanger networks. *Chemical Engineering Research and Design*, *85*(6), 839–851.

Rohsenow, W. M., & Hartnett, J. P. (1973). *Handbook of heat transfer*. New York: McGraw-Hill.

Saffar-Avval, M., & Damangir, E. (1995). A general correlation for determining optimum baffle spacing for all types of shell and tube exchangers. *International Journal of Heat and Mass Transfer*, *38*(13), 2501–2506.

Selbas, R., Kizilkan, O., & Reppich, M. (2006). A new design approach for shell-and-tube heat exchangers using genetic algorithms from economic point of view. *Chemical Engineering and Processing*, *45*, 268e275.

Shah, R. K. (1991). In E. A. Foumeny, & P. J. Heggs (Eds.), *Compact heat exchanger technology and applications, in heat exchanger engineering, Vol. 2, compact heat exchangers: Techniques for size reduction*. London: Ellis Horwood.

Shah, R. K., & Bell, K. J. (2000). *CRC handbook of thermal engineering*. Florida: CRC Press.

Shinde, S., & Chavan, U. (2017). Numerical and experimental analysis on shell side thermo-hydraulic performance of shell and tube heat exchanger with continuous helical FRP baffles. *Thermal Science and Engineering Progress.*. https://doi.org/10.1016/j.tsep.2017.11.006.

Sinnott, R. K. (1993). *Chemical engineering design. Vol. 6* (2nd ed.). Pergamon.

Srinivasan, M. (2005). Heat exchanger fouling of some Canadian crude oils. *Heat Transfer Engineering*, *7–14*.

Taghizadeh-Tabari, Z., Zeinali Heris, S., Moradi, M., & Kahani, M. (2016). The study on application of TiO2/water nanofluid in plate heat exchanger of milk pasteurization industries. *Renewable and Sustainable Energy Reviews*, *58*, 1318–1326.

TEMA. (1999). *Standards of TEMA* (8th ed.). New York: Tubular Exchanger Manufacturers Association.

Tkachuk, V., et al. (2019). *IOP Conference Series: Materials Science and Engineering*, *589*, 012007.

Walker, G. (1990). *Industrial heat exchangers: A basic guide* (2nd ed.). Washington, DC: Hemisphere Publishing.

Wang, Q. W., Chen, G. D., Xu, J., & Ji, Y. P. (2010). Second-law thermodynamic comparison and maximal velocity ratio design of shell-and-tube heat exchangers with continuous helical baffles. *Transactions of the ASME, Journal of Heat Transfer*, *132*(10), 101801–101809.

Yu, C., Ren, Z., & Zeng, M. (2018). Numerical investigation of shell-side performance for shell and tube heat exchangers with two different clamping type anti-vibration baffles. *Applied Thermal Engineering*, *133*, 125–136.

Zeinali Heris, S., Noie, S. H., Talaii, E., & Sargolzaei, J. (2011). Numerical investigation of Al2O3/water nanofluid laminar convective heat transfer through triangular ducts. *Nanoscale Research Letters*, *6*(1), 179–189.

Further reading

Shah, R. K., & Sekulic, D. P. (2003). *Fundamentals of heat exchangers*. John Wiley and Sons.

Direct heat exchangers in the food industry

7

Chi-Ching Lee[a] *and İbrahim Agah Taştemir*[b]
[a]Istanbul Sabahattin Zaim University, Faculty of Engineering and Natural Sciences, Department of Food Engineering, Istanbul, Turkey, [b]Istanbul Sabahattin Zaim University, Faculty of Engineering and Natural Sciences, Department of Interior Architecture and Environmental Design, Istanbul, Turkey

7.1 Introduction

Heat refers to a type of energy as the molecules of a substance move in lockstep. The absolute temperature is proportional to the substance's mean kinetic energy per molecule. Because of the temperature difference among different materials, thermal energy starts to move and transfer from one to another. The spontaneous, irreversible flow of heat from hotter (higher temperature) substances to the colder (lower temperature) substances is known as heat transfer. This field of study focuses on the analysis and quantification of heat transfer. The science of heat flow and its relationship to mechanical work is known as thermodynamics. The principle of energy conservation is well known as the first law of thermodynamics. Except in subatomic processes, energy cannot be created or destroyed and it can only be transformed from one form to another. The terms "work" and "heat" are interchangeable. Unless there is an energy storage component, the energy output must match the energy input. Moreover, the second law of thermodynamics is that heat transmission may only occur in one direction, which is from a hotter substance to a colder one or from a higher to a lower energy state, just like water flow which takes place only downhill. Heat does not flow from a low-temperature substance to a high-temperature one spontaneously. A device can only provide heat in the reverse direction if it receives external energy. Any device or equipment that performs work must have an energy source and operation units, which means that energy must flow through the machine. However, only a portion of this flow may be converted into work (Savulescu, Sorin, & Smith, 2002).

7.2 Direct heat-transfer mechanism

Heat can be transferred from one system to another resulting in a temperature change. Thermodynamic research examines the amount of heat transfer that occurs as a system moves from one equilibrium state to another. The study of heat transfer focuses on understanding the mechanism of this energy exchange. Heat flows from a higher-temperature substance into a lower-temperature substance. When both substances reach the same temperature, heat transfer ends. There are three ways of heat transfer;

Thermal Processing of Food Products by Steam and Hot Water. https://doi.org/10.1016/B978-0-12-818616-9.00008-0

they are conduction, convection, and radiation. All forms of heat transmission need a temperature difference. Conduction, convection, and radiation use indirect heat transfer to cause temperature change, whereas conduction and convection are usually direct heat-transfer mechanisms (Cengel & Ghajar, 2020; Welty, Rorrer, & Foster, 2020). Conduction heat transfer is a status of molecular-level kinetic energy transfers between solids, liquids, and gases. In solids, heat conduction and electrical conduction have a tight relationship. Conduction heat moves in the direction of low temperature. Convection heat transfer is a result of larger-scale motions of a fluid, either liquid or gas. The rate of convection heat transfer increases as the fluid flow velocity increases. Radiation heat transfer is the energy transition with electromagnetic waves. Radiation can occur without any substance present other than the surface exchanging energy. The presence of two surfaces with different temperatures is the only requirement for the occurrence of radiation heat transfer. To generate internal energy, radiation must be absorbed by matter. For example, solar energy is transmitted from the sun to earth by means of radiation (Baehr & Stephan, 2008; Bergman, Incropera, DeWitt, & Lavine, 2011).

Conduction is a direct heat-transfer mechanism in which the energy transfer from a substance occurs from more energetic particles to nearby less energetic particles because of particle interactions. Conduction can occur in solids, liquids, and gases. The geometry, the thickness, the property of a medium, and the temperature difference between the media determine the rate of heat conduction. High molecular energies are usually involved in high temperatures. When nearby molecules collide, energy must be transferred from the more energetic to the less energetic molecules. Energy transfer via conduction must proceed in the direction of a lower temperature in the presence of a temperature gradient. Molecular collisions help with energy transmission, and energy diffusion is a net energy transfer caused by random molecule motion (Holman, 2008; Szokolay, 2012). When two substances with different temperatures come into physical contact, direct contact heat transfer can occur. The implication is that the two substances do not have a barrier between them. Indirect heat transfer occurs when there is a surface wall between the two streams (Kreith & Boehm, 2013; Rajput, 2019).

Applications of direct contact heat exchangers in the food industry have been in different products, such as dairy foods, beverages, oils, and innovative products (Demirci, Lee, Çavuş, & Çağlar, 2020). Direct contact heat transfer can occur at the interface of two continuous streams. For example, a dispersed spray injected into a gaseous or water vapor might contain the condensation of the water vapor with the spray droplets. Besides, a hot gas or fuel might be vaporized and combusted while a gas drifts across a thin liquid sheet. Direct contact heat transfer can occur at the interface of two continuous fluid streams. For example, a hot gas may be vaporized or the fuel can be combusted when the gas flows over a thin liquid film. Furthermore, a dispersed spray injected into a gaseous or vapor stream could cause vapor condensation on droplets inside the spray. Another example is the cooling of small liquid droplets that are solidifying, such as during the production of metal shot or glass beads. Many bioactive compounds are sensitive to many external factors, such as oxygen, light, and heat during the process of heat exchange (Ashaolu et al., 2021; Lee, Tomas, & Jafari,

2020). Chemical interactions between the mass streams can occur in many circumstances of direct-contact heat transfer, and one of the gas streams may be entirely devoured by the other. The sensible heat transfer can occur within two immiscible liquids. When mass fluids contain at least one kind of stream, it can be either laminar flow or turbulent (streamline) flow. Because turbulence might cause difficulties with the mass stream or the dynamic changes in the bulk fluid, turbulence should not allow in many industrial applications (Jacobs, 1995).

Industrial direct heat exchangers have been classified based on the following:

1) Phase characteristics of the contacting fluids
2) The relative flow directions
3) The dynamic forces of the flows
4) Process purposes
5) Heat-transfer surfaces
6) Constructions
7) Phase change mechanisms

These classifications are mentioned in Table 7.1. This chapter will discuss different types of direct heat exchangers for more details according to phase characteristics of the contacting fluids, process purpose, heat-transfer surfaces, and phase change mechanisms (Shah & Sekulic, 2003; Thulukkanam, 2000).

Table 7.1 Classifications of direct heat-transfer processes.

Classified by	Types	Remarks
Phase characteristics of the contacting fluids	Liquid-liquid Solid-gas Gas-liquid *Immiscible fluid* Liquid-vapor	Heat is allowed to transfer in direct contact between any combinations of substance fluids effectively. These phases contain liquids (water), gases (air), vapor, solids, and *immiscible fluids*
The relative flow directions	Parallel flow Counter flow Cross flow	The selection of a specific flow direction is affected by many parameters, including streamflow pathways, device designs, thermal stresses, heat exchanging efficiency, and required temperature
The dynamic forces of the flows	Gravity Pressure Centrifugal forces	Gravity is the most common driving factor for the separation of two immiscible fluids from the other. However, when a force other than gravity is used, pressure or centrifugal forces may be involved in this situation
Process purposes	Steam-driven jet heat exchangers Cooling towers Scrubbers	For different functions and applications in the industry, a variety of heat exchangers are designed and developed. For example, the fluid in the steam-driven jet heat exchanger with a greater pressure at high speed is expelled from the nozzle. The

Continued

Table 7.1 Continued

Classified by	Types	Remarks
	Hybrid condensers	low-pressure fluid is injected into the mixing chamber, where it comes into direct contact with the jet fluid for heat transfer and then flows into the diffuser together. The output of the diffuser is delivered to the user after it achieves the same pressure and temperature
Heat-transfer surfaces	Drop- or spray-type Jet- or sheet-type Film type Bubble- or pool-type	These types of heat exchangers are classified with the contact surface between two fluids during heat exchange. For example, a film-type heat exchanger allows one kind of fluid to exchange heat with another type of liquid in the form of a liquid film
Constructions	Tubular or shell heat exchangers Plate heat exchangers Extended surface heat exchangers Regenerators	The type of exchanger is designed as a recuperator to allow the heat transfer from the hot stream to the cold stream through a separating wall. The tubular or shell type contains a double pipe and coiled tube. The plate-type contains lamella, panel coils, brazed types, and spiral types. The extended surface type includes tubes and plate fins
Phase change mechanisms	Condenser Evaporator	This mechanism of heat transfer between two fluids usually occurs in two-phase convection. Condensers cool down the gas (air) or liquid (water). Besides, evaporators turn the liquid (water) form into a gas (vapor) form

7.3 Direct heat exchangers classified by process purpose

This section will mention various direct heat exchangers based on their process purposes. Condensers, evaporators, cooling towers, flash evaporators, and scrubbers are the examples.

7.3.1 Condensers

The condenser is a type of heat exchanger and also a component of the refrigeration system. The heat exchanging system is an exothermic process because it can transform a gas or vapor into a liquid and transfer the heat inside the tube to the surrounding air near the tube very efficiently. This is the reason that there is a relatively high temperature in the condenser and its nearby environment. In the refrigeration industry, the refrigerants ammonia and freon are commonly manufactured by condensers. During the distillation process, the function of a condenser is to convert a vapor to a liquid by removing the heat from the gas and steam. Direct-contact condensers have been designed and applied in various industries for more than 100 years. In 1903, Hausbran described the commercial design of barometric condensers theoretically in his book "Evaporating Condensing and Cooling Apparatus" (Hausbrand, 1919). Afterward, the design of direct-contact condensers in 1956 and 1972 appeared and was published (Fair, 1972; How, 1956). Most of the work described is more suitable for cooling gases than steam condensation. In 1977, the report "Direct Contact Condensers-A Literature Survey" was published from the U.S. Department of Energy (Jacobs & Fannar, 1977). Based on the interface formation of steam and the coolant liquid, a direct contact condenser was first introduced into four types: drop- or spray-type, jet- or sheet-type, film type, and bubble- or pool-type. These classifications are still used currently and also in this chapter. The application of the direct contact condenser on deaerators and their potential in geothermal systems was reported (Oliker, 1977; Oliker & Permyakov, 1971). Twenty-one companies in western Europe and the United States produced direct contact equipment, and 13 of them built condensers (Goldstick, 1981). Nowadays, direct contact condensers can be found in a variety of applications, including nuclear reactors; solar energy, geothermal, or space power plants; water desalination; ocean thermal energy conversion systems; and all kinds of chemical or food processing industries (Ashoor, Mansour, Giwa, Dufour, & Hasan, 2016; Bharathan, Parsons, & Althof, 1988; Fisher & Wright, 1984; Najafabadi, 2015). This is because the advantages of the application of direct contact condensers contain simple design and construction, a high efficiency of heat transfer, nonfouling and nonscaling situations, low cost for built-in equipment, and easy maintenance (Hewitt, Shires, & Bott, 1994).

7.3.1.1 Drop- or spray-type condensers

The condensation in drop-type or spray-type condensers occurs when a liquid coolant is injected with spraying or dropping into a steam chamber full of a gas-vapor mixture. The drops can be presumed sphere-shaped, and the transient conduction would direct the heat transfer inside the drops (Kutateladze & Borishanskiĭ, 1966). The spherical surface of drops was rapidly contacted to the hot steam at the saturation temperature. Its instant heat transfer can be exhibited by the experimental equations of transient conduction (Ford & Lekic, 1973). In the drop-type or spray-type condenser, cold water is injected from the upper nozzle, and steam enters the side inlet. After the steam contacts the cold water, it is condensed into water. At the same time, the condensed liquid flows down the pipe, and the noncondensable vapor would also be exhausted. The drop-type or spray-type condenser includes a condensation chamber with a heat exchanger. The bottom of the condensation chamber is provided with a spray water collection tank. A spray pump is installed on the side of the spray water collection tank. The spray pump is connected to a water pipe. The water pipe leads into the condensing chamber. The water pipe is equipped with a spray device on a condensing chamber. The upper end of the heat exchanger is provided with a refrigerant vapor

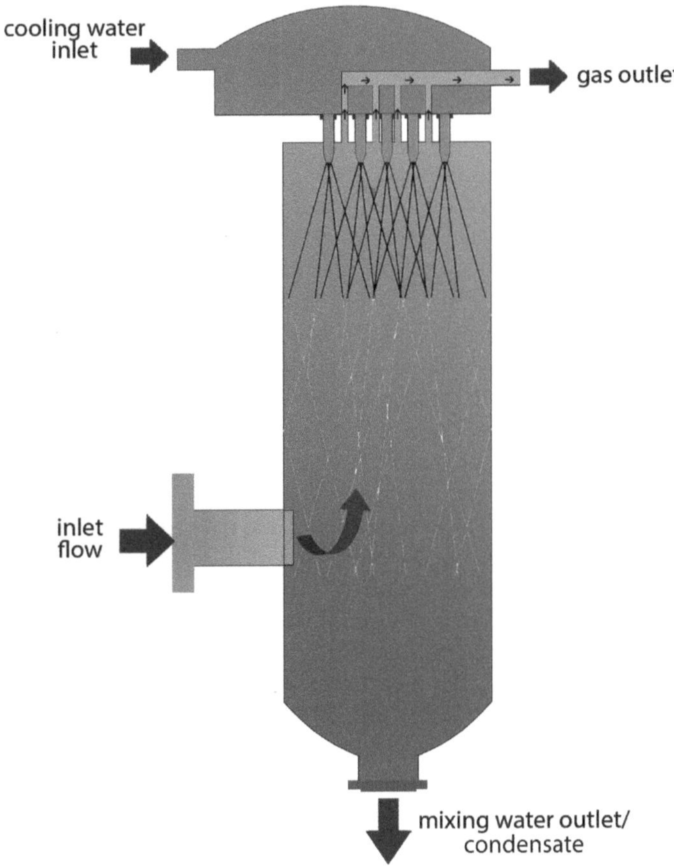

Fig. 7.1 Spray-type condenser.

inlet, and the lower end is provided with a refrigerant vapor outlet. A schematic representation of a spray-type condenser could be seen in Fig. 7.1. In some modern spray-type condensers, a splash-proof water inlet style grille is designed between the heat exchanger and the spray collection tank, and the top of the condensing chamber contains a water collector. Besides, the fluid is driven by the main circulating pump, and the fan is replaced with a spray device. When the circulating cooling water is sprayed, the outside air is pumped in without pressure. The water and air undergo heat exchange during the mixing process (Apriyanti, Adriansyah, Abdurrachim, & Pasek, 2018; Tissot et al., 2014).

7.3.1.2 Jet- or sheet-type condensers

There is direct contact between exhaust vapor and cooling water in jet- and sheet-type condensers used broadly for decades commercially. The cooling water and exhaust vapor are mixed up in a chamber. The temperatures of cooling water and the condensate are the same while being released from the condensers. The condensate would not be recycled and reused to the boilers as feed water. Because the vapor and cooling water are mixed up very well in the chamber of a jet-type condenser, it requires less amount of cooling water for the condensation of the vapor to achieve the high-efficiency process of condensation. Moreover, the advantages of jet-type condensers include simplicity in construction, building space saving, and low cost. The jet condensers are classified into three types:

1) Low-level jet condensers including the parallel flow type and counter flow type.
2) High-level jet condenser or barometric jet condenser.
3) Ejector condenser.

The major components of the jet condenser include nozzles or distributors of the condensing water, a steam inlet, a hot well, and a mixing chamber. If the steam and water flow in the same direction in the mixing chamber before condensation, it is a parallel flow-type jet condenser. However, the steam and water move in the opposite direction in a counter-flow-type chamber. The construction and elements of the sheet-type condenser are very similar to those of the jet-type condenser. The difference is that the coolant liquid flows into the steam chamber as sheet forms instead of a liquid jet. The purpose of both types of condensers is to increase the contact area between cooling water and steam (Bakay & Jaszay, 1978; Kreith & Boehm, 2013). The direct contact jet- and sheet-type condensers are shown schematically in Figs. 7.2 and 7.3 individually.

7.3.1.3 Film-type condensers

The structure of the film-type condenser is more intricate than the two previous types. In the chamber of the film-type condenser, there is a solid surface composed of plats, baffles, or particles. The condensate attaches the surface to form a thin liquid film that slumps down because of the gravity effect. This design offers a huge surface area for heat transfer in the packed bed. In addition, film condensation could lead to low heat-transfer rates because the heat transfer is hindered by the condensate film (Chung, Kim, & Ahmadinejad, 2008). The heat-transfer rate between the steam and surface is decreased by the thermal resistance caused by the film. Many factors can affect the thickness of film formation, such as liquid viscosity, condensation rate, and the orientation of the surface (Camaraza-Medina et al., 2019; Del Col, Parin, Bisetto, Bortolin, & Martucci, 2017).

Film-type condensers are also called packed bed condensers because the coolant flows over a solid substrate to form a thin film. After the steam enters the side pipe, the cooling water is sprayed from the upper side of the contact chamber of the condenser, which is filled with ceramic ring packing. After the packing is wetted by water, the contact area between cold water and steam is increased. The steam is condensed to water and discharged out along the lower pipeline. The noncondensable gas within the upper pipeline is pumped out by the vacuum pump to ensure a certain degree of vacuum in the condenser.

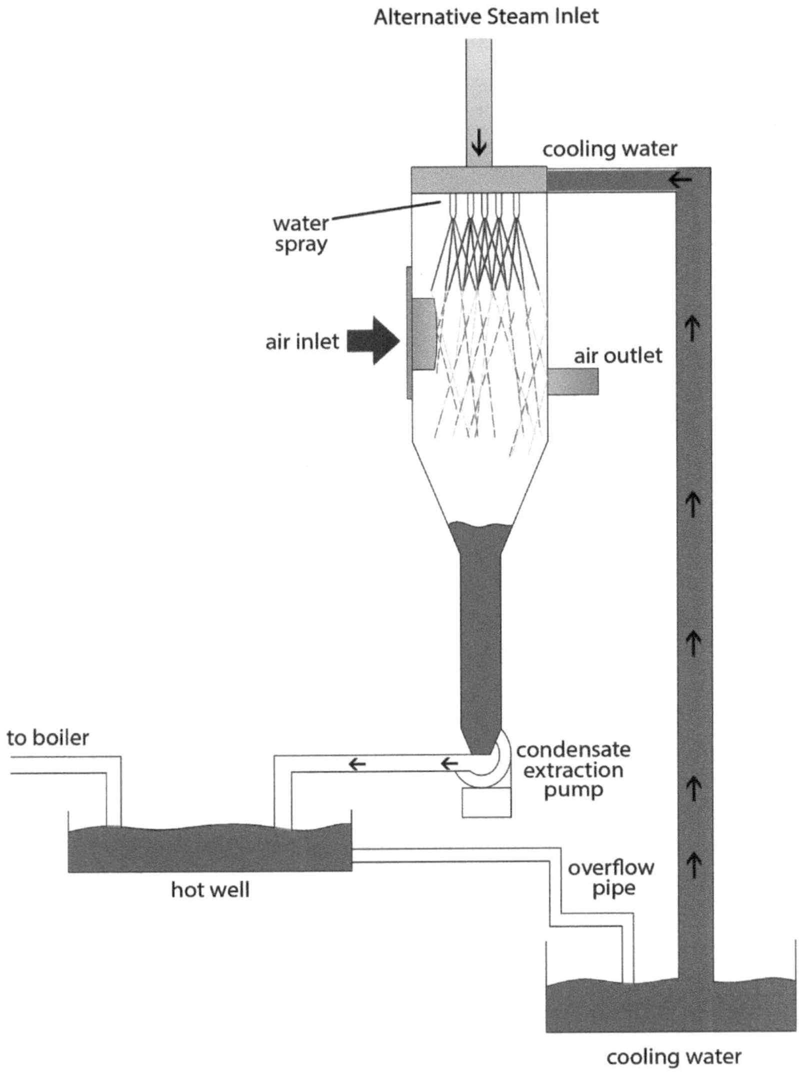

Fig. 7.2 Jet-type condenser.

7.3.1.4 Bubble- or pool-type condensers

In the bubble-type condenser, steam as a jet or bubble is injected into the continuous fluid or a pool of the cooling liquid. In the comparison of the three types of condensers mentioned above, their cooling water is injected into the chamber filled with vapor. The pool-type condenser is designed as a horizontal device. The vapor in the high-pressure injector enters the condenser under the action of the high-speed flow of liquid ammonia in the high-pressure ejector. Mixing, condensation, and possible chemical reaction could occur. The heat generated by the reaction or by gas-liquid condensation

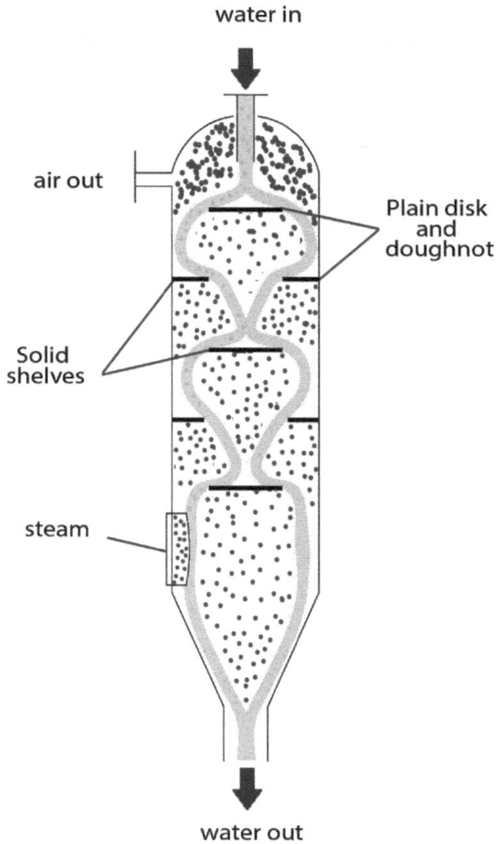

Fig. 7.3 Sheet-type condenser.

is absorbed by the steam condensate in the heat exchanger tube. Afterward, the low-pressure steam is produced, and it is used for the follow-up reheating, desorption, and vacuuming. The pool-type condenser is designed with a horizontal baffle, a vertical baffle, and a stripping gas distributor. After the vapor enters the bottom of the pool-type condenser, it is evenly distributed in the pool-type condenser through the distributor. This structure significantly increases the gas-liquid contact surface, and the incoming gas has a strong stirring effect in the continuous liquid phase, which improves the mass- and heat-transfer rate. Because the side of the pool-type condenser has a particular volume, the gas and liquid in this chamber are thoroughly mixed (Chan & Lee, 1982; Kakac, Liu, & Pramuanjaroenkij, 2012).

7.3.2 Evaporators

The evaporator is a kind of recuperative heat exchanger, which is an essential part of the four significant parts of refrigeration, including a compressor, a condenser, an expansion device, and an evaporator. The low-temperature condensed liquid passes through the evaporator to exchange heat with the outside air, where the gas absorbs heat to achieve the cooling effect. The medium to be cooled is usually water or air. For this reason, evaporators can be classified into two categories, evaporators for the cooling liquid, which is usually water or saltwater, and evaporators for cooling air. A common evaporator consists of a heating chamber, tubes, fins, a header, and an evaporative tank. The heating chamber offers the heat needed for evaporation to the liquid for boiling and vaporizing the liquid. The evaporative tank completely separates the gas and liquid phases. The vapor generated in the heating chamber has a large amount of liquid foam. After reaching the evaporative tank, these liquids can be separated from the vapor by self-condensation or by using a defoamer. The defoamer is usually located at the top of the evaporation chamber (Ribeiro Jr & Lage, 2005).

The header is a refrigerant distributor to make the refrigerant distributed in the pipes and placed between the expansion valve and the evaporator. The header is the larger pipe that connects to all circuits and supplies a specific amount of refrigerant to each circuit in the evaporator. The refrigerant flows inside the finned tube when air is released from the tube and fin surfaces. By increasing the total heat-transfer surface, the fin design improves the heat transfer from the gas in the evaporator pipes into the high-temperature refrigerant. The total heat-transfer surface of the evaporator is increased to increase the heat transfer from the air in the evaporator tube to the boiling refrigerant. Rectangular flat fin plates with the systematical arrangement are usually pierced by plenty of tubes or pipes an elliptical or circular cross-section (Huang, Xiao, Liu, & Wang, 2019). In the food industry, evaporators use heat to remove water from the foods, such as processed milk, vegetables, pastes, concentrates, coffee, fruit juices, and so on. The boiling point of the foods increases with concentration of the product. The quality of thermos-labile foods, such as juices and milk, can be maintained at low temperatures while the evaporation process occurs at low pressure. A variety of types of evaporators are applied in the food industry, such as natural circulation, rising-film, falling-film, batch-type, and agitated thin-film evaporators (Singh & Heldman, 2014).

The application of evaporation can be observed widely in various industries, such as the caustic chlorine industry, pulp, and paper production, phosphate industry, and food processing. During evaporation, solids and liquids are separated to recover solute from the solvent, usually the water. The unit operation of evaporation is similar to fractional distillation for the separation of the mixture with various volatile liquids. Vacuum evaporation is usually applied in single or multiple operations of each continuous stage. It can use the heat energy carried out from steam condensation in the previous stages. The advantage of multiple-stage evaporation includes the enhancement of process efficiency and the cost reduction of the operation in heat energy. The combination of low-temperature and high-vacuum techniques in the final stage of evaporation is used to dry the heat-liable product in the food and pharmaceutical industry. The main applications of the evaporation include air cooling in heating, ventilation, and air conditioning (HVAC) systems, heat rejection in spray ponds and cooling towers, the vapor production of flash evaporation in ocean thermal energy conversion (OTEC) and desalination, solute extraction, and petroleum distillation (Kreith & Boehm, 2013).

An evaporator with a direct contact heat transfer is also used in many types of industrial production. It mixes fuel (usually a gas and oil) with air and burns in a combustion chamber immersed in the solution. The high-temperature flame and flue generated are directly sprayed into the evaporated solution through a nozzle in the lower part of the combustion chamber. The high-temperature gas is in direct contact with the solution; heat transfer is carried out to evaporate and vaporize the water, and the generated water vapor and waste flue gas are discharged from the top of the evaporator together. The immersion depth of the combustion chamber in the solution is generally 0.2–0.6 m in some equipment, and the temperature of the gas exiting the combustion chamber can reach above 1000 °C. Because of the direct contact heat transfer, its heat transfer is highly effective, and the heat utilization is comparatively high. Because it does not require a fixed heat-transfer wall, the structure is simple, especially suitable for the evaporation of materials that are easy to be crystallized, are affected by scaling, and corrosive. It has been broadly used in the hazardous waste treatment of acid solutions and evaporation of ammonium sulfate solution. However, this type of evaporator is generally not applicable if the evaporated liquid is not allowed to be contaminated by the vaporized flue gas. Moreover, because of a large amount of flue gas, the utilization of secondary steam is limited. In addition, the nozzle is easily damaged because it is immersed in a high-temperature liquid (Kakac et al., 2012; Mori, 1991).

There are two basic designs of the direct contact evaporator with the immersed heating tank (Fig. 7.4) and with the external heating tank (Fig. 7.5). The water vapor production was enhanced with the initial droplet velocity in a single-stage system during the droplet evaporation. The size reduction of feed droplets significantly increased the droplet evaporation operation (Chen et al., 2016). Subsequently, a multistage model of direct contact spray evaporation and condensation system was reported in comparison with a single-stage system. A significant increase in water vapor production and thermal efficiency was displayed. For a 14-stage desalination plant, this technology has a better performance ratio of 6.5, which is defined as the ratio of equivalent heat of the distillate to heat input (Chen, Li, & Chua, 2016). A multistage low-temperature desalination system was powered with thermal 10 million watts concentrating solar power in cogeneration with thermal 7 million W diesel engines to desalinate 500 m^3 per day of fresh water. The cogeneration was predicted with the electrical power of 2.2 million W in the concentrating solar power plant to produce 520 m^3 per day of fresh water. Moreover, the output production would also be influenced by a variety of process parameters, including temperature, salinity load, and cogeneration fraction (Wellmann, Meyer-Kahlen, & Morosuk, 2018; Wellmann, Neuhäuser, Behrendt, & Lehmann, 2015).

7.3.3 Cooling towers

The cooling tower has been broadly used for decades as one of the major direct contact heat exchangers in a variety of industries, including crude oil refining and petrochemical operations, power stations, refrigeration, and air conditioning processes. In the cooling tower, water is used as a circulating coolant for absorbing heat from a system and discharging it into the environment to reduce the temperature of the loading hot

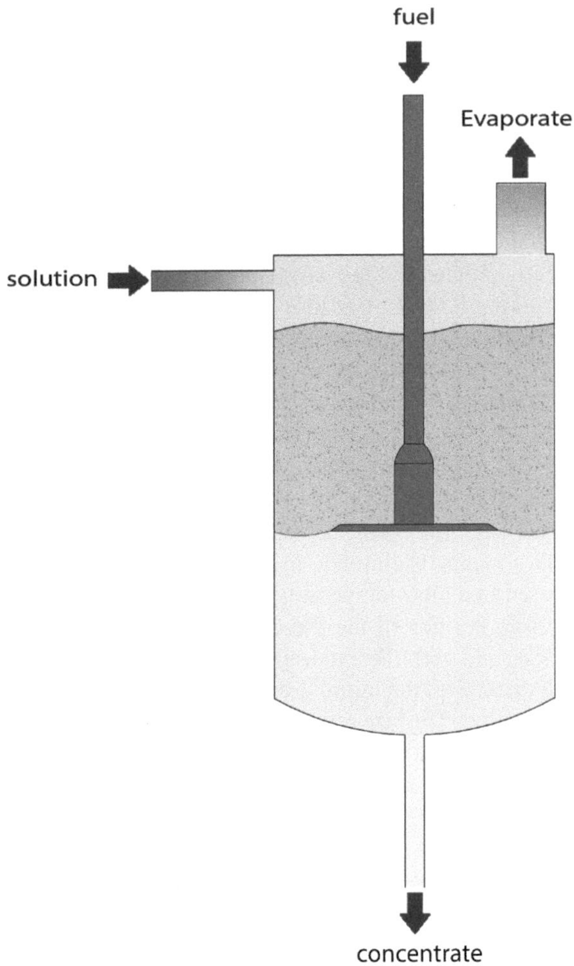

Fig. 7.4 Direct contact evaporator with the immersed heating tank.

water. The cooling tower uses water to directly contact air under heat exchanging, resulting in cooling down water and producing volatilized steam. The evaporation of steam takes away the heat to achieve heat dissipation. Besides, the convection and radiation of heat transfer disperse the waste heat generated in the industrial refrigeration and air conditioning to decrease the water temperature for maintaining the regular operation in the whole system (Naphon, 2005). The cooling tower can be classified into open (direct), closed (indirect), and hybrid circuit cooling towers (adiabatic cooling towers). In an open cooling tower, there is direct contact to decrease the water temperature with air. Direct heat transfer occurs during the interaction between discharged water and air. Besides, the evaporation with a small amount of cooled water makes the temperature of the cooling system decline (Lebrun & Silva, 2002; Milosavljevic & Heikkilä, 2001).

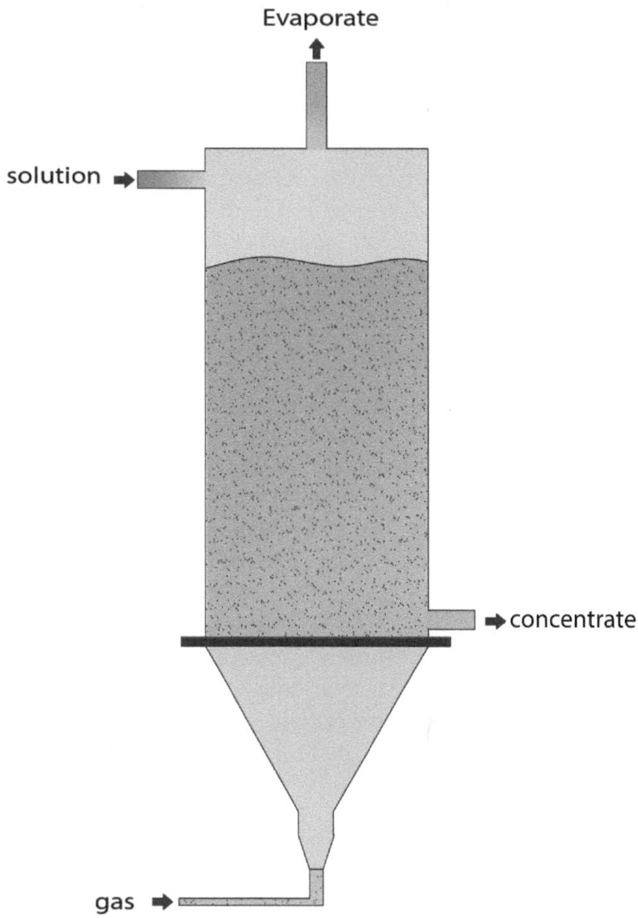

Fig. 7.5 Direct contact evaporator with the external heating tank.

7.3.3.1 Direct or open-circuit cooling towers

A direct or open-circuit cooling tower is a sealed structure with nozzles installed on the upper top of the cooling tower. The water that needs to be cooled down is introduced to these nozzles, and the circulating water is sprayed onto the packed bed full of glass fiber as the filler. It provides a large contact surface leading to the heat exchange effect achieved through water and air contact. The air circulation is introduced into the whole packed bed by a fan with blowing or extracting surrounding air. The fan drives the air circulation inside the tower to cool down the water by removing the heat with main evaporation and the convection between cool air and hot water. After heat exchange, the hot saturated humid air would be released from the tower to decrease the water temperature. The cooled water can be collected and recycled into the production process. The packed bed filling has multiple layers with a mainly vertical structure. The sprayed water or lateral splash water was spread for wetting the surface to create numerous small water droplets with a large surface area and several cascade layers of thin films (Lemouari, Boumaza, & Kaabi, 2009). In natural draft dry cooling towers, the performance of direct and indirect cooling for the heat rejection of a supercritical carbon dioxide Brayton cycle was evaluated. The comparison of direct and indirect cooled supercritical carbon dioxide Brayton cycles was estimated through a variety of ambient temperatures for concentrated solar power plants. In addition, the comparison of direct and indirect natural draft dry cooling towers in isolation was investigated through a range of input conditions (Duniam, Jahn, Hooman, Lu, & Veeraragavan, 2018; Hooman, 2010; Murray, 2009).

7.3.3.2 Indirect or closed-circuit cooling towers

In the indirect or closed-circuit cooling tower, direct contact is not involved between air and the water that needs to be cooled down. Nevertheless, an additional heat exchanger is used with piping and a plate heat exchanger. Unlike a direct cooling tower, there are two independent fluid circuits in an indirect cooling tower. One is that the water in the external circuit is in the secondary pathway, which is the hot fluid connected to the external loop of the tube bundle that is cooled and returned in a closed circuit. The air is driven through water circulation outside the entire heat pipe to provide a similar evaporative cooling. The heat flows from the internal fluid circuit, through the coil tube wall and the external circuit, and then heated by some evaporation of air and water into the ambient environment. In this process, the coolant is in a closed circuit, not directly exposed to external circulating water (Stabat & Marchio, 2004; Xia, Chen, & Wang, 2011).

7.3.3.3 Hybrid circuit or adiabatic cooling towers

The operation of the hybrid circuit or adiabatic cooling tower is similar to that of the dry cooling tower, except for the additional precooling pads inside the chamber of the adiabatic cooling tower. The adiabatic cooling system uses wet cooling but is equipped with a dry cooling unit. This design can use the environmental air to cool down the water temperature before entering the wet segment. Flowing water through precooling pads and then forcing air over the pads compress the dry bulb of the entering air to allow more extra heat rejection during operation. The major components of an adiabatic cooling tower contain a water basin, fan elements, dry cooling exchangers, cooling pads, mixing units, drift eliminators, a water distribution system, and a tower supporting structure (Shim, Sarker, Moon, Lee, & Yoon, 2010). The outside of the hybrid circuit cooling tower is usually made of different materials, including concrete, fiber-reinforced plastic, and stainless steel. Because consumption of water is decreased in the adiabatic system than in conventional evaporators, it can be used in high-temperature and low-moisture environments very efficiently.

Compared with dry condensers or coolers, the advantages of adiabatic systems include taking up a small amount of space for cooling ability and saving operative energy by reducing fan motor watt. Besides, the steam plume generated during operation is diminished by mixing the humid air from the wet segment and high-temperature air from the dry cooling unit. It can cause less potential icing visually in the atmosphere, especially favorable in the urban area or any surroundings close to the densely populated area (Sarker, Kim, Moon, & Yoon, 2008; Vitkovic, Storch, Puncochar, & Stodulka, 2016).

7.3.4 Flash evaporators

When high-pressure saturated water enters a relatively low-pressure container, the saturated water becomes part of the saturated steam and saturated water under the container's pressure owing to the abrupt reduction in pressure. The major use of flash evaporation is the recovery of boiler drainage in thermal power plants and geothermal power generation. The flash tank's primary role is to allow the higher-pressure stream to enter into the lower-pressure tank, depressurize the high-pressure stream, and evaporate the lower-boiling point components in the stream to accomplish the effect of separation. The water recovery of boiler drainage in thermal power plants and geothermal power generation is the major usage of flash evaporation. The primary function of a flash evaporator is to transfer the higher-pressure stream into the lower-pressure tank, where it is rapidly depressurized. The components in the stream with lower boiling points are evaporated to accomplish the effect of separation. A flashing jet with rapid phase alteration might result from an inadvertent leak of pressurized liquid from supply pipelines or liquid storage containers. This quick phase transition can result in a rapid combination with an oxidant such as air. It can cause severe consequences such as explosions, causing damage to industrial equipment while the liquid is flammable.

Besides, the advantage of flash evaporation in industrial applications has the potential to have a significant impact on the performance of a variety of devices, such as injectors and reactors (Mansour & Müller, 2019). Heat transport data are supplied for direct contact evaporator and condenser designs suited for open-cycle ocean thermal energy conversion (OC-OTEC) applications. Under normal operational circumstances, falling turbulent jets and films were evaluated. Because of the breaking of the jets (or films) into sprays, the flash evaporator performance was rather consistent over the range of circumstances tested, with efficiencies as high as 95%. The Reynolds number of the jet and the steam air content has an impact on condenser performance. Jets produced a condenser heat-transfer coefficient of 27 kW/m^2 °C, which was greater than those recorded with films. After it was demonstrated that none of the existing correlations discovered in the literature could correlate all of the data patterns detected, an empirical correlation was established for the condenser data (Sam & Patel, 1984). A conventional flash distillation system indicated that the significant single component of capital cost and thermodynamic irreversibility is related to the transfer of heat from condensing steam via tubes into the brine. The heat energy available in flashing brine is used to raise brine against gravity. A vertical configuration allows gravity to be utilized to force brine to flow through a countercurrent heat-transfer system employing direct contact with a fluid. In addition, the work done may be favorably used as potential energy for pumping. Experiments on vertical flashing brine flow and direct contact heat transfer are presented. Besides, the theoretical design approaches were considered for both fluid and metallic heat-transfer surfaces (Walker, Newson, & Johnson, 1967).

The direct contact spray flash evaporation and condensation systems have been studied for many years (Alrowais et al., 2020). For the prediction of distillate production during spray flash evaporation, a superheated water jet was driven by nozzles into a cooling tank where the feed temperatures were between 40 and 80 °C to determine empirical equations (Miyatake, Tomimura, Ide, & Fujii, 1981; Miyatake, Tomimura, Ide, Yuda, & Fujii, 1981). The flash evaporation performance on saline water was estimated. Both computational and experimental approaches were used at vacuum pressures less than 2.40 kPa and low feed temperatures from 26 to 32 K (Muthunayagam, Ramamurthi, & Paden, 2005). Based on experimental results of upward jet flash evaporation, a comparison analysis of a spray flash desalination method with the injection direction is determined. The superheated stream at 24.0, 30.0, and 40.0°C was pumped upward into a depressurized container over a stainless-steel cylinder nozzle to compare the occurrences of a downward jet flash evaporation technique. A series of tests were performed to estimate the influence of injection direction on the spray flash evaporation (Ikegami, Sasaki, Gouda, & Uehara, 2006). The flash evaporation system containing upward jets with larger nozzles was also studied. The results showed that the power of operation in the flash evaporation system could be enhanced with the superheat stream's degree and initial liquid temperature (Mutair & Ikegami, 2009, 2010). The flashing operation using tap water at various flow rates was demonstrated with the inlet fluid temperature between 40 and 70°C. Meanwhile, the pressure of stream injection was 6 bar, and the temperature of superheated fluid was from 6 to 18 K. The results indicated that the production of water steam and the efficiency of the flash process are inversely proportional to the tank size (El-Fiqi, Ali, El-Dessouky, Fath, & El-Hefni, 2007).

7.3.5 Scrubbers

The scrubber is widely used in various industries to clean and purify the gas vapor for different purposes, such as using a liquid to absorb specific components in the gas mixture, removing dust from the gas, humidifying or drying the gas, and so on. Industrial scrubbers are used to improve air quality by removing contaminant particles and harmful dust in gases from industrial exhaust streams to maintain air quality. They can also protect employers and their employees from combustion and provide a safe and healthy working environment. Therefore, the scrubber plays an important role in air pollution control devices (APCDs). In 1858, the first version of an air scrubber was designed and developed in Spain by Spanish engineer Narcís Monturiol i Estarriol. It was built to remove carbon dioxide from the submarine by driving the air through a vessel of calcium hydroxide to allow the submarine to *stay* and operate underwater for a very long time.

The process of direct contact heat transfer occurs between cold water and hot flue gases in scrubbers. Both substances under different temperatures have heat exchange to initiate the interaction of molecules by heat conduction on the basis of a temperature gradient. The efficiency of heat recovery from flue gases in scrubbers is influenced by convection, condensation, evaporation, and heat conduction during heat and mass transfer (Blumberga et al., 2020). A total of more than 15 variable factors are involved in these heat recovery processes (Cao, 2010). According to the methods of scrubbing pollutants out of the exhaust, the industrial scrubbers are classified into three major types—wet scrubber, semidry scrubber, and dry scrubber.

The wet scrubber is particularly used to treat the water-soluble pollutants in the exhaust gas as shown in Fig. 7.6. The coarse dust and specific pollutants in the fuel gas can be removed by spraying a chemical liquid through the gas. The exhaust gas flows into the packing layer in the tower. The exhaust gas and the scrubbing liquid produce a chemical absorption reaction in the packing layer, which can effectively control the volatile organic compounds, sulfuric acid, hydrochloric acid, hydrofluoric acid, nitric acid, and ammonia contained in the exhaust gas, because they are soluble compounds in water solution. Direct heat transfer between the gas bubbles and their surrounding scrubbing liquid results in several correlations for the heat-transfer coefficient and free convection between gas bubbles and liquids. The wet scrubber has the advantages of low initial cost, easy maintenance, and simple operation (Abdulwahid, Situ, Brown, & Lin, 2020). In some wet scrubbing methods, the combusted flue gas is first treated by heat recovery before being passed into the scrubber. The aqueous solution with the adsorbent is sprayed into the scrubber to contact the organic exhaust gas, and the gas is dissolved into the liquid phase through diffusion. The chemical reaction with the absorbent in the liquid phase can achieve the purpose of waste removal. The final product is removed from the bottom of the tower. Because the final discharged substance contains high moisture, it is called the wet scrubbing method.

On the other hand, semidry scrubbing is also called spray scrubbing. The theory is similar to that of the wet scrubbing method. The difference is that the water content in the final exhaust materials is reduced, and the droplets are completely removed inside the chamber. As a result, the final product is in dry powder form, and the granular collector can gather it in the latter stage to remove the pollutant gas (Fan, Yang, & Yuan, 2009). The dry scrubbing method is to pass the exhaust gas through the packed bed containing a solid adsorbent. The toxicity of the exhaust gas is removed by adsorption reaction. Solid adsorbents mainly include activated carbon, activated aluminum oxide, zeolite, and so on. The dry scrubbing does not produce secondary wastewater, has simple design, can be easily maintained, has a low cost, and does not require additional energy. Besides, their adsorption material can be regenerated after saturation, and a high removal efficiency for toxic gases can be provided. Based on the above advantages, many enterprises use dry scrubbers to deal with the problem of toxic gas emissions (Wang, Taricska, Hung, Eldridge, & Li, 2004).

7.4 Phase characteristics of the contacting fluids

This section will mention various direct heat exchangers based on their phase characteristics of the contacting fluids. Immiscible fluid, gas-liquid, liquid-liquid, and liquid-vapor heat exchangers will be discussed.

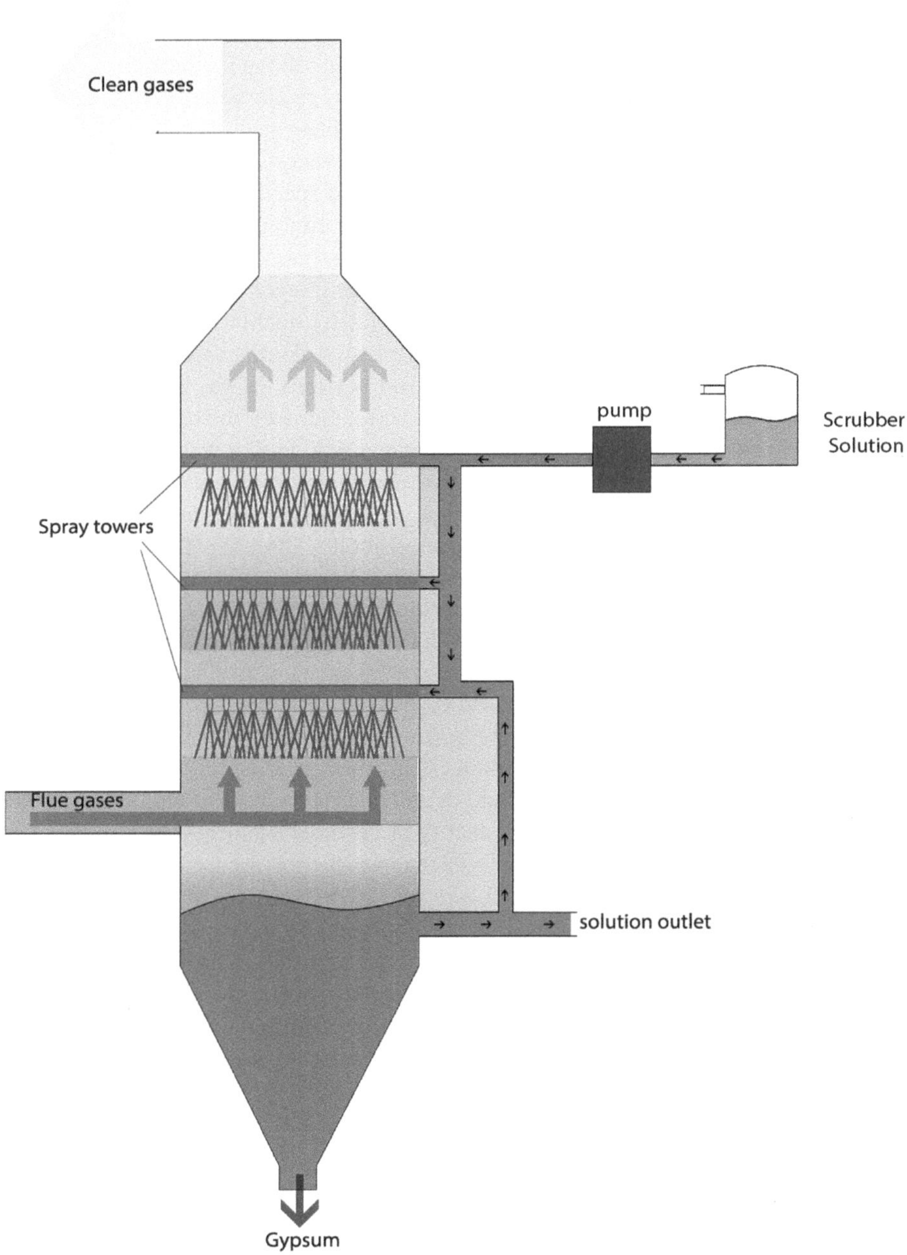

Fig. 7.6 Wet scrubber.

7.4.1 Immiscible fluid heat exchangers

There is direct contact between two immiscible fluids in the immiscible fluid exchanger, and this type of heat transfer has been studied for decades. Both dispersed and continuous streams provide scientific and practical significance in the fundamental properties of heat transfer. The single-phase or multiphase fluids may be used in evaporation or condensation during processing. The typical examples are oil vapors and organic vapors condensing with water or air (Prakash, 1966; Shah & Sekulic, 2003). Understanding heat-transfer processes between drops and continuous fluids and precise prediction of transfer coefficients for a particular system have developed well in decades.

Nonetheless, the prior experience can offer a hint at the transfer mechanism to be encountered, allowing the required coefficients to be estimated. In the water desalination process, the benefits of direct contact heat transfer with metallic transfer surfaces have been studied and utilized for decades. The heat-transfer mechanism is complicated when drops enter a constant variable temperature field and a constant-temperature field. Through a simultaneous phase transition, heat is transmitted to droplets and bubbles. The entirely mixed drop, the drop with internal circulation, and the rigid drop are the three models considered when addressing a constant-temperature field. With a simultaneous phase transition, heat is transmitted to droplets and bubbles. When addressing a constant-temperature field, three models are considered: stiff drop, entirely mixed drop, and drop with internal circulation. Low-cost, efficient water desalination devices and equipment are widely used in the principle of direct heat exchangers. Multiphase heat transfer is involved in latent heat exchanging across immiscible fluids. This multiphase exchanger has been successfully employed in direct contact freezing elements where a disseminated volatile fluid evaporates in saline water while a portion of the water freezes simultaneously (Marschewski & Stojunoff, 1983; Sideman, 1966).

Heat exchanging between two immiscible fluid phases in turbulent flow is important because this type of heating system can improve the compactness, residence time, and energy cost. The high-efficiency vortex unit employed as a direct contact heat exchanger is a general multifunctional exchanger/reactor in which wall tabs create longitudinal vortices that increase phase dispersion, macromixing, and rapid temperature uniformity in a turbulent flow. The experimental thermal data in the direct-contact heat exchanger are used to verify an algebraic one-dimensional thermal model that accounts for the axial evolution of phase temperatures associated with drop breakup. This model needs to understand the turbulent flow under single-phase circumstances to apply to various flow geometries for engineering design. Regardless of the phase separation at the output, the direct contact heat exchanger has a higher global Nusselt number than a double-jacketed heat exchanger. Besides, the heat transfer of a high-efficiency vortex requires less energy than the other direct contact heat exchangers with the same capacity of heat exchanging (Lemenand, Durandal, Della Valle, & Peerhossaini, 2010).

7.4.2 Gas-liquid heat exchangers

There is direct contact heat transfer between the gas and a low-pressure liquid in a gas-liquid exchanger, where the gas and liquid are usually air and water. Partial liquid evaporation and vapor removal by the gas occur in the operation of water cooling and air humidification. Because of the evaporation of the liquid, mass transfer accounts for most of the energy transfer in these exchangers, with convective heat transfer being a minor factor that takes less than 10% of total energy transfer. The application of gas-liquid exchangers can be found in a variety of industrial units, such as cooling towers with forced- or natural-draft ventilation, spray dryers or ponds, and air-conditioning spray vessels (Shah & Sekulic, 2003). The energy and mass-transfer coefficients in the gas-liquid phase and the relationship of interfacial layers are exhibited in various devices, such as packed columns in co- and countercurrent flow, packed bubble columns, plate columns with bubble cap and sieve trays, stirred tanks, spray towers, and bubble columns (Charpentier, 1981).

Energy may be recovered from hot gas discharge streams by transferring it straight to a coolant liquid in one of many gas-liquid direct contact units. The heat energy may be used in the evaporation of a Rankine cycle fluid or in a fluid stream during the process of petroleum and chemical reaction. Heat exchangers with direct gas-liquid contact are built along the same tube to gas absorbers. The diameter of tubes should be wide enough to avoid excessive entrapment of the liquid in a flowing gas and flooding. Based on the design of the direct contact unit, the maximum velocity of the fluid stream varies. The dimensions of the heat exchanger should allow the appropriate temperature to be reached between the entering water and the discharged air. Many designs of direct contact devices include sprays, trays, structured packing, and random packing used as in gas absorbers (Fair, 1990).

The heat-transfer coefficient in a packed column was estimated for the nonvolatile oil or cooling water. Mass and energy transfer of the gas-liquid packed column shows a high efficiency in distillation, fascination, and shedding. A number of applications of packed columns use the direct contact between the gas and liquid phases to achieve heat exchanging effectively. Meanwhile, the mass transfer also occurs simultaneously. The process applications are the consequence of the direct contacting mode's uncomplicatedness and low cost and the challenges with surface fouling that may be expected in shell-and-tube equipment. While the methods for estimating pressure drop and flooding are similar to those used in mass-transfer applications such as distillation and gas absorption, both need empirical methodologies. The methods for predicting heat-transfer rates were studied and developed (Huang & Fair, 1989).

The gas-liquid heat exchanger involving energy-efficient polymers was studied. The compression process used on offshore platforms to liquefy natural gas creates more heat commonly dispersed using seawater-cooled plate heat exchangers. Gasket plate heat exchangers are widely used as pasteurization units in the food industry to deactivate microorganisms for foodborne illness prevention (Lee, Chen, & Frank, 2015). A conductive polymer packed with a carbon fiber thermally was examined for the possibility of using commercially available material to improve thermal conductivity by order of magnitude or more. According to the research, the total coefficient of performance is roughly twice that of aluminum, and a total coefficient 10 times that of titanium might be reached. A thermally improved polymer heat exchanger uses 70% of its total lifetime energy in manufacturing, compared to 97% and 85% for titanium and aluminum heat exchangers, respectively. In terms of thermal performance and life cycle energy consumption, the study revealed the advantages of thermally enhanced polymer heat exchangers over traditional ones (Luckow, Bar-Cohen, Rodgers, & Cevallos, 2010).

7.4.3 Liquid-liquid heat exchangers

The heat transfer in the liquid-liquid heat exchangers occurs in a heat-transfer plate or from a liquid loop to another fluid stream. A heat exchanger is integrated to enhance the liquid's surface area for rejecting or absorbing heat and transporting it to a secondary fluid stream. The heat load is transported from the drive to the heat exchanger through the drive loop. The supply loop collects the heat load and delivers it to the user's cooling system liquid source through the heat exchanger. The user is responsible for using and selecting a heat load disposal system (Abu-Hamdeh, 2002; Bar-Cohen, Rodgers, & Cevallos, 2008; Omelchuk, Stambouli, & Chagnes, 2018). The phase of the working fluid, the type and size of the contactor, and operating circumstances are all critical considerations in designing an industrial liquid-liquid process. The end product is inexpensive in thermal operations where the liquid-liquid contactor is utilized for heating or cooling, such as in desalination or power plants. Equipment and operating costs must be kept to a minimum in such procedures. It is important to keep the equipment clean and free of particles on the heat-transfer surfaces. However, the fundamental problem in such systems is the working fluid losses induced by dissolution and entrainment. The device is intended to prevent entrainment, and insoluble organic solutions are used to reduce dissolving. In other thermal procedures, one of the liquids is used to heat or cool the other liquid to alter precipitation, dissolution, or reaction. The primary goal may be the product's qualities rather than the cheapest contacting liquids or the simplest equipment. Extraction techniques are examined in the same way. The characteristics of a more costly product may require a more soluble contacting liquid or a smaller contactor. Solvent losses and recovery costs as well as equipment maintenance may be acceptable in such operations. Besides, the availability of a low-cost insoluble solvent and simple equipment may lead to the deployment of extraction techniques that would be overlooked (Letan, 1988).

7.4.4 Liquid-vapor heat exchangers

Direct contact heat transfer between two immiscible liquids has been applied in many industrial operations, including water desalination (Snyder & Spiegler, 1966), crystallization (Byrd & Mulligan, 1986; Tao, Chunxiang, & Ziqiu, 1997), and solar and geothermal power (Jacobs, 1988). This design provides many advantages, including less scaling and fouling issues, a relatively simple structure, and a higher heat-transfer coefficient, which is 20–100 times more than the single-phase stream, no corrosion occurring in metallic heat-transfer surfaces, and operation at relatively small thermodynamic driving forces. The heat transfer with a gas-liquid-liquid three-phase system in a direct-contact parallel flow exchanger was studied. The influence of operational factors on the volumetric heat-transfer coefficient was also investigated and developed. The evaporation of the continuous phase fluid into the dispersed phase, possible coalescence, and break-up of the two-phase droplets were evaluated (Peng, Yiping, Cuili, & Kun, 2001).

Evaporation and condensation in well-designed heat exchangers can obtain significantly high heat-transfer coefficients to improve liquid-vapor heat exchange. A two-phase stream is operated in the enhanced surface. In addition, evaporation and condensation in channel geometries are exhibited along with correlations. The design correlations for predicting two-phase pressure loss are presented in these surfaces, including offset strip fin, perforated fin, and plate-fin geometries and tubes with twisted-tape inserts as well as ribbed or grooved walls. Last, the designs of evaporators and condensers were developed and exhibited (Carey & Shah, 1988). The heat exchangers in pasteurization use heat at high temperature in a moist environment for microorganism inhibition instead of using chemicals (Lee, Chen, & Frank, 2016).

The liquid-vapor heat exchangers are commonly found in condensation or evaporation units. With direct contact between two phases, the waste stream can usually heat up water or steam condensed by cooling water. The remaining noncondensable steam and hot water are released in the discharged stream. A deaerator and an open direct contact feed water heater in many power plants are two-phase types of heat exchangers. Furthermore, the desuperheating system also uses liquid-vapor heat transfer with direct contact to restore superheated steam to its saturated condition or lower the superheated temperature (Jacobs, 1988).

The direct contact heat exchanger with three-phase vapor-liquid-liquid is designed and developed in recent years. Its volumetric heat-transfer coefficient with direct contact in a three-phase condenser was investigated. Although the heat-transfer coefficient increased, the continuous phase mass flow rate and the holdup ratio also increased. Besides, the optimal value of the continuous phase mass flow rate increased when the vapor dispersed phase mass flow rate was also enhanced. Nevertheless, the temperature obtained by the continuous phase has a significant influence (Mahood, Campbell, Thorpe, & Sharif, 2018).

7.5 Conclusion

Direct heat transfer is usually considered heat transfer from one fluid to the other fluid directly without intervening plate-like or sheet-like walls. Direct heat transfer may occur between two or more mass fluids, which can be concurrent flow (parallel flow), counter flow, and even cross flow. In addition, these fluids can be miscible, immiscible, partly miscible, and partly immiscible. According to the phase characteristics of the contacting fluids, direct heat transfer between the two fluids includes solid-gas, gas-liquid, liquid-vapor, liquid-liquid, immiscible fluid, and even solid-solid. For the typical heat exchanging system, the liquid is commonly water or sometimes an organic chemical solution. The gas is usually air or steam during the heat-transfer process.

One of the main challenges of designing an effective heat conversion system is that efficient heat transfer at a specific temperature is capable of maximizing the potential of the heat source thermodynamically in the designated system. In a conventional heat exchanger, heat transfer occurs between the hot and cold streams in the heating surface. Therefore, the maximum potential of thermodynamic ability in a conventional heat exchanger is restrained because of built-in energy losses of the intervening solid heating surface that divides two different fluid streams. According to this type of design, the heat-transfer efficiency has degenerated significantly because the heat-transfer coefficients decrease over time. It is caused by the accumulation of undesirable materials deposits on heating surfaces for the duration of the heating and cooling process. The fouling is found among most industrial heat exchangers to influence the heat-transfer effectiveness, fluid resistance, and pressure decline. Under thermal processing, high temperature leads to a variety of deteriorations on the wall of the heating surface by corrosion regardless of the two-fluid steams between gases, liquids, vapors, and solids. In the direct heating system, two fluid streams are allowed to contact directly in order to alter the final temperature in the materials. On the other hand, the indirect heating system utilizes conduction, convection, or radiation to reach energy balance instead of involving any form of contact and exposure between solids, liquids, gases, or any other type of media. The application of direct heat exchangers can be predicted to develop rapidly in a variety of industries because of all the advantages discussed in this chapter.

Acknowledgments

The authors would like to appreciate Muhammed Emin AKYÜREK from Istanbul Sabahattin Zaim University, Department of Interior Architecture and Environmental Design.

References

Abdulwahid, A., Situ, R., Brown, R., & Lin, W. (2020). Bubble dynamics in a diesel exhaust wet scrubber. In *Conference: 22nd Australasian fluid mechanics conference AFMC2020*.

Abu-Hamdeh, N. H. (2002). Control of a liquid–liquid heat exchanger. *Heat and Mass Transfer*, *38*(7), 687–693.

Alrowais, R., Qian, C., Burhan, M., Ybyraiymkul, D., Shahzad, M. W., & Ng, K. C. (2020). A greener seawater desalination method by direct-contact spray evaporation and condensation (DCSEC): Experiments. *Applied Thermal Engineering*, *179*, 115629.

Apriyanti, V., Adriansyah, W., Abdurrachim, & Pasek, A. D. (2018 July). A review of direct contact condensation of steam on water droplets. In *AIP conference proceedings (Vol. 1984, No. 1, p. 020035)*AIP Publishing LLC.

Ashaolu, T. J., Samborska, K., Lee, C. C., Tomas, M., Capanoglu, E., Tarhan, Ö., et al. (2021). Phycocyanin, a super functional ingredient from algae: Properties, purification characterization, and applications. *International Journal of Biological Macromolecules*, *193*(6).

Ashoor, B. B., Mansour, S., Giwa, A., Dufour, V., & Hasan, S. W. (2016). Principles and applications of direct contact membrane distillation (DCMD): A comprehensive review. *Desalination*, *398*, 222–246.

Baehr, H. D., & Stephan, K. (2008). *Heat and mass transfer* (6. new rev. ed.). Waerme-und Stoffuebertragung.

Bakay, A., & Jaszay, T. (1978). High performance jet condensers for steam turbines. In *International heat transfer conference digital library* Begel House Inc.

Bar-Cohen, A., Rodgers, P., & Cevallos, J. G. (2008). Application of thermally enhanced thermoplastics to seawater-cooled liquid-liquid heat exchangers. In *Proceedings 5th European thermal-sciences conference.*

Bergman, T. L., Incropera, F. P., DeWitt, D. P., & Lavine, A. S. (2011). *Fundamentals of heat and mass transfer*. John Wiley & Sons.

Bharathan, D., Parsons, B. K., & Althof, J. A. (1988). *Direct-contact condensers for open-cycle OTEC applications: Model validation with fresh water experiments for structured packings*. No. SERI/TR-252-3108 Golden, CO, USA: Solar Energy Research Inst.

Blumberga, D., Priedniece, V., Rumba, R., Kirsanovs, V., Ņikitenko, A., Lavendelis, E., et al. (2020). Mathematical modeling of heat and mass processes in a scrubber: The Box–Wilson optimization method. *Energies*, *13*(9), 2170.

Byrd, L. W., & Mulligan, J. C. (1986). A population balance approach to direct-contact secondary refrigerant freezing. *AIChE Journal*, *32*(11), 1881–1888.

Camaraza-Medina, Y., Hernandez-Guerrero, A., Luviano-Ortiz, J. L., Mortensen-Carlson, K., Cruz-Fonticiella, O. M., & García-Morales, O. F. (2019). New model for heat transfer calculation during film condensation inside pipes. *International Journal of Heat and Mass Transfer*, *128*, 344–353.

Cao, E. (2010). *Heat transfer in process engineering* (p. 147). New York: McGraw-Hill.

Carey, V. P., & Shah, R. K. (1988). Design of compact and enhanced heat exchangers for liquid-vapor phase-change applications. In *Two-phase flow heat exchangers* (pp. 909–968). Dordrecht: Springer.

Cengel, Y., & Ghajar, A. J. (2020). *Heat and mass transfer: Fundamentals and applications*. McGraw-Hill Education.

Chan, C. K., & Lee, C. K. B. (1982). A regime map for direct contact condensation. *International Journal of Multiphase Flow*, *8*(1), 11–20.

Charpentier, J. C. (1981). Gas-liquid absorptions and reactions. In *Advances in chemical engineering*. New York: Academic Press.

Chen, Q., Li, Y., & Chua, K. J. (2016). On the thermodynamic analysis of a novel low-grade heat driven desalination system. *Energy Conversion and Management*, *128*, 145–159.

Chen, Q., Thu, K., Bui, T. D., Li, Y., Ng, K. C., & Chua, K. J. (2016). Development of a model for spray evaporation based on droplet analysis. *Desalination*, *399*, 69–77.

Chung, B. J., Kim, M. C., & Ahmadinejad, M. (2008). Film-wise and drop-wise condensation of steam on short inclined plates. *Journal of Mechanical Science and Technology*, *22*(1), 127–133.

Del Col, D., Parin, R., Bisetto, A., Bortolin, S., & Martucci, A. (2017). Film condensation of steam flowing on a hydrophobic surface. *International Journal of Heat and Mass Transfer*, *107*, 307–318.

Demirci, M., Lee, C. C., Çavuş, M., & Çağlar, M. Y. (2020). Oleogels for food applications. In *Biopolymer-based formulations* (pp. 781–811). Elsevier.

Duniam, S., Jahn, I., Hooman, K., Lu, Y., & Veeraragavan, A. (2018). Comparison of direct and indirect natural draft dry cooling tower cooling of the sCO2 Brayton cycle for concentrated solar power plants. *Applied Thermal Engineering*, *130*, 1070–1080.

El-Fiqi, A. K., Ali, N. H., El-Dessouky, H. T., Fath, H. S., & El-Hefni, M. A. (2007). Flash evaporation in a superheated water liquid jet. *Desalination*, *206*(1–3), 311–321.

Fair, J. R. (1972). Designing direct-contact coolers/condensers. *Chemical Engineering, 79* (13), 91.

Fair, J. R. (1990). Direct contact gas-liquid heat exchange for energy recovery. *Journal ofSolarEnergyEngineering, 112*, 216–222.

Fan, F., Yang, L., & Yuan, Z. (2009). Properties of water vapor supersaturation under spray scrubbing conditions. *Journal of the Chemical Industry and Engineering Society of China, 7*.

Fisher, E. M., & Wright, J. D. (1984). *Direct-contact condensers for solar pond power production.* NASA STI/Recon Technical Report N. 85 (p. 21771).

Ford, J. D., & Lekic, A. (1973). Rate of growth of drops during condensation. *International Journal of Heat and Mass Transfer, 16*(1), 61–64.

Goldstick, R. J. (1981). *Survey of flue gas condensation heat recovery systems.* Final report February–October 1981 (no. PB-84-215425; KVB-13-45400-1289) Irvine, CA, USA: KVB Inc.

Hausbrand, E. (1919). *Evaporating, condensing and cooling apparatus.* Scott.

Hewitt, G. F., Shires, G. L., & Bott, T. (1994). *Process heat transfer* (pp. 725–727). Begel House.

Holman, J. P. (2008). *Heat transfer (SI units) Sie.* Tata McGraw-Hill Education.

Hooman, K. (2010 December). Natural draft dry cooling tower modelling. In *17th Australasian fluid mechanics conference 2010* (pp. 154–159).

How, H. (1956). How to design barometric condensers. *Chemical Engineering, 63*, 174–182.

Huang, C. C., & Fair, J. R. (1989). Direct-contact gas-liquid heat transfer in a packed column. *Heat Transfer Engineering, 10*(2), 19–28.

Huang, J., Xiao, Q., Liu, J., & Wang, H. (2019). Modeling heat transfer properties in an ORC direct contact evaporator using RBF neural network combined with EMD. *Energy, 173*, 306–316.

Ikegami, Y., Sasaki, H., Gouda, T., & Uehara, H. (2006). Experimental study on a spray flash desalination (influence of the direction of injection). *Desalination, 194*(1–3), 81–89.

Jacobs, H. R. (1988). Direct-contact heat transfer for process technologies. *Journal of Heat Transfer, 110*(4B), 1259–1270.

Jacobs, H. R. (1995). Direct contact heat transfer. In *Heat exchanger design handbook* Begell Press.

Jacobs, H. R., & Fannar, H. (1977). *Direct contact condensers: A literature survey.* United States: Department of Mechanical Engineering, University of Utah.

Kakac, S., Liu, H., & Pramuanjaroenkij, A. (2012). Condensers and evaporators. In *Heat exchangers: Selection, rating, and thermal design* (3rd ed., pp. 491–541). CRC Press.

Kreith, F., & Boehm, R. F. (Eds.). (2013). *Direct-contact heat transfer* Springer Science & Business Media.

Kutateladze, S. S., & Borishanskiĭ, V. M. (1966). *A concise encyclopedia of heat transfer.* Pergamon.

Lebrun, J., & Silva, C. A. (2002). Cooling tower—Model and experimental validation. *ASHRAE Transactions, 108*, 751.

Lee, C. C., Chen, J., & Frank, J. F. (2015). Role of cellulose and colanic acid in attachment of Shiga toxin-producing Escherichia coli to lettuce and spinach in different water hardness environments. *Journal of Food Protection, 78*(8), 1461–1466.

Lee, C. C., Chen, J., & Frank, J. F. (2016). Influence of extracellular cellulose and colanic acid production on the survival of Shiga toxin-producing Escherichia coli on spinach and lettuce after chlorine treatment. *Journal of Food Protection, 79*(4), 666–671.

Lee, C. C., Tomas, M., & Jafari, S. M. (2020). Optical analysis of nanoencapsulated food ingredients by color measurement. In *Characterization of nanoencapsulated food ingredients* (pp. 505–528). Academic Press.

Lemenand, T., Durandal, C., Della Valle, D., & Peerhossaini, H. (2010). Turbulent direct-contact heat transfer between two immiscible fluids. *International Journal of Thermal Sciences, 49*(10), 1886–1898.

Lemouari, M., Boumaza, M., & Kaabi, A. (2009). Experimental analysis of heat and mass transfer phenomena in a direct contact evaporative cooling tower. *Energy Conversion and Management, 50*(6), 1610–1617.

Letan, R. (1988). Liquid-liquid processes. In *Direct-contact heat transfer* (pp. 83–118). Springer, Berlin, Heidelberg.

Luckow, P., Bar-Cohen, A., Rodgers, P., & Cevallos, J. (2010). Energy efficient polymers for gas-liquid heat exchangers. *Journal of Energy Resources Technology, 132*(2).

Mahood, H. B., Campbell, A. N., Thorpe, R. B., & Sharif, A. O. (2018). Measuring the overall volumetric heat transfer coefficient in a vapor-liquid-liquid three-phase direct contact heat exchanger. *Heat Transfer Engineering, 39*(3), 208–216.

Mansour, A., & Müller, N. (2019). A review of flash evaporation phenomena and resulting shock waves. *Experimental Thermal and Fluid Science, 107*, 146–168.

Marschewski, W. W., & Stojunoff, C. G. (1983, August). Liquid-liquid direct-contact, immiscible fluids heat exchanger analysis. In *Proc. Intersoc. Energy Convers. Eng. Conf. (United States) (Vol. 4, No. CONF-830812-)*. Boulder City, NV: University of Nevada System, Desert Research Institute, Energy Systems Center.

Milosavljevic, N., & Heikkilä, P. (2001). A comprehensive approach to cooling tower design. *Applied Thermal Engineering, 21*(9), 899–915.

Miyatake, O., Tomimura, T., Ide, Y., & Fujii, T. (1981). An experimental study of spray flash evaporation. *Desalination, 36*(2), 113–128.

Miyatake, O., Tomimura, T., Ide, Y., Yuda, M., & Fujii, T. (1981). Effect of liquid temperature on spray flash evaporation. *Desalination, 37*(3), 351–366.

Mori, Y. H. (1991). An analytic model of direct-contact heat transfer in spray-column evaporators. *AIChE Journal, 37*(4), 539–546.

Murray, C. (2009). Natural draft dry cooling tower. In *Australian geothermal energy conference 2009*.

Mutair, S., & Ikegami, Y. (2009). Experimental study on flash evaporation from superheated water jets: Influencing factors and formulation of correlation. *International Journal of Heat and Mass Transfer, 52*(23–24), 5643–5651.

Mutair, S., & Ikegami, Y. (2010). Experimental investigation on the characteristics of flash evaporation from superheated water jets for desalination. *Desalination, 251*(1–3), 103–111.

Muthunayagam, A. E., Ramamurthi, K., & Paden, J. R. (2005). Modelling and experiments on vaporization of saline water at low temperatures and reduced pressures. *Applied Thermal Engineering, 25*(5–6), 941–952.

Najafabadi, A. (2015). Geothermal power plant condensers in the world. In *Proceeding world geothermal congress 2015. Australia, Melbourne*.

Naphon, P. (2005). Study on the heat transfer characteristics of an evaporative cooling tower. *International Communications in Heat and Mass Transfer, 32*(8), 1066–1074.

Oliker, I. (1977 August). Application of direct contact heat exchangers in geothermal systems. In *American Institute of Chemical Engineers and American Society of Mechanical Engineers, heat transfer conference*.

Oliker, I. I., & Permyakov, V. A. (1971). *Termicheskaya deaeratsiya vody na teplovykh elektrostantsiyakh [Thermal deaeration of water at heat power stations]*. Leningrad: Energiya.

Omelchuk, K., Stambouli, M., & Chagnes, A. (2018). Investigation of aggregation and acid dissociation of new cationic exchangers for liquid-liquid extraction. *Journal of Molecular Liquids*, *262*, 111–118.

Peng, Z., Yiping, W., Cuili, G., & Kun, W. (2001). Heat transfer in gas–liquid–liquid three-phase direct-contact exchanger. *Chemical Engineering Journal*, *84*(3), 381–388.

Prakash, C. B. (1966). *Direct contact heat transfer between two immiscible liquids during vaporization*. Doctoral dissertation University of British Columbia.

Rajput, R. K. (2019). *A textbook of heat and mass transfer. SI units*. S. Chand Publishing.

Ribeiro, C. P., Jr., & Lage, P. L. (2005). Gas-liquid direct-contact evaporation: A review. *Chemical Engineering & Technology: Industrial Chemistry-Plant Equipment-Process Engineering-Biotechnology*, *28*(10), 1081–1107.

Sam, R., & Patel, B. (1984). An experimental investigation of OC-OTEC direct-contact condensation and evaporation processes. *Journal ofSolarEnergyEngineering*, *106*(1), 120–127.

Sarker, M. M. A., Kim, E., Moon, C. G., & Yoon, J. I. (2008). Performance characteristics of the hybrid closed circuit cooling tower. *Energy and Buildings*, *40*(8), 1529–1535.

Savulescu, L. E., Sorin, M., & Smith, R. (2002). Direct and indirect heat transfer in water network systems. *Applied Thermal Engineering*, *22*(8), 981–988.

Shah, R. K., & Sekulic, D. P. (2003). Classification of heat exchangers. In *Fundamentals of heat exchanger design* (pp. 1–77). John Wiley & Sons.

Shim, G. J., Sarker, M. M. A., Moon, C. G., Lee, H. S., & Yoon, J. I. (2010). Performance characteristics of a closed-circuit cooling tower with multiple paths. *Heat Transfer Engineering*, *31*(12), 992–997.

Sideman, S. (1966). Direct contact heat transfer between immiscible liquids. In *Vol. 6. Advances in chemical engineering* (pp. 207–286). Academic Press.

Singh, R. P., & Heldman, D. R. (2014). Evaporation. In *Introduction to food engineering* (pp. 565–592).

Snyder, A. E., & Spiegler, K. S. (1966). Freezing methods. In *Principles of desalination*. New York: Academic Press.

Stabat, P., & Marchio, D. (2004). Simplified model for indirect-contact evaporative cooling-tower behaviour. *Applied Energy*, *78*(4), 433–451.

Szokolay, S. (2012). *Introduction to architectural science*. Routledge.

Tao, Q., Chunxiang, L., & Ziqiu, S. (1997). Vaporizing cooling heat transfer of the alcohol liquid drops in sucrose solution and crystallizaton process. *Journal of Chemical Engineering of Chinese Universities*, *1*.

Thulukkanam, K. (2000). Heat exchangers—Introduction, classification, and selection. In *Heat exchanger design handbook* (pp. 17–42). CRC Press.

Tissot, J., Boulet, P., Trinquet, F., Fournaison, L., Lejeune, M., & Liaudet, F. (2014). Improved energy performance of a refrigerating machine using water spray upstream of the condenser. *International Journal of Refrigeration*, *38*, 93–105.

Vitkovic, P., Storch, V., Puncochar, J., & Stodulka, J. (2016 September). Measurement of thermal performance of hybrid cooling tower. In *Vol. 1768 (1). AIP conference proceedings* (p. 020023). AIP Publishing LLC.

Walker, P. T., Newson, I., & Johnson, K. D. B. (1967). Development of direct contact flash distillation concepts. *Desalination*, *2*(2), 196–206.

Wang, L. K., Taricska, J. R., Hung, Y. T., Eldridge, J. E., & Li, K. H. (2004). Wet and dry scrubbing. In *Air pollution control engineering* (pp. 197–305). Totowa, NJ: Humana Press.

Wellmann, J., Meyer-Kahlen, B., & Morosuk, T. (2018). Exergoeconomic evaluation of a CSP plant in combination with a desalination unit. *Renewable Energy, 128*, 586–602.

Wellmann, J., Neuhäuser, K., Behrendt, F., & Lehmann, M. (2015). Modeling an innovative low-temperature desalination system with integrated cogeneration in a concentrating solar power plant. *Desalination and Water Treatment*, *55*(12), 3163–3171.

Welty, J., Rorrer, G. L., & Foster, D. G. (2020). *Fundamentals of momentum, heat, and mass transfer*. John Wiley & Sons.

Xia, Z. Z., Chen, C. J., & Wang, R. Z. (2011). Numerical simulation of a closed wet cooling tower with novel design. *International Journal of Heat and Mass Transfer*, *54*(11 – 12), 2367–2374.

Section C

Different thermal processing operations in the food industry

Blanching in the food industry

8

Felipe Richter Reis
Instituto Federal de Educação, Ciência e Tecnologia do Paraná,
Campus Colombo, Colombo, PR, Brazil

8.1 Introduction

Among the thermal processes used for food that use steam and hot water, blanching can be categorized as a pretreatment from a sequential viewpoint in the processing line and as an intermediate treatment from a time–temperature intensity viewpoint. This is justified by the fact that blanching is applied to vegetables prior to canning, freezing, and drying and that there are milder and faster thermal processes than blanching such as HTST (high temperature short time) pasteurization and more heat intense and longer thermal processes than blanching such as sterilization in autoclaves. Blanching usually takes place at temperatures between 85°C and 100°C for 1 to 5 min, which makes it an HTST process (Gowen, Abu-Ghannam, Frias, & Oliveira, 2007; Guiamba & Svanberg, 2016; Guiamba, Svanberg, & Ahrné, 2015; Ngobese & Workneh, 2018; Xanthakis, Gogou, Taoukis, & Ahrné, 2018). However, there are reports of the use of LTLT (low temperature long time) blanching, which can be described as a thermal treatment that takes place between 60°C and 75°C for 10–40 min (Gamboa-Santos, Montilla, Soria, & Villamiel, 2012; Munyaka, Oey, Loey, & Hendrickx, 2010; Munyaka, Oey, Verlinde, Loey, & Hendrickx, 2009; Ngobese, Workneh, & Siwela, 2017). Additionally, other studies were conducted under conditions of time and temperature that extrapolate or combine the conditions described above, such as Reis, Masson, and Waszczynskyj (2008) (50–85°C for 2–10 min), Oner and Walker (2011) (60°C for 10–20 min, followed by 98°C for 1–10 min), and Zhao et al. (2016) (40–90°C for 10–30 min).

There are several published articles dedicated to investigating the effect of blanching on food quality. Most of them were collected in a book (ed. Richter Reis, 2017), which reviews blanching from a product quality viewpoint and shows that blanching affects food physical, chemical, sensory, and nutritional quality; accelerates drying; and is slowly being substituted by new methods of enzymatic inactivation that cause less sensory changes and nutritional loss. In the context of this chapter, the term "blanching" will be used to describe hot water blanching processes, and when blanching is performed with steam, the term "steam blanching" will be used, except from the cases where the two processes are presented together, when the term "hot water blanching" will be used to differentiate the two processes.

Thermal Processing of Food Products by Steam and Hot Water. https://doi.org/10.1016/B978-0-12-818616-9.00007-9

8.2 Blanching equipment

The conventional blanching methods are hot water blanching and steam blanching, for which there are commercially available equipment from different manufacturers. Hot water blanching consists in immersing food pieces in hot water for a suitable time that varies according to the food under treatment, followed by cooling, which is necessary to inhibit thermophilic microorganisms and to avoid the product presenting a cooked taste (ed. Richter Reis, 2017). Steam blanching is carried out by passing food pieces through an environment of saturated steam and results in enhanced retention of nutrients as cooling is made by cold air (Fellows, 2000).

Novel blanching methods have been proposed in order to overcome the drawbacks associated with conventional blanching methods. The equipment used for these methods is not usually commercially available, being restricted to academic purposes. In this sense, microwave blanching is a technology by means of which molecules of high polarity presenting an electric dipole moment rotate as to align in an electromagnetic field, increasing the kinetic energy and consequently the temperature of food (Ranjan, Dasgupta, Walia, Thara Chand, & Ramalingam, 2017). Microwave equipment of a frequency between 915 and 2450 MHz operating at a power density between 1 and 14 W/g has been currently used for microwave blanching (Delfiya, Mohapatra, Kotwaliwale, & Mishra, 2018; Fiore et al., 2019; Kachhadiya, Kumar, & Seth, 2018; Liburdi, Benucci, & Esti, 2019; Srimagal, Mishra, & Pradhan, 2017; Tang et al., 2019; Xu, Wang, & Wang, 2019).

Infrared blanching has been successfully used to inactivate enzymes in different foods and comprises a technique in which heat is radiated to the food by means of electromagnetic waves of wavelengths between 0.78 and 1000 μm (Krishnamurthy, Khurana, Jun, Irudayaraj, & Demirci, 2008). A distance from the infrared emitter varying from 5 to 20 cm has been currently used for infrared blanching (Jamali, Kashaninejad, Amirabadi, Aalami, & Khomeiri, 2018; Lara, Wilson, Chen, & Atungulu, 2019; Li et al., 2019). Ohmic blanching can be defined as a blanching technique in which the heating is obtained by means of the passing of an alternating electric current through the food (Knirsch, Santos, Vicente, & Penna, 2010). Ohmic heaters of voltages between 163 and 220 V and frequencies between 50 and 60 Hz have been currently used for Ohmic blanching (Bhat, Saini, & Sharma, 2017; Gomes, Sarkis, & Marczak, 2018). Radiofrequency blanching is a novel technology that uses electromagnetic energy at a frequency range of 3 kHz–300 MHz with advantages such as a higher penetration depth and a much faster heating rate (Zhang, Guo, et al., 2018). Equipment presenting between 3.5 and 6.0 kW of power and 27.12 MHz of frequency has been used for radio frequency blanching (Manzocco, Anese, & Nicoli, 2008; Zhang, Guo, et al., 2018; Zhang, Wang, et al., 2018).

High-humidity hot air impingement blanching (HHAIB), previously known as superheated steam impingement blanching (SSIB), is a blanching technique that uses humid hot air at a high velocity. The results are process uniformity, fastness, energy efficiency, reduced water-soluble nutrient losses, and reduced generation of wastewater (Xiao, Bai, Sun, & Gao, 2014). An air velocity between 10 and 14.0 ± 0.5 m/s, a temperature of 110°C, and a hot air relative humidity between 35% and 50% have been currently used for HHAIB (Deng et al., 2018; Wang, Fang, et al., 2017; Wang, Yang, et al., 2018).

The equipment commercially available for blanching are mainly rotary drum blanchers, screw blanchers, and belt/conveyor blanchers. In the websites of commercially available rotary drum blanchers, such as https://lycomfg.com/ and https://www.dodman.com/, there are descriptions of the equipment's mode of operation available. One of them states that the product moves into the blancher through a gravity chute or a water-cushioned in-feed flume and passes through the cook zone that blanches, very gently agitating the product as it passes through it, and after blanching, the product then transfers to the cool zone and gently discharges from the machine (Lyco Manufacturing, Inc., 2020). Fig. 8.1 shows a photograph and a drawing of the above-mentioned equipment.

In another model, the product is fed into the blancher and then gently fed through a drum suspended and rotating in hot water. After the required dwell time, lifters fitted in the form of a Ferris wheel at the end of the drum discharge the product into the outfeed system. Temperatures used can reach 98°C (Dodman Limited, 2019). This model of equipment can be used for virtually any vegetable.

Screw blanchers mode of operation are also described by manufacturers. One of them mentions that in its equipment, which operates continuously, the product is distributed uniformly inside the cooking chamber through the feed conveyor and is moved by a screw to the discharge zone. The product moves along the cooking chamber rotating vigorously in the transverse plane. Intensive water circulation ensures uniform heating and efficient heat transfer as well as ensures that the product will not stick. This model can be used for fruits and vegetables, seafood, mushrooms, and so on (EMA EUROPE S.R.O., 2020).

Finally, there are belt or conveyor blanchers, which can operate with both hot water and vapor. They are used for delicate products as product movement inside the equipment is not turbulent as in rotary drum and screw blanchers. One example of the product blanched in this kind of equipment is soft seafoods. According to a manufacturer, its conveyor blancher consists of an area for loading the product, a working area with

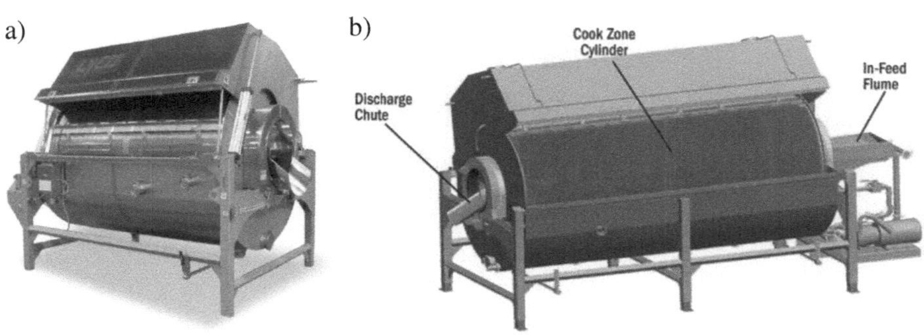

Fig. 8.1 Photograph (A) and drawing (B) of a rotary drum blancher
Credit: Lyco Manufacturing, Inc. (2020) *Cooker-cooler*. Available from: https://lycomfg.com/equipment/blanchers/cooker-cooler. [19 March 2020].

heating, and an unloading area. The equipment can be supplied with a section to turn the product for uniform heating and with a system for supplying steam or hot water from the lower part of the conveyor belt (EMA EUROPE S.R.O., 2020). The website https://www.cabinplant.com/ describes a model of a belt steam blancher (Fig. 8.2) in which the product is fed to the blanching zone through a horizontal screw conveyor and discharged onto a continuously moving belt. According to the manufacturer, this solution reduces steam losses. Steam is circulated by means of fans located inside the equipment. There are two cooling steps: the first is made with ambient air (evaporative cooling) and includes a small amount of water, which prevents drip loss, and the second is made with mechanically chilled air, yielding a product final temperature of about 10°C (Cabinplant A/S, 2020).

a)

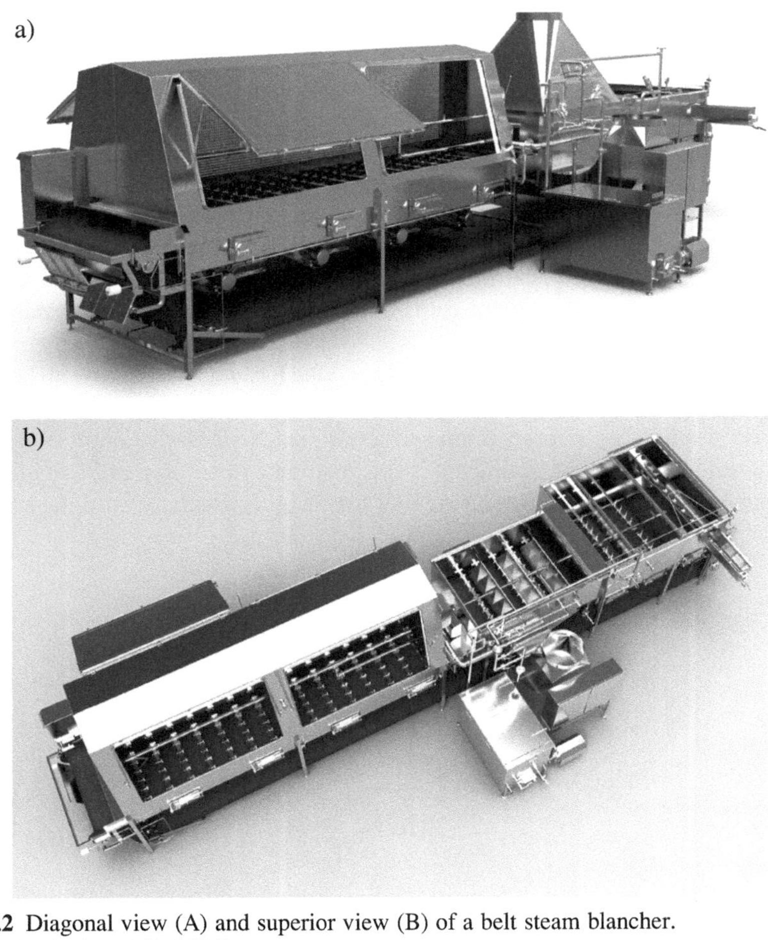

Fig. 8.2 Diagonal view (A) and superior view (B) of a belt steam blancher. Credit: Cabinplant A/S (2020).

8.3 Mathematical modeling of blanching as a method of enzymatic inactivation

Enzymes are catalyzers of the reactions that occur in biological systems. They present an extraordinary catalytic specificity, usually much higher than that of synthetic catalyzers, and a high level of specificity for their substrates and accelerate specific chemical reactions (Lehninger, Nelson, & Cox, 1995). Enzymes such as peroxidase, polyphenol oxidase, lipoxygenase, and polygalacturonase present deleterious effects if they are not inactivated in vegetables during processing.

Peroxidase (POD) promotes undesirable changes in texture, flavor, and color of vegetables (Collins, Biles, Perkins-Veazie, & Maness, 1996). It is used as an index of blanching efficiency in a way that if POD is inactivated, the blanching process is considered efficient. Sometimes, the minimization of POD activity, that is, 10% of enzymatic activity, is enough to assure the desired shelf-stability while maintaining an acceptable nutritional value (Agüero, Ansorena, Roura, & del Valle, 2008) as it is a very heat-resistant enzyme. Polyphenol oxidase (PPO) catalyzes the oxidation of phenols to quinones, which polymerize to melanin (brown) and can cause important losses to the processors of light-colored pulp vegetables, especially potatoes. Blanching can efficiently inactivate PPO and avoid the so-called enzymatic browning (Reis et al., 2008). Lipoxygenase (LOX) is claimed to release intracellular gases of the plant tissue and cause off-flavor in chilies (Ismail & Revathi, 2006), enzymatic generation of volatile compounds in bell peppers (Wampler & Barringer, 2012), and beany off-flavor in soybeans (Mozzoni et al., 2009). Polygalacturonase (PG) is a target in the tomato processing industry as it can reduce tomato juice viscosity by hydrolysis of pectic acid and cause tomato sauce rejection by the customer (ed. Fennema, 1996). Its inactivation can be achieved by blanching during tomato paste processing with a process known as "hot break."

Several studies have proposed equations that fit the effect of blanching time on the activity of deleterious enzymes, which can be used to build graphs of thermal inactivation. This is achieved through nonlinear regression by using enzymatic activity and blanching time data, the equation of choice, and a suitable piece of software. Agüero et al. (2008) combined a monophasic first-order model with the Arrhenius equation to yield a model (Eq. 8.1) that describes the effect of blanching time and blanching temperature on butternut squash POD activity for blanching temperatures above 70°C.

$$RA = exp\left\{-k_{ref}\, exp\left[-\frac{E_a}{R}\left(\frac{1}{T}-\frac{1}{T_{ref}}\right)\right]t\right\} \tag{8.1}$$

where RA is the residual POD activity (dimensionless), k_{ref} is the inactivation rate constant (min^{-1}) at a reference temperature (T_{ref}), E_a is the activation energy for POD inactivation (J mol^{-1}), R is the universal gas constant ($8.314\,\text{J mol}^{-1}\,\text{K}^{-1}$), T is the temperature (K), t is the time (min), and T_{ref} is 358.15 K. This model considers that only the heat-resistant fraction of POD was active. For blanching temperatures below 70°C, both the heat-labile and the heat-resistant fractions of butternut squash POD were active, which makes a biphasic model (Eq.8.2) more suitable for fitting the effect of blanching conditions on POD activity, as follows:

$$RA = f_L\, exp\left\{-k_{L,ref}\, exp\left[-\frac{E_{aL}}{R}\left(\frac{1}{T}-\frac{1}{T_{ref}}\right)\right]t\right\}$$
$$+ f_R\, exp\left\{-k_{R,ref}\, exp\left[-\frac{E_{aR}}{R}\left(\frac{1}{T}-\frac{1}{T_{ref}}\right)\right]t\right\} \quad (8.2)$$

where f_L and $k_{L,ref}$ (min^{-1}) are the initial fraction and the inactivation rate constant of the labile isoenzyme at the reference temperature, respectively, and f_R and $k_{R,ref}$ (min^{-1}) are those of the heat-resistant fraction; E_{aL} and E_{aR} are the activation energies for the inactivation of the heat-labile fraction and of the heat-resistant fraction ($J\,mol^{-1}$), respectively; T_{ref} is 338.15K; and RA, R, T, and t are the same parameters presented in Eq. (8.1).

Fante and Noreña (2012) fitted the inactivation kinetics of POD and PPO in garlic slices by using a biphasic first-order model (Eq. 8.3) proposed by Ling and Lund (1978):

$$y = a_L\, exp^{-k_L t} + b_R\, exp^{-k_R t} \quad (8.3)$$

where y is the residual enzymatic activity (g^{-1} fresh sample); k_L and k_R are the velocity constants of the heat-labile and heat-resistant components (min^{-1}), respectively; a_L and b_R are the initial concentrations of the labile and resistant components, respectively; and t is the blanching time (min).

Bai, Gao, Xiao, Wang, and Zang (2013) used several mathematical models to describe the thermal inactivation of PPO in apple quarters, among which the zero-order kinetic model and the first-order kinetic model are presented below:

$$P = P_0 - kt \quad (8.4)$$

$$P = P_0 \exp(-kt) \quad (8.5)$$

where P is the enzymatic activity at blanching time t (min), P_0 is the initial enzymatic activity, and k is the rate constant (min^{-1}). Their results suggest that high humidity air impingement blanching (HHAIB) is able to inactivate PPO in apple quarters when applied under conditions between 90°C for 7min and 120°C for 5min.

Goyeneche, Di Scala, and Roura (2013) investigated the thermal inactivation of PPO from radish using Eq. (8.3) to fit the effect of blanching time on PPO activity and the Arrhenius law to model the effect of blanching temperature on the rate constant, as follows:

$$k = k_0\, exp\left(\frac{-E_a}{RT}\right) \quad (8.6)$$

where k is the rate constant (min^{-1}), k_O is the pre-exponential factor, E_a is the activation energy ($kJ\,mol^{-1}$), R is the universal gas constant ($8.314\,kJ\,mol^{-1}\,K^{-1}$), and T is the absolute temperature (K). They observed that radish PPO presents a heat-labile and a heat-resistant component in a way that blanching at 90°C causes an inactivation of 90% after 2 min and then PPO activity persists over 5 min of blanching, not falling into zero (Fig. 8.3).

Lopes, Toralles, and Rombaldi (2014) studied the blanching of peach puree aiming at evaluating the inactivating effect of heat on PPO and POD, finding that blanching at 80°C and 90°C were highly effective in this sense. They used a first-order kinetic model (Eq. 8.5) to fit the effect of blanching time on enzymatic activity and Eq. (8.6) to fit the effect of blanching temperature on rate constant. Additionally, they calculated the half-life value of inactivation and the decimal reduction time by means of the following equations:

$$t_{1/2} = \frac{\ln(2)}{k} \tag{8.7}$$

$$D = \frac{\ln(10)}{k} \tag{8.8}$$

where $t_{1/2}$ is the half-life value of inactivation (min), which is the time necessary for the enzymatic activity to decrease to a half of its original activity, D is the decimal reduction time, defined as the time that it takes for the initial enzymatic activity decrease by 90%, and k is the rate constant (min^{-1}). PPO was found to be more

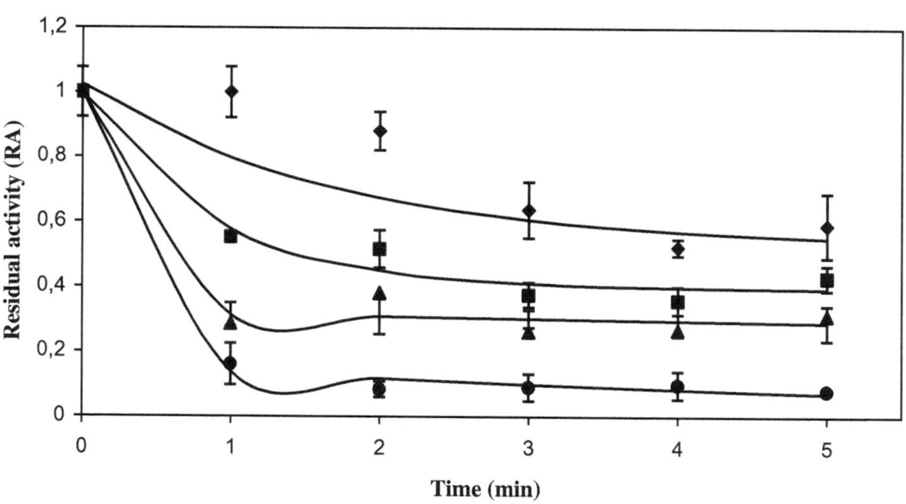

Fig. 8.3 Loss of activity for the radish PPO as a function of blanching time and temperature ($n=3$). Values are mean ± s.d., $n=3$; ◆, 60°C; ▪, 70°C; ▲, 80°C; ●, 90°C; –, Eq. (8.3). Credit: Goyeneche, R., Di Scala, K., & Roura, S. (2013). Biochemical characterization and thermal inactivation of polyphenol oxidase from radish (Raphanus sativus var. sativus). *LWT—Food Science and Technology*, *54*(1), 57–62.

heat-stable than POD, with a higher inactivation rate constant, a higher half-life value of inactivation, and a lower value of activation energy.

Niu, Li, Liu, and Liu (2015) studied the inactivation of lipoxygenase (LOX) extracted from sweetcorn and waxy corn by means of thermal treatment because a grassy, green, and beany flavor of processed vegetables has been mainly ascribed to the LOX pathway (Salas, Sánchez, García-González, & Aparicio, 2005). They discovered that a monophasic first-order model (Eq. 8.5) was suitable for fitting the effect of blanching time on LOX activity for the range of temperatures from 50°C to 70°C and that waxy corn LOX presented less thermal stability than sweet corn LOX. Deylami, Rahman, Tan, Bakar, and Olusegun (2016) successfully explained the effect of blanching time on mangosteen pericarp PPO by means of a first-order kinetic model (Eq. 8.5), besides calculating the half-life value of PPO inactivation by means of Eq. (8.7). They found that after 12 min of blanching at 90°C and 100°C, 23% and 14% of residual PPO activity were detected, respectively, which led them conclude that mangosteen pericarp PPO is highly thermostable. Gomes et al. (2018) used several mathematical models for fitting the effect of blanching time on POD activity. Among them, the multicomponent first-order model (Fujikawa & Itoh, 1996) is presented below:

$$A = \frac{\{\exp(-k_1 t) + r(\exp(-k_2 t)\}}{(1 + r)} \tag{8.9}$$

where A is the ratio C/C_0 that represents the total POD activity per unit volume of the sample at times t and 0, respectively (dimensionless), k_1 is the inactivation rate constant of the first component (min^{-1}), k_2 is the inactivation rate constant of the second component (min^{-1}), t is the blanching time (min), and r is the ratio of C_{02} to C_{01}, that is, the ratio of the POD activity per unit of the sample for second and first components, respectively, at time t (dimensionless). Additionally, other models used by those authors, namely, the series model (Henley & Sadana, 1985) (Eqs. 8.10 and 8.11), the fractional conversion model (Rizvi & Tong, 1997; Rudra Shalini, Shivare, & Basu, 2008) (Eq. 8.12), and the Weibull model (Rudra Shalini et al., 2008) (8.13), are presented below:

$$E \xrightarrow{k_1} E_1 \xrightarrow{k_2} E_2 \tag{8.10}$$

$$\begin{aligned} a = \alpha_2 &+ \left[1 + \frac{\alpha_{1k_1}}{k_2 - k_1} - \frac{\alpha_2 k_2}{k_2 - k_1}\right] \exp(-k_1 t) \\ &- \left[\frac{\alpha_{1k_1}}{k_2 - k_1} - \frac{\alpha_2 k_1}{k_2 - k_1}\right] \exp(-k_2 t) \end{aligned} \tag{8.11}$$

$$A = A_r + (A_0 - A_r) \exp(-kt) \tag{8.12}$$

$$s = s_0 \exp(-bt^n) \tag{8.13}$$

where E, E_1, and E_2 are the specific enzymatic activities, k_1 and k_2 are the first-order deactivation rate coefficients, α_1 and α_2 are the ratios of specific enzymatic activities E_1/E *and* E_2/E, respectively, and t is the inactivation process time. A is the POD activity at time t (min), A_r is the POD activity at equilibrium or the nonzero maximum retainable enzymatic activity, A_o is the initial POD activity, k is the inactivation rate coefficient, and t is the heating time. s is the residual POD activity, s_0 is the initial residual POD activity, b is scale factor (min^{-n}), n is the shape factor, and t is the heating time (min).

Jamali et al. (2018) used infrared heating for blanching pumpkin and fitted POD activity to blanching time by means of a first-order kinetic model (Eq. 8.5), calculated the decimal reduction time by means of Eq. (8.8), and also calculated the parameter called z-value (Aghajanzadeh, Ziaiifar, Kashaninejad, Maghsoudlou, & Esmailzadeh, 2016), defined as the temperature increase required to promote a 90% decrease in the decimal reduction time, as follows:

$$z = \frac{T_2 - T_1}{\log D_1 - \log D_2} \tag{8.14}$$

where z is the number of degrees Celsius (°C), and T_2 and T_1 are the temperatures corresponding to D_2 and D_1, respectively.

Li et al. (2019), when studying the infrared-steam blanching of flowers of *Chrysanthemum indicum* L., used a first-order kinetic model (Eq. 8.5) to adjust the effect of blanching time on POD activity and an Arrhenius-type equation to adjust the effect of blanching temperature on the parameter k, typical of the first-order kinetic model, as follows:

$$ln\left(\frac{k_0}{k_T}\right) = \frac{E_a}{RT}\left(\frac{1}{T_2} - \frac{1}{T_1}\right) \tag{8.15}$$

k_0 is the pre-exponential constant (min^{-1}), k_T is the inactivation rate constant at T, T is the experimental temperature (K), and R_T is the universal gas constant (J mol^{-1} K^{-1}) and E_a is the activation energy (J mol^{-1}).

Moscetti et al. (2019) assessed the effectiveness of hot water and steam blanching at 80–90°C of 1 cm thick potato slices regarding PPO inactivation. They used a first-order kinetic model (Eq. 8.5) to fit the effect of blanching time on potato PPO activity, Eq. (8.7) to calculate the half-life value of PPO inactivation, and Eq. (8.8) to calculate the PPO decimal reduction time. Even though both treatments increased lightness, inactivated PPO, and reduced its substrate specificity, hot water blanching was more effective than steam blanching in this sense.

In conclusion, there are several equations available in the literature for modeling the blanching process, most of which are used to fit the blanching time to the enzymatic activity. The first-order kinetic model is the most used equation to describe the thermal enzymatic inactivation, even though it is usual that vegetable enzymes present a heat-labile and a heat-resistant fraction, which makes a biphasic model more suitable for this purpose. Different forms of the Arrhenius equation are extensively used for describing the effect of blanching temperature on the rate constant (k). Finally, parameters such as the half-life value of inactivation, the decimal reduction value, and z-value can be calculated in order to evaluate the thermal resistance of the enzymes that are the target of blanching treatments by means of specific equations.

8.4 Material and energy balances in a blanching process

In a continuous blanching process operating at steady state, the overall material balance is given by (Smith, 2011):

$$Mp_{in} + Mw_{in} = Mp_{out} + Mw_{out} \tag{8.16}$$

where Mp_{in} is the mass flow rate of the product entering the blancher (kg s^{-1}), Mw_{in} is the mass flow rate of hot water entering the blancher (kg s^{-1}), Mp_{out} is the mass flow rate of the product leaving the blancher (kg s^{-1}), and Mw_{out} is the mass flow rate of hot water leaving the blancher (effluent) (kg s^{-1}).

The energy balance for a blanching process is given by (Smith, 2011)

$$Q = mc_p(T_w - T_P) \tag{8.17}$$

where Q is the rate of heat addition required for blanching (kW), m is the mass flow rate of the product (kg s^{-1}), c_p is the product heat capacity at constant pressure (kJ kg^{-1} K^{-1}), T_W is the temperature of the blanching water, and T_p is the product temperature.

Consider the following problem: (1) Potato cubes are blanched in hot water at a mass flow rate (output) of 3.5 kg s^{-1}, while blanching water enters the blancher at a mass flow rate of 8.7 kg s^{-1} and leaves the blancher (effluent stream) at a mass flow rate of 9.4 kg s^{-1}. The product initial temperature is 21°C, and the blanching temperature is 88°C. Find the other terms of the material and energy balances. Consider the potato heat capacity as being 3.52 kJ kg^{-1} K^{-1} (Smith, 2011).

The equation for material balance is

$$Mp_{in} + 8.7 = 3.5 + 9.4$$

and therefore

$$Mp_{in} = 4.2\ kg\ s^{-1}$$

The observed result can be attributed to the loss of soluble compounds from the product to the blanching water by leaching. The equation for energy balance is

$$Q = 4.2x3.52x(361.15 - 294.15)$$

from which

$$Q = 990.53\ kW$$

Consider the following problem: (2) Peas were blanched in a continuous blancher and the following data were obtained: the mass flow rate of peas leaving the blancher of $2.5\,kg\,s^{-1}$, the mass flow rate of hot water entering the blancher of $6.0\,kg\,s^{-1}$, and the mass flow rate of the product entering the blancher of $3.0\,kg\,s^{-1}$. The peas' initial temperature is 25°C, and the blanching temperature is 94°C. Find the other terms of the material and energy balances. Consider the potato heat capacity as being $3.31\,kJ\,kg^{-1}$ -K^{-1} (Smith, 2011).

From Eq. (8.16), the material balance is

$$3.0 + 6.0 = 2.5 + Mw_{out}$$

and thus

$$Mw_{out} = 6.5\ kg\ s^{-1}$$

From Eq. (8.17), the energy balance is

$$Q = 3.0.x3.31x(367.15 - 298.15)$$

which gives

$$Q = 685.17\ kW$$

Some of the data used in the problems were extracted from the study published by Arroqui, Lopez, Esnoz, and Virseda (2003).

An equation for calculating heat capacity was proposed by Heldman and Singh (Smith, 2011), as follows:

$$cp = 4.187x_w + 0.837x_a + 1.424x_c + 1.549x_p + 1.675x_f \quad (8.18)$$

where x_a, x_c, x_p, and x_f are the mass fractions of ash, carbohydrate, protein, and fat, respectively.

8.5 Current research in blanching

As stated in Section 8.1, most of the research on blanching conducted until the end of year 2015 was compiled in a previously published book (ed. Richter Reis, 2017). Therefore, it is worthwhile to present to the reader the advances in blanching since then.

8.5.1 Current research on the influence of blanching on food physical and sensory properties

Table 8.1 shows the current reports on the effects of blanching on physical and sensory properties of foods.

Palmyra palm (*Borassus flabellifer*) flesh discoloration was found to be insignificant when the flesh was blanched without peel in distilled water at 75–85°C, while blanching at 95°C in tap water containing more than 100 ppm of alkalinity caused a red-violet discoloration in the flesh with some peel left on, which was attributed to the presence of anthocyanins in the peel (Chamchong et al., 2016). Yam slices blanched at 70°C for 15 min were used for making yam flour, which was analyzed for its functional properties, that is, properties related to its technological use (Wahab et al., 2016). Blanching was found to increase yam flour wettability, resulting in slower reconstitution properties, reduce foam capacity, reduce pasting temperature, which is the minimum temperature required to cook the flour, and reduce breakdown, which

Table 8.1 Current reports on the effects of blanching on physical and sensory properties of foods and the respective physical and sensory properties and products affected.

Physical/sensory property	Product	Report
Color	Palmyra palm flesh	Chamchong, Tatidin, and Misayan (2016)
Wettability Foam capacity Pasting temperature Breakdown	Yam flour	Wahab et al. (2016)
Hardness	Lotus rhizome discs	Zhao et al. (2016)
pH Soluble solids Lightness Yellowness Redness Apparent viscosity Storage modulus (G') Loss modulus (G'')	Apricot sauce	Levent and Alpaslan (2018)
Lightness Redness Yellowness	Potato chips	Barreto, Tribuzi, Marsaioli Junior, Carciofi, and Laurindo (2019)
Greenness Crunchiness Stickiness	*U. pinnatifida* seaweed	Hamid et al. (2020)

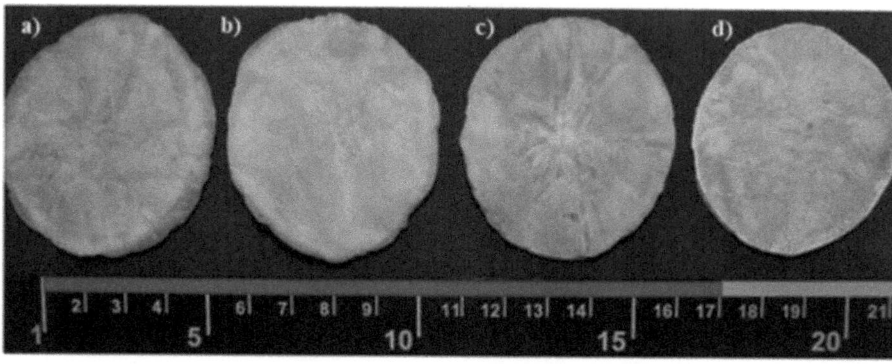

Fig. 8.4 Photograph of potato chips obtained by (A) hot water blanching (BL(A)) followed by MWVD, (B) hot water blanching (BL(A)) followed by MWMFD, (C) vacuum impregnation and microwave blanching (BL(B)) followed by MWMFD, and (D) vacuum impregnation and hot water blanching (BL(C)) followed by MWMFD.
Credit: Barreto, I.M.A., Tribuzi, G., Marsaioli Junior, A., Carciofi, B.A.M., Laurindo, J.B. (2019). Oil–free potato chips produced by microwave multiflash drying. *Journal of Food Engineering*, *261*, 133–139.

is the ease with which the swollen starch granule can be disintegrated. Lotus rhizome discs presented low hardness when blanched at 90°C and high hardness when blanched at 40°C and immersed in a $CaCl_2$ solution; the results that were attributed, respectively, to the inactivation of PME, and to the activation of PME, claimed to hydrolyze methyl ester linkages and promote the formation of a calcium pectate network (Zhao et al., 2016). Apricot sauce blanched at variable temperatures (85–95°C) and times (5–10 min) presented lower pH as the blanching temperature increased and higher soluble solids content as the blanching time increased (Levent & Alpaslan, 2018). In addition, the report showed that an increase in blanching temperature and time decreased sauce lightness and yellowness and increased its redness, besides decreasing its apparent viscosity, storage modulus (G'), and loss modulus (G''). Oil-free potato chips obtained by microwave multiflash drying (MWMFD) presented a golden yellow color when preceded by blanching (Fig. 8.4), as denoted by relatively high lightness values combined with simultaneous high redness and high yellowness values (Barreto et al., 2019).

Blanching of leaves and sporophylls of *Undaria pinnatifida* edible seaweed was studied by Hamid et al. (2020), who discovered that blanching in boiling water for 20–40 s was ideal for its greenness, crunchiness, and stickiness.

8.5.2 Current research on the influence of blanching on food nutritional properties

Table 8.2 shows the current reports on the effects of blanching on nutritional properties of foods.

Table 8.2 Current reports on the effects of blanching on nutritional properties of foods and the respective nutritional properties and products affected.

Nutritional property	Product	Report
Moisture Vitamin C Antioxidant capacity Total phenolics Total chlorophyll	Kale slices	Araújo, Oliveira, Ramos, Brandão, and Silva (2016)
Dry matter Vitamin B_1 Vitamin B_3 L-ascorbic acid	*A. bisporus* mushrooms	Bernaś and Jaworska (2016)
Glucosinolates Total phenolics Chlorophyll Antioxidant capacity	Broccoli florets	Cai et al. (2016)
Phenolics Flavonoids Antioxidant activity	Peanut skins	Larrauri, Zunino, Zygadlo, Grosso, and Nepote (2016)
Ash Fiber Protein Sugar	Carrot slices	Barroca, Guiné, Calado, Correia, and Mendes (2017)
Total phenolics Total flavonoids Antioxidant activity	Broccoli and collard	Radošević et al. (2017)
As, Cd, Co, Cu, Mn, Ni, Pb, Rb, Tl Ba, Cr, U, V, Zn, Ag, Sr	*A. fulva* mushroom	Drewnowska et al. (2017)
Hg	*C. cibarius* and *A. fulva* mushrooms	Falandysz and Drewnowska (2017)
Insoluble protein	Soymilk	Wang, Xing, Wang, and Guo (2017)
Sulforaphane	Broccoli florets	Mahn, Saavedra, and Paz Rubio (2018)
Chlorophyll a Chlorophyll b	Curry leaf puree	Meher, Keshav, Mazumdar, and Keshav (2018)
Moisture Protein Oil	Potato strips	Ngobese and Workneh (2018)
Phenolics Carotenoids Antioxidant activity	Celery roots	Priecina, Karklina, and Kince (2018)
Total phenolic Total flavonoid Total anthocyanin Antioxidant capacity	Onion slices	Ren et al. (2018)

Table 8.2 Continued

Nutritional property	Product	Report
Volatile basic nitrogen ammonia	Abalone	Tsai, Shiau, and Sung (2018)
Reducing sugars Asparagine Acrylamide	Potato chips	Zhang, Kahl, Bizimungu, and Lu (2018)
3-Monochloropropane-1,2-diol fatty acid esters.	French fries	Arisseto et al. (2019)

Kale slices that were blanched by saturated steam or hot water at 98°C, sometimes in combination with chemical dipping treatments, and dried by hot air at 55°C presented a different quality when compared to unblanched samples (Araújo et al., 2016). For example, steam blanched kale presented the lowest final moisture content, whereas steam blanched/sodium metabisulphite (5 g/L) pretreated kale presented the highest final moisture content. Concerning vitamin C retention, steam blanching was considered as the best pretreatment, yielding a 70.1% retention of this nutrient in dried kale. Total antioxidant capacity assays showed that hot water blanching produced the highest losses in the antioxidant capacity of dried kale. On the other hand, steam blanching and steam blanching/sodium metabisulphite dipping led to increases in the total antioxidant capacity of 10% and 5% of dried kale, respectively, a result that they attributed to the retention of vitamin C and to the antioxidant action of sulfur. The total phenolic content decreased between 44% and 76% in dried kale pretreated by blanching, results that they attributed to the inaccessibility of polyphenols to the analytical method used. Finally, the study showed that the total chlorophyll (a + b) retention in dried kale pretreated by steam blanching was found to be 5% superior to that of kale not subjected to pretreatment.

Blanching of *Agaricus bisporus* mushrooms strips in sodium metabisulphite (0.2% w/w) and citric acid (0.5% w/w) solution for 3 min at 96–98°C resulted in increased dry matter, vitamin B_3, and L-ascorbic acid contents but also in reduced levels of vitamin B_1 (Bernaś & Jaworska, 2016), possibly because of leaching, once this is a water-soluble vitamin. Broccoli florets were assessed for their antioxidant capacity and bioactive compound content after cutting, washing, blanching, cooling, and freezing (Cai et al., 2016). They concluded that blanching in boiling water for 90 s reduced the concentration of glucosinolates, which are health-promoting compounds typically found in broccoli, and caused an increase in antioxidant capacity, total phenolics, and chlorophyll content. They attributed the results to the dissolution of glucosinolates into blanching water, as cell rupture occurred during blanching, to the improvement of the extractability of bioactive compounds with the resulting increase in antioxidant activity and to the inactivation of the chlorophyllase enzyme, delaying the degradation of chlorophyll.

Peanut skins obtained from blanching and roasting processes were used to obtain extracts that were characterized for their bioactive features in a study whose results showed that when compared to roasted extracts, blanched extracts were higher in phenolic content but lower in flavonoid content and showed higher activities (lower IC_{50}) toward hydroxyl radicals, ferrous ions and superoxide anions, and similar DPPH radical-scavenging activity (Larrauri et al., 2016). Hot air-dried carrot slices previously blanched at 95°C during 4 min were compared to chemically pretreated slices in terms of their chemical composition, with the former presenting lower ash and protein content but higher fiber and sugar content than the latter (Barroca et al., 2017). Blanching of *Amanita fulva* caps at 100°C for 15 min decreased their As, Cd, Co, Cu, Mn, Ni, Pb, Rb, and Tl concentrations by 57%–86% and their Ba, Cr, U, V, and Zn by 24%–40%, whereas their Ag concentration became 70% higher and their Sr concentration became 160% higher (Drewnowska et al., 2017).

Fresh *Cantharellus cibarius* and *Amanita fulva* mushrooms blanched in boiling water presented 56% and 15% decreases in the concentration of Hg, respectively, whereas frozen blanched *A. fulva* presented a 35% decrease, with the rates of mercury leaching similar when using potable and deionized water (Falandysz & Drewnowska, 2017). A higher antitumor activity was detected in extracts prepared by blanching of broccoli and collard at 60°C for 5 min when compared to extracts prepared from fresh broccoli and fresh collard, and additionally, the vegetables' total phenolic and total flavonoid content and antioxidant activity were increased by blanching (Radošević et al., 2017). The results also showed that the remarkable antitumor activity of blanched extracts was well correlated with their phenols content and antioxidant capacity.

Even though soymilk made from blanched soybeans is free from unpleasant beany flavor, a study showed that they present an increase in the concentration of precipitable protein because of blanching, remarkably when it is conducted above 3 min/60°C, which ultimately leads to poor product stability (Wang, Xing, et al., 2017). They attributed the results to protein denaturation by blanching heat, which increased the amount of insoluble protein. Broccoli florets blanched at 57°C for 13 min presented degradation of sulforaphane, a bioactive compound, during frozen storage for 83 days, which was not observed for unblanched samples in which there was the formation of this compound (Mahn et al., 2018). Chlorophyll a and chlorophyll *b* were better retained in curry leaf puree when the leaves where blanched in MgO/KCl solutions at 80°C for 5 min, followed by cooling in potassium metabisulphite solution (Meher et al., 2018). Additional results from the study showed that concentration of both chlorophylls decreased as blanching proceeded, especially at lower pH, following a first-order kinetics. They attributed the good retention of chlorophyll during blanching in salt solutions to the presence of MgO, which could have risen the pH of the blanching medium.

Most of the potato varieties tested by Ngobese and Workneh (2018) presented a reduced moisture content after blanching at 75° for 10 min and at 85°C for 5 min. Additionally, the results showed that the latter blanching conditions led to higher protein leaching and to higher oil content in potatoes after frying when compared to the former. They attributed the moisture loss to evaporation during blanching. Steam blanching for 1.5 min before microwave-vacuum drying was suggested for enhancing the extractability of phenolics compounds and carotenoids along with the antioxidant activity of celery as measured by means of the ABTS method (Priecina et al., 2018).

They justified their results by the damaging effect of blanching on cell walls, which leads to more accessible bioactive compounds, which could interfere in the antioxidant activity.

Onion slices blanched at 60°C for 3 min and at 70°C for 1 min showed higher total phenolic, flavonoid, and anthocyanin contents, along with higher antioxidant capacity as measured by the FRAP and DPPH methods when compared to unblanched samples (Ren et al., 2018). They affirmed that the mechanisms responsible for the observed effects might be the inactivation of oxidative enzymes, in the case of phenolic compounds, the better extractability as a result of cell disruption, in the case of flavonoids, the inactivation of degradation enzymes, in the case of anthocyanins, and the increase in the individual phenolic compounds content, in the case of antioxidant activity.

The shelf life of fresh abalone, a marine mollusk, was extended from 3 to 6 days using a blanching treatment at 100°C for 1 min, as denoted by reduced levels of aerobic plate counts, volatile basic nitrogen, ammonia levels, and *K*-value, defined as ratio of the sum of inosine and hypoxanthine to the total concentration of all ATP-related compounds (Tsai et al., 2018). Optimization of potato chips blanching as a function of reducing sugars, asparagine, and acrylamide reduction showed that the optimal process conditions were 8.8–9.7 min and 68.7–75.0°C, which led to reductions of 64.2%, 49.8%, and 61.3%, respectively (Zhang, Kahl, et al., 2018). Additional results showed that an increase in blanching time was more effective than an increase in blanching temperature for reducing the concentration of the selected compounds. Blanching at 98°C for 3 min followed by dipping in a 4% pectin/0.1% $CaCl_2$ solution reduced the concentration of 3-monochloropropane-1,2-diol fatty acid esters, possible human carcinogens, and the oil uptake by French fries fried in palm olein (Arisseto et al., 2019). They attributed the results to the dipping treatment, more specifically to the formation of a Ca-pectate complex, which resulted in lower oil uptake and consequently in lower contamination of potatoes by 3-monochloropropane-1,2-diol fatty acid esters.

8.5.3 Current research on the influence of blanching on food drying and dehydration performance

Table 8.3 shows the current reports on the effects of blanching on food drying and dehydration performance.

Ginger cubes blanched at 100°C for 1 min and osmotically dehydrated in sucrose solutions of a concentration of 35–65°Brix at 40–70°C presented superior water loss, solids gain, effective diffusivity of water, and effective diffusivity of solids, but at the same time, they contained a reduced total polyphenol content and lost antioxidant capacity and displayed a color more different from fresh ginger when compared to unblanched samples (García-Toledo et al. 2016). According to the authors, blanching opened channels or pores in the product tissue, allowing faster moisture diffusivity. The weight loss and solids gain were increased by blanching during osmotic dehydration of bitter gourd slices in sodium chloride solutions, whereas firmness was decreased, and additionally, optimization analyses showed that the ideal process

Table 8.3 Current reports on the effects of blanching on food drying and dehydration performance and the respective drying/dehydration processes used and products affected.

Drying/dehydration process	Product	Report
Osmotic dehydration	Ginger cubes	García-Toledo et al. (2016)
Osmotic dehydration	Bitter gourd slices	Nambi, Behera, Saharan, and Meena (2016)
Hot air drying	Meat pieces	Hernández, Claramount, Kučerová, and Banout (2017)
Hot air drying	Peppers	Weil, Shum Cheong Sing, Méot, Boulanger, and Bohuon (2017)
Hot air drying	Chili fruits	Kamal et al. (2019)
Osmo-convective drying	Cranberries	Wiktor et al., 2019
Microwave and hot air drying	Celery stalks	Chen, Song, Li, Chen, and Jin (2020)
Hot air drying and freeze-drying	Broccoli wastes	Ferreira et al., 2020

conditions were blanching at 80°C for 5.26 min, followed by osmotic dehydration in 15% sodium chloride solutions for 5 h (Nambi, Behera, et al., 2016).

Meat pieces that were blanched at 90°C for 2 min in the presence of different concentrations of oregano essential oil dried faster because of blanching but presented inferior sensory quality, whereas hot air blanching, that is, drying at 35°C for 10 min, was also carried out in the presence of oregano essential oil and yielded good sensory quality but was not as effective as hot water blanching for improving drying kinetics (Hernández et al., 2017). They attributed the effect of blanching on drying kinetics to softening of the tissue. Blanching of *Piper borbonense* peppers greatly increased their drying rate, reducing the drying time, besides preserving high product lightness, yellowness, and redness, but the two latter were drastically reduced after hot air drying at 60°C (Weil et al., 2017). They attributed the results to the partial destruction of pepper cell wall during blanching, which facilitated water transfer during drying, and to the decrease in enzymatic activity during blanching with the consequent limitation of browning, with a significant impact on color.

Blanching of chili fruits in 2% acetic acid solution at 100°C for 2 min followed by soaking in 0.3% sodium metabisulfite/1% $CaCl_2$ solution for 10 min accelerated drying at 60°C until a final moisture content of around 8% and affected the properties of the chili powders obtained (Kamal et al., 2019). In this sense, the powders obtained from pretreated fruits presented high chlorophyll retention and acceptable retention of β-carotene and phenolics, along with higher lightness, yellowness, chroma, and pungency when compared to the powder obtained from untreated fruits. Blanching of cranberries prior to osmo-convective drying was found to reduce the drying time, increase the cranberry effective diffusion coefficient, and reduce the dried cranberry sucrose concentration while maintaining acceptable levels of bioactive compounds and desirable color (Wiktor et al., 2019).

Drying of celery stalks was accelerated by blanching, which was also positive for attenuating color changes caused by drying (Chen et al., 2020). Nevertheless, the authors affirmed that blanching was not a suitable pretreatment for celery drying because of the damage it caused to the plant tissue and because of its deleterious effect on the retention of aromatic compounds. Different drying processes were preceded by blanching during the processing of broccoli wastes, and different results were obtained depending on the drying technique used (Ferreira et al., 2020). For example, hot air drying was accelerated by blanching, but at the same time, a negative impact of blanching on the concentration of chlorophylls was observed, a result that they attributed to the diffusion of these pigments into the blanching water. On the contrary, freeze-dried blanched samples presented increased carotenoid extractability and increased recovery of phenolic compounds, a result that they attributed to cell disruption during blanching.

8.5.4 Current research on the influence of blanching on food miscellaneous properties

Table 8.4 shows the current reports on the effects of blanching on food miscellaneous properties.

Yacon (*Smallanthus sonchifolius*) slices of 5 mm thickness had their PPO and their POD activity inactivated in more than 95% by both steam blanching and hot water blanching in boiling water for 6 min (Campos et al., 2016). The report showed that blanching of 5 mm thickness slices in boiling solutions of ascorbic acid and $CaCl_2$ caused nonsignificant losses of fructooligosaccharides, which are prebiotic components of yacon tubers, besides yielding a light color in the dried flour obtained by hot air drying, especially when drying was conducted at 80°C. In this case, it is well known that enzyme-driven color changes in light-colored vegetables can be stopped by blanching, which inactivates PPO and POD to produce foods with light color. Additionally, the addition of chemicals with inhibitory activity on PPO like ascorbic acid and $CaCl_2$ to the blanching water potentializes the inactivating effect of blanching.

Mango puree presented 0% of PPO activity and 8% of ascorbic acid oxidase activity, an enzyme that destroys ascorbic acid by converting it into dehydroascorbic acid, after being subjected to blanching at 90°C for 4 min (Guiamba & Svanberg, 2016). In addition, the report showed that blanching preserved mango purée vitamin C but reduced its all-*trans*-β-carotene retention in more than 35%, a result that they attributed to *trans-cis*-isomerization of all-*trans*-β-carotene. Blanched mushrooms were found to lose phenolic compounds and antioxidant activity, besides presenting darker, redder, and yellower colors when compared to fresh mushrooms, but their chemical

Table 8.4 Current reports on the effects of blanching on food miscellaneous properties, the respective properties, and products affected.

Property	Product	Report
PPO activity POD activity Fructooligosaccharides Color	Yacon flour	Campos, Aguilar-Alvez, and Pedreschi (2016)
PPO activity Ascorbic acid oxidase activity Vitamin C All-*trans*-β-carotene	Mango puree	Guiamba and Svanberg (2016)
Phenolic compounds Antioxidant activity Color	Mushrooms	Jabłońska-Ryś, Sławińska, Radzki, and Gustaw (2016)
LOX activity POD activity Vitamin C Color Texture	Green peas	Kar, Hallsworth, and Singhal (2016)
Phenolic compounds Vitamin C Antioxidant activity Color Firmness	Green peas Cubes of eggplants Cubes of beetroot Cuboids of green pepper	Nambi, Gupta, Kumar, and Sharma (2016)
PPO activity Microbial counts Texture Color	Potato strips	Amaral, Achaerandio, Benedetti, and Pujola (2017)
PPO activity POD activity Myrosinase activity Anthocyanins Antioxidant activity	Red cabbage	Bernstein and Noreña (2017)
Dripping after thawing Color Firmness Resistant starch	Potato cubes	Icier, Cokgezme, and Sabanci (2017)
Volatile composition	Tomato-onion puree	Koutidou, Grauwet, Loey, and Acharya (2017)
Total aerobic count Molds and yeasts count	Edible insects	Megido et al. (2017)
POD activity Phenolic compounds Anthocyanins Antioxidant activity	Tamarillo fruits/ sweet product	Preciado-Iñiga, Amador-Espejo, and Bárcenas (2018)

Table 8.4 Continued

Property	Product	Report
Aerobic mesophilic bacteria count Molds and yeasts count Total soluble solids Titratable acidity Firmness Flavor (electronic nose)	Tomatoes	Xu et al. (2018)
LOX activity Thiobarbituric reactive species Hexanal	Peas	Yonny, Medina, Nazareno, and Chaillou (2018)
Phenolics Flavonoids Antioxidant activity PPO activity POD activity Aerobic bacteria count Molds and yeasts count Microstructure	*Aruncus dioicus* plants	Kim et al. (2019)
PPO activity POD activity Protein Total phenolic Total starch Amylose Color Antioxidant activity Swelling power Water solubility	*L. lancifolium* bulbs/ bulbs flour	Liu et al. (2019)
Physical stability Color Phenolics	Lettuce waste juice	Plazzotta and Manzocco (2019)
Redness Antioxidant activity 2,4,6,3′,4′-pentahydroxybenzophenone-2-O-β-D-glucopyranoside	*G. xanthochymus* fruits	Wickramasinghe et al. (2020)

and physical quality was comparable to that of the fermented mushrooms developed in the study (Jabłońska-Ryś et al., 2016). They attributed the decrease in phenolic compounds and antioxidant activity to high temperature and to the leaching of water-soluble compounds during blanching.

Green peas blanched in a solution of glycine betaine (10% w/v) for 60 s at 100°C presented a minimal LOX activity, which is necessary for a good pea flavor and a minimal POD activity, which is indicative of a well-designed blanching process (Kar et al., 2016). Other benefits of using glycine betaine in the blanching solution, according to the study, were a decrease by more than half in the loss of vitamin C with blanching when compared to tap water, along with slower deterioration of vitamin C, green color, and texture during 90 days of frozen storage.

The impact of blanching on bioactive compounds, antioxidant activity, color, and firmness of green peas, cubes of eggplant and beetroot, and cuboid pieces of green pepper was assessed by Nambi, Gupta, et al. (2016), who discovered that blanching caused a decrease in the concentration of phenolic compounds and vitamin C, a decrease in antioxidant activity (except from beetroot), and changes in color and firmness (softening), with this effect being more pronounced for a higher blanching time and a higher blanching temperature. They attributed the results to the breakdown of phenolic compounds and dissolution of polyphenols into the blanching water to the leaching of vitamin C into the blanching water and to the solubilization and depolymerization of pectic polysaccharides, responsible for softening.

Potato strips were treated by alginate coating, sonication combined or not with citric acid dipping, and blanching, in a report whose results showed that potatoes blanched at 85°C for 3 min presented undetectable PPO activity and low microbial counts but, at the same time, a softened texture and an undesirable color (Amaral et al. 2017). Steam blanching of red cabbage was found to be superior to hot water blanching regarding simultaneous inactivation of POD and PPO, preservation of myrosinase, an enzyme responsible for producing bioactive compounds, and manutention of acceptable levels of anthocyanins and antioxidant activity, as predicted by first-order (Eq. 8.5) and biphasic first-order (Eq. 8.3) kinetic models (Bernstein & Noreña, 2017). Potato cubes blanched at 92°C for 3.5–4.5 min frozen faster and thawed slower than unblanched samples, besides presenting lower dripping after thawing, a lighter color, lower firmness, and a lower resistant starch concentration (Icier et al., 2017). The hypotheses of the authors for the longer thawing times were losses of ionic compounds and changes in textural properties of potatoes.

The use of blanching at 95°C for 6 min for tomatoes and onions for preparing a tomato-onion puree resulted in a decrease in the level of volatile compounds responsible for the puree aroma, as measured by means of gas chromatography-olfactometry techniques (Koutidou et al., 2017). Blanching of edible insects was shown to be a necessary step before their consumption as it reduced their total aerobic count below the high limit for fresh minced meat established by European authorities (6.70 log CFU/g), besides preventing yeast and mold development (Megido et al., 2017). Tamarillo (*Cyphomandra betacea*) fruits blanched at 93°C for 12 min presented insignificant POD activity, a 14.6% decrease in phenolic compound content and insignificant changes in anthocyanin content and in antioxidant activity when compared to unblanched fruits (Preciado-Iñiga et al., 2018). Additional results from the study showed that a sweet made from blanched tamarillo fruits presented lower aerobic mesophilic bacteria and molds and yeast counts over 63 days of storage when compared a sweet made from unblanched fruits.

The impact of blanching at 50°C and 100°C for 1 min on total soluble solids, titratable acidity, firmness, and flavor of tomatoes was assessed by Xu et al. (2018), who discovered that blanching increased soluble solids content, decreased firmness, and changed the flavor, especially with the treatment conducted at 100°C, which was suggested for processing tomatoes into sauces. Blanching in *Ilex paraguariensis* solutions was found to enhance the oxidative stability of frozen peas during storage, as confirmed by low LOX activity and low thiobarbituric reactive species and hexanal levels (Yonny et al., 2018). The impact of blanching on bioactive compounds, antioxidant activities, PPO and POD inactivation, aerobic bacteria and yeast and mold counts, and microstructures of *Aruncus dioicus* plants was evaluated by Kim et al. (2019), who concluded that blanching at 98°C for 30 s increased total and selected phenolics, total flavonoids, and antioxidant activity, as measured by the DPPH, ABTS, and FRAP methods, respectively, besides inactivating PPO and POD and reducing microbial counts with only slight changes in microstructure.

Lilium lancifolium bulbs were blanched by hot water and steam and evaluated for their PPO and POD activity, whereas the flour made by hot air drying and freeze-drying followed by milling was assessed for its protein, total phenolics, total starch and amylose content, color, antioxidant activity, swelling power, and water solubility, among other analyses (Liu et al., 2019). The residual activity of PPO was 0.94% after steam blanching and 0.34% after blanching in boiling water for 60 s, whereas the residual activity of POD was 0.78% after blanching in boiling water for the same time. Blanching presented no effect on total starch but reduced the flour amylose and protein content, results which the authors justified by the inactivation of enzymes such as α-amylase and β-amylase and by the leaching of soluble proteins into the blanching water. Furthermore, blanching pretreatment inhibited the decrease in flour lightness caused by hot air drying and led to a significant decrease in redness and increase in yellowness. Additionally, the phenolic content and antioxidant activity of flours made from blanched bulbs were higher when compared to regular flours, especially those made from steam blanched bulbs, results that they attributed to a lower loss of antioxidant compounds during steam blanching when compared to hot water blanching. Finally, swelling power and water solubility of flours made from blanched bulbs were lower than those of regular flours, results that they attributed to an increase in insoluble amylose content in these samples.

High-pressure homogenization of lettuce waste juice preceded by blanching at 90°C for 30 s was a suitable method for producing a product of high physical stability, a bright green color, and an acceptable phenolic content, with blanching being crucial for avoiding phase separation (Plazzotta & Manzocco, 2019). The authors attributed the results to the effect of blanching on pectolytic enzymes, in the case of physical stability, and to its effect on oxidative enzymes responsible for the oxidation of polyphenols and for the degradation of chlorophyll, in the case of the green color. Both low-temperature (60°C) blanching and high-temperature (100°C) blanching of *Garcinia xanthochymus* fruits for 3 min, followed by hot air drying and grinding, were favorable for improving redness, DPPH radical scavenging activity, and preservation of 2,4,6,3′,4′-pentahydroxybenzophenone-2-*O*-β-D-glucopyranoside, a compound with anti-inflammatory properties, in a beverage made with the fruits (Wickramasinghe et al., 2020).

8.5.5 Current research on innovations in blanching technology

Table 8.5 shows the current reports on innovations in blanching technology.

Table 8.5 Current reports on innovations in blanching technology, the technology used, and the product assessed.

Technology	Product	Report
Blanching + Sonication	Bottle gourd cubes	Bhat and Sharma (2016)
Ohmic blanching	Bottle gourd cubes	Bhat et al. (2017)
Plasma	Tomato slices	Khani, Shokri, and Khajeh (2017)
Microwave blanching	*L. edodes* mushrooms	Li et al. (2017)
Microwave blanching	Chestnut fruits	Liu, Chang, Hu, Brennan, and Guo (2017)
Microwave blanching	Bitter gourd slices	Srimagal et al. (2017)
HHAIB	Peppers	Wang, Fang, et al. (2017)
Microwave blanching	Carrot slices and shreds	Delfiya et al. (2018)
HHAIB	De-pitted halves of apricots	Deng et al. (2018)
Microwave blanching	Sweet corn	Kachhadiya et al. (2018)
Pulsed electric fields + Blanching	Carrot discs	Leong, Du, and Oey (2018)
HHAIB	Red peppers	Wang et al. (2018)
Radiofrequency blanching	Potato cuboids	Zhang, Guo, et al. (2018)
Radiofrequency blanching	Potato cuboids and puree	Zhang, Wang, et al. (2018)
Microwave blanching	Mackerel fillets	Fiore et al. (2019)
Infrared blanching	Edamame	Lara et al. (2019)
Microwave blanching	Potato and cabbage, various geometries	Liburdi et al. (2019)
Microwave blanching	Water dropwort	Tang et al. (2019)
Radiofrequency blanching and Microwave blanching	Cabbage	Xu et al. (2019)

Blanching of bottle gourd (*Lagenaria siceraria*) cubes at 90°C for 5 min, followed by sonication of its juice, was found to positively affect juice quality parameters, namely, the increase in the extractability of total carotenoids and total phenolics, the decrease in particle size, the increase in total soluble solids, and the decrease in total plate count and in yeast and mold count (Bhat & Sharma, 2016). The only parameter impaired by the hybrid treatment was ascorbic acid content. The authors stated that the mechanism behind the results is the occurrence of sono-chemical reactions during ultrasonication, facilitating better extractability of bioactive compounds. The study concluded that even though blanching was effective for improving juice quality, the combination between blanching and sonication proved to be even better in this sense.

Ohmic blanching of bottle gourd cubes was found to be superior to hot water blanching in regard of phenolics extractability and color retention (Bhat et al., 2017). The results of the study also showed that the best product quality was attained by means of

Ohmic blanching at 80°C for 4 min and hot water blanching at 90°C for 5 min, namely, the increase in the concentration of total phenolic compounds, the lowest values of browning index, and an acceptable green color. Plasma, defined as "a high-energy and electrically conductive environment composed by a large number of positive and negative ions and high-energy electrons," was the technology used in a gliding arc system designed to inactivate POD in tomato slices, which was compared to hot water blanching, steam blanching, and microwave blanching systems in terms of their effectiveness for POD inactivation (Khani et al., 2017). The results produced in the study show that blanching in boiling water for 3 min, steam blanching for 9 min, microwave blanching at 2.45 GHz and 650 W for 1 min, and plasma treatment (air plasma, a flow of 3 L/min) for 7 min yielded residual POD activities of 10.51%, 13.84%, 7.71%, and 7.32%, respectively.

Blanching of *Lentinula edodes* mushrooms in boiling water for 60 s and in microwave at 300 W for 90 s caused a marked increase in aldehyde, hydrocarbon, and ester concentration, whereas the sulfide and ketone concentration decreased (Li et al., 2017). They attributed the results on chemical composition to the destruction of cellular integrity and to the increase in the interaction between enzymes and their respective substrates. Chestnut fruit phenolics and flavonoid content and antioxidant activity were higher after steam blanching for 30 min than after microwave blanching at 1150 W for 1.5 min and blanching in boiling water for 20 min, and in addition, steam-blanched fruits showed a bioactive content and an antioxidant activity superior to that of fresh chestnut fruits (Liu et al., 2017). They explained the increase in phenolics and flavonoids by steam blanching based on the degradation of enzymes, with the resulting increase in antioxidant activity in chestnut fruit.

Hot air drying of bitter gourd slices was accelerated by both microwave blanching at 600 W for 135 s and hot water blanching at 80°C for 3 min, especially by the former when preceded by dipping in 2% ethyl oleate and 4% potassium carbonate solutions for 1 min at room temperature (Srimagal et al., 2017). They attributed the accelerating effect of blanching on drying to alterations in cell wall structure and the formation of pores in the tissues allowing increased water diffusion. High-humidity hot air impingement blanching (HHAIB) was used as a pretreatment for drying peppers (Wang, Fang, et al., 2017). Minimal PPO activity, proper color, and texture and reduced drying time (Fig. 8.5) were observed when HHAIB was carried out for 120 s. They attributed the reduced drying time to removal of air from the tissue, enhancement of tissue permeability, and the formation of microcracks during HHAIB, which facilitated mass-transfer during drying.

For dried carrot slices and shreds, longer microwave blanching times as carried out at a 2450 MHz frequency and a 900 W power (a power density of 3.8 W/g) were associated with lower final moisture, higher drying rates, higher rehydration capacity, lower color and texture change, and higher Beta carotene content (Delfiya et al., 2018). They attributed the obtained results to, respectively, an increase in the formation of pores in carrot because of the effect of microwaves, resulting in a lower final moisture content; the opening of cell walls, allowing quick evaporation and movement of water during drying; increased cell disruption and formation of pores in carrots, which allowed better rehydration; the inhibition of enzymatic browning, yielding

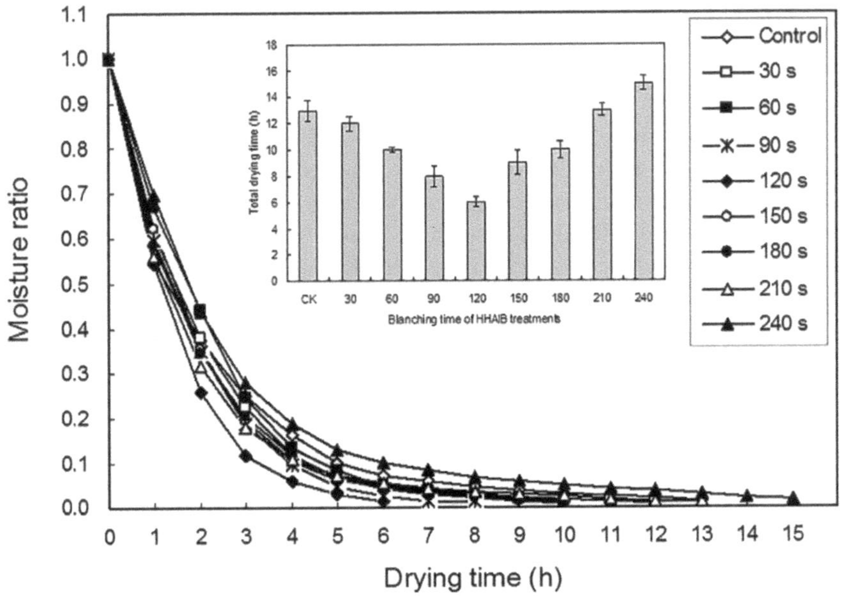

Fig. 8.5 Effect of different HHAIB time on drying kinetics of red pepper samples with 65°C as the constant drying temperature. The total drying time of each treatment is also shown. Credit: Wang, J., Fang, X.M., Mujumdar, A.S., Qian, J.Y., Zhang, Q., Yang, X.H., Liu, Y.H., Gao, Z.J., Xiao, H.W. (2017). Effect of high-humidity hot air impingement blanching (HHAIB) on drying and quality of red pepper (Capsicum annuum L.). *Food Chemistry*, *220*, 145–152.

lower color change; the inactivation of POD, avoiding changes in firmness caused by alterations in mechanical properties of cell walls; and enhancement in the availability of carotenoids, caused by the breakdown of crystalline carotenoid complexes and breakdown of pectin in carrot cell wall during blanching.

Deng et al. (2018) elucidated the effect of HHAIB on drying kinetics, hardness, and pectic fraction concentration of apricots, finding that HHAIB at 110°C reduced the drying time and increased drying rates and caused a decrease in hardness, effects which were more pronounced as the blanching time increased from 30 to 120 s. Regarding pectic fractions, the conversion of insoluble pectin into soluble pectin by the thermal effect of HHAIB was observed, which was positively correlated with the apricot hardness decrease. The optimal blanching times for hot water (100°C), steam, and microwave (2450 MHz, 900 W, 100 g of the sample) blanching of sweet corn were established in 120, 90, and 60 s, respectively, resulting in 95% of POD inactivation (Kachhadiya et al., 2018). The report also showed that microwave blanching was superior to the other blanching techniques, as denoted by a higher inactivation rate constant, a lower half-life value of inactivation, and a lower decimal reduction time, along with higher product total sugar and ascorbic acid content.

The use of pulsed electric fields in 300 ppm $CaCl_2$ solutions as a pretreatment for blanching at 100°C for 5 min of carrots could improve the hardness of blanched carrots (Leong et al., 2018). They mentioned that the mechanism responsible for hardness improvement was that pulsed electric fields facilitated mass transfer of calcium ions into carrot cells, which led to a better accessibility of cell walls to calcium ions to aid in the firmness achievement in the tissues by means of the formation of additional ionic linkages with pectins. HHAIB of red peppers at 110°C for 120 s was the most favorable condition for red pigment and ascorbic acid retention and for high antioxidant activities in both blanched and dried peppers, as measured immediately after processing and after 6-month storage (Wang et al., 2018).

Radiofrequency blanching was used for improving heating uniformity in potato cuboids (Zhang, Guo, et al., 2018). The results showed that the best heating uniformity was obtained by using a sample height of 60 mm, an electrode gap of 120 mm, and a NaCl solution with an electrical conductivity of $0.5\,\mathrm{s\,m^{-1}}$. The effect of radiofrequency blanching on quality features of potatoes was evaluated by Zhang, Wang, et al. (2018), who found that using an electrode gap of 120 mm was the most favorable condition for PPO inactivation under the same final temperature. In addition, for the same electrode gap, lightness, yellowness, and hardness values decreased with increasing temperature, whereas redness slightly increased. They justified the decrease in color values by cell collapse and liquid release, which led to a lower reflectance, and the decrease in hardness to water loss during radiofrequency blanching, which reduced turgor pressure.

Marinating of mackerel (*Scomber scombrus*) fillets, followed by hot water blanching, microwave blanching, or high-pressure processing, resulted increased protein, ash, and solids and decreased lipid and polyunsaturated fatty acid content (Fiore et al., 2019). Additionally, microwave blanching (900 W/1 min) and high-pressure processing (500 MPa/5 min/4°C) were the best treatments for preserving mackerel omega-3 fatty acids, whereas microwave blanching was the best treatment for its sensory quality. The authors explained the variations in composition as affected by the treatments by the significant solubilization of collagen, in the case of protein increase, by the NaCl penetration during the marinating process, in the case of ash increase, by the water loss from the tissue induced by the processing, in the case of solids increase, and by the loss of liquefied lipids through the muscle and lipid hydrolysis, in the case of lipid and polyunsaturated fatty acid decrease.

The effect of infrared heating at variable heating intensities and heating times on edamame weight loss, texture, and green color intensity was assessed by Lara et al. (2019), who discovered that the higher the heating intensity and heating time, the higher the weight loss. Furthermore, a higher infrared heating time caused more edamame softening and intensification of the green color. Microwave blanching of potatoes and cabbages was optimized having a blanching time and microwave power as independent variables and POD activity as dependent variable (Liburdi et al., 2019). The results showed that when optimal microwave blanching conditions were used, potatoes mimicking the most common geometries, namely, cylinder, cube, and parallelepiped of two sizes, presented a POD residual activity of less than 10%, whereas savoy cabbage and white cabbage mimicking a slab presented higher POD residual activity even after increasing the microwave power and the blanching time (Fig. 8.6).

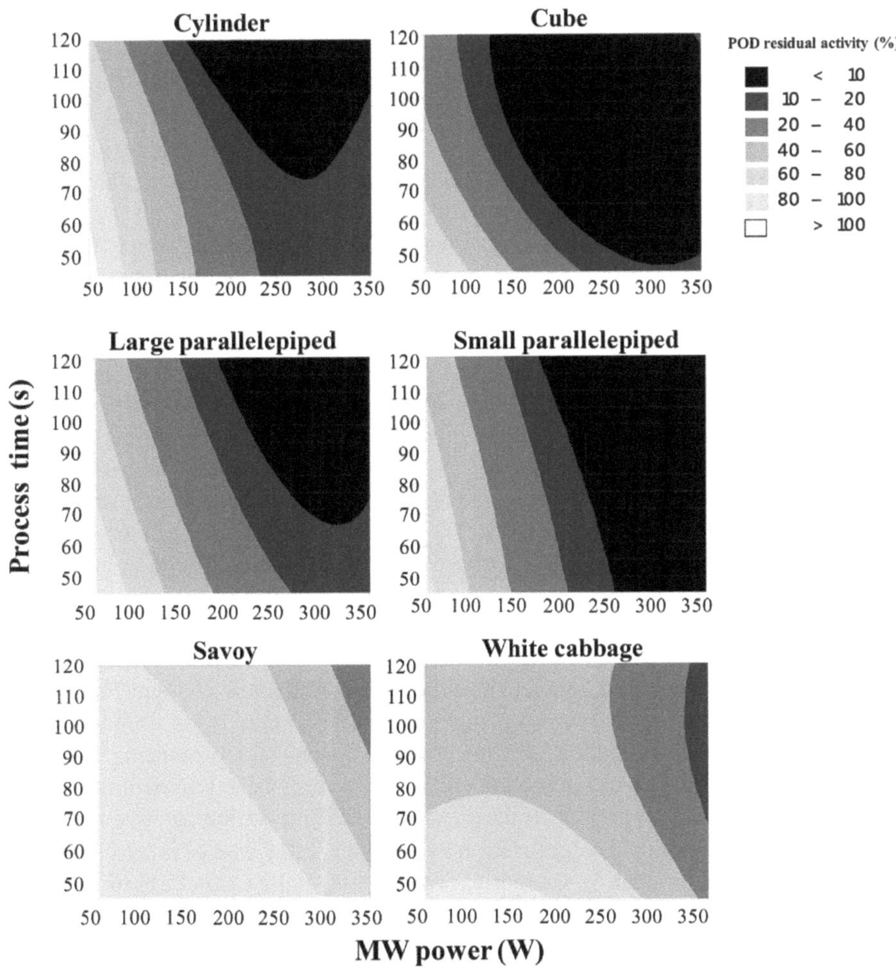

Fig. 8.6 Contour plots showing the microwave (MW) power and process time effect on the peroxidase (POD) residual activity (%) for potatoes and leafy vegetables (savoy cabbage and white cabbage), mimicking the most common geometries (cylinder, cube, parallelepiped, and slab, respectively).

Credit: Liburdi, K., Benucci, I., Esti, M. (2019). Effect of microwave power and blanching time in relation to different geometric shapes of vegetables. *LWT—Food Science and Technology, 99*, 497–504.

Water dropwort blanching through hot water and microwave showed that more than 90% of PPO can be inactivated by using temperatures of 80–90°C and power densities of 9–14 W/g for 1–2 min (Tang et al., 2019). Additional data showed that increasing the microwave power reduced softening, a result that they attributed to dehydration and shrinkage during microwave blanching. Finally, hot water blanching was found to have more potential to reduce unpleasant odor in water dropwort than microwave blanching, a result that they attributed to the leaching of volatile compounds into the blanching water. The effect of different blanching techniques on quality features of cabbage was evaluated by Xu et al. (2019), who found that hot water blanching, microwave blanching at 2450 and 915 MHz, and radiofrequency blanching were able to reduce POD activity to 10% in 40, 50, 90, and 100 s, respectively. In addition, all blanching treatments reduced cabbage chlorophyll and vitamin C content and enhanced the cabbage odor, whereas radiofrequency blanching was especially favorable for preserving the cabbage volatile compounds.

8.6 Conclusion and future outlook

Along the years, blanching processes have been giving an important contribution to the obtaining of canned, frozen, and dried foods of an extended shelf life, especially vegetables. Hot water and steam blanching are still extensively used in commercial applications, but the limitations associated with these technologies stimulate new research on the field, which ultimately led to the development of modern blanching technologies. HHAIB is receiving increasing attention from the food industry because of advantages such as a high energy efficiency and reduced effluent generation. Microwave blanching is very consolidated in academic studies, whereas Ohmic blanching, infrared blanching, and radiofrequency blanching are gaining notability. Nevertheless, it is uncertain whether these technologies will someday completely substitute hot water and steam blanching as there are issues related to scale-up that could impose difficulties to the use of technologies that look promising at the laboratory scale.

References

Aghajanzadeh, S., Ziaiifar, A. M., Kashaninejad, M., Maghsoudlou, Y., & Esmailzadeh, E. (2016). Thermal inactivation kinetic of pectin methylesterase and cloud stability in sour orange juice. *Journal of Food Engineering*, *185*, 72–77.

Agüero, M. V., Ansorena, M. R., Roura, S. I., & del Valle, C. E. (2008). Thermal inactivation of peroxidase during blanching of butternut squash. *LWT—Food Science and Technology*, *41*(3), 401–407.

Amaral, R. D. A., Achaerandio, I., Benedetti, B. C., & Pujola, M. (2017). The influence of edible coatings, blanching and ultrasound treatments on quality attributes and shelf-life of vacuum packaged potato strips. *LWT—Food Science and Technology*, *85*(Part B), 449–455.

Araújo, A. C., Oliveira, S. M., Ramos, I. N., Brandão, T. R. S., & Silva, C. L. M. (2016). Influence of pretreatments on quality parameters and nutritional compounds of dried galega kale (*Brassica oleracea* L. var. *acephala*). *Food and Bioprocess Technology*, *9*, 872–881.

Arisseto, A. P., Silva, W. C., Marcolino, P. F. C., Scaranelo, G. R., Berbari, S. A. G., Miguel, A. M. R. O., et al. (2019). Influence of potato cultivar, frying oil and sample pre-treatments on the contamination of French fries by 3-monochloropropane-1,2-diol fatty acid esters. *Food Research International*, *124*, 43–48.

Arroqui, C., Lopez, A., Esnoz, A., & Virseda, P. (2003). Mathematical model of heat transfer and enzyme inactivation in an integrated blancher cooler. *Journal of Food Engineering*, *58* (3), 215–225.

Bai, J. W., Gao, Z. J., Xiao, H. W., Wang, X. T., & Zang, Q. (2013). Polyphenol oxidase inactivation and vitamin C degradation kinetics of Fuji apple quarters by high humidity air impingement blanching. *International Journal of Food Science and Technology*, *48*(6), 1135–1141.

Barreto, I. M. A., Tribuzi, G., Marsaioli Junior, A., Carciofi, B. A. M., & Laurindo, J. B. (2019). Oil–free potato chips produced by microwave multiflash drying. *Journal of Food Engineering*, *261*, 133–139.

Barroca, M. J., Guiné, R. P. F., Calado, A. R. P., Correia, P. M. R., & Mendes, M. (2017). Artificial neural network modelling of the chemical composition of carrots submitted to different pre-drying treatments. *Food Measure*, *11*, 1815–1826.

Bernaś, E., & Jaworska, G. (2016). Vitamins profile as an indicator of the quality of frozen *Agaricus bisporus* mushrooms. *Journal of Food Composition and Analysis*, *49*, 1–8.

Bernstein, A., & Noreña, C. P. Z. (2017). Kinetics of enzymatic inactivation and loss of anthocyanins and antioxidant activity in red cabbage blanched under different conditions. *Journal of Food Biochemistry*, *41*(3), e12340.

Bhat, S., & Sharma, H. K. (2016). Combined effect of blanching and sonication on quality parameters of bottle gourd (*Lagenaria siceraria*) juice. *Ultrasonics Sonochemistry*, *3*, 182–189.

Bhat, S., Saini, C. S., & Sharma, H. K. (2017). Changes in total phenolic content and color of bottle gourd (*Lagenaria siceraria*) juice upon conventional and ohmic blanching. *Food Science and Biotechnology*, *26*(1), 29–36.

Cabinplant A/S. (2020). *Heat treatment: Industrial blanchers*. Available from: https://cabinplant.com/products/blanchers. Accessed on 19 March 2020.

Cai, C., Miao, H., Qian, H., Yao, L., Wang, B., & Wang, Q. (2016). Effects of industrial pre-freezing processing and freezing handling on glucosinolates and antioxidant attributes in broccoli florets. *Food Chemistry*, *210*, 451–456.

Campos, D., Aguilar-Alvez, A., & Pedreschi, R. (2016). Stability of fructooligosaccharides, sugars and colour of yacon (*Smallanthus sonchifolius*) roots during blanching and drying. *International Journal of Food Science and Technology*, *51*(5), 1177–1185.

Chamchong, M., Tatidin, Y., & Misayan, A. (2016). Violet-red discolouration of canned Palmyra palm during processing. *Journal of Food Science and Technology*, *53*(1), 631–639.

Chen, Y., Song, C., Li, Z., Chen, H., & Jin, G. (2020). Effects of hot air and combined microwave and hot air drying on the quality attributes of celery stalk slices. *Journal of Food Processing and Preservation*, *44*(1), e14310.

Collins, J. K., Biles, C. L., Perkins-Veazie, P., & Maness, N. (1996). Flavour qualities of frozen sweetcorn are affected by genotype and blanching. *Journal of the Science of Food and Agriculture*, *72*(4), 425–429.

Delfiya, A., Mohapatra, D., Kotwaliwale, N., & Mishra, A. K. (2018). Effect of microwave blanching and brine solution pretreatment on the quality of carrots dried in solar-biomass hybrid dryer. *Journal of Food Processing and Preservation*, *42*(2), e13510.

Deng, L. Z., Mujumdar, A. S., Yang, X. H., Wang, J., Zhang, Q., Zheng, Z. A., et al. (2018). High humidity hot air impingement blanching (HHAIB) enhances drying rate and softens texture of apricot via cell wall pectin polysaccharides degradation and ultrastructure modification. *Food Chemistry*, *261*, 292–300.

Deylami, M. Z., Rahman, R. A., Tan, C. P., Bakar, J., & Olusegun, L. (2016). Effect of blanching on enzyme activity, color changes, anthocyanin stability and extractability of mangosteen pericarp: A kinetic study. *Journal of Food Engineering*, *178*, 12–19.

Dodman Limited. (2019). *Rotary drum blancher*. Available from: https://www.dod (web archive link, 19 March 2020) man.com/product/rotary-drum-blancher. [19 March 2020].

Drewnowska, M., Falandysz, J., Chudzińska, M., Hanć, A., Saba, M., & Barałkiewicz, D. (2017). Leaching of arsenic and sixteen metallic elements from amanita fulva mushrooms after food processing. *LWT—Food Science and Technology*, *84*, 861–866.

EMA EUROPE S.R.O. (2020). *NEAEN ThermoScrew continuous screw cooker/blancher*. Available from: https://neaen.com/neaen-thermoscrew-continuous-screw-cooker-blancher (19 March 2020).

EMA EUROPE S.R.O. (2020). *NEAEN ContiBlanch conveyor blancher/cooker*. Available from: https://neaen.com/neaen-contiblanch-conveyor-blancher-cooker. [19 March 2020.

Falandysz, J., & Drewnowska, M. (2017). Cooking can decrease mercury contamination of a mushroom meal: *Cantharellus cibarius* and *Amanita fulva*. *Environmental Science and Pollution Research*, *24*, 13352–13357.

Fante, L., & Noreña, C. P. Z. (2012). Enzyme inactivation kinetics and colour changes in garlic (*Allium sativum* L.) blanched under different conditions. *Journal of Food Engineering*, *108*(3), 436–443.

Fellows. (2000). *Food processing technology: Principles and practice*. Abington: Woodhead Publishing.

Fennema, O. R. (Ed.). (1996). *Food chemistry*. New York: Marcel Dekker.

Ferreira, S. S., Monteiro, F., Passos, C. P., Silva, A. M. S., Wessel, D. F., Coimbra, M. A., et al. (2020). Blanching impact on pigments, glucosinolates, and phenolics of dehydrated broccoli by-products. *Food Research International*, *132*, 109055.

Fiore, A. G., Palmieri, L., Orlando, I., de Giorgi, A., Derossi, A., & Severini, C. (2019). Comparison of different combined processes of preservation on the nutritional and sensory changes of "ready to eat" mackerel fillets. *Journal of Food Processing and Preservation*, *43*(3), e13886.

Fujikawa, H., & Itoh, T. (1996). Characteristics of a multicomponent first-order model for thermal inactivation of microorganisms and enzymes. *International Journal of Food Microbiology*, *31*(1–3), 263–271.

Gamboa-Santos, J., Montilla, A., Soria, A. C., & Villamiel, M. (2012). Effects of conventional and ultrasound blanching on enzyme inactivation and carbohydrate content of carrots. *European Food Research and Technology*, *234*, 1071–1079.

García-Toledo, J. A., Ruiz-López, I. I., Martínez-Sánchez, C. E., Rodríguez-Miranda, J., Carmona-García, R., Torruco-Uco, J. G., et al. (2016). Effect of osmotic dehydration on the physical and chemical properties of Mexican ginger (*Zingiber officinale* var. Grand Cayman). *CyTA—Journal of Food*, *14*(1), 27–34.

Gomes, C. F., Sarkis, J. R., & Marczak, L. D. F. (2018). Ohmic blanching of Tetsukabuto pumpkin: Effects on peroxidase inactivation kinetics and color changes. *Journal of Food Engineering*, *233*, 74–80.

Gowen, A., Abu-Ghannam, N., Frias, J., & Oliveira, J. (2007). Modelling the water absorption process in chickpeas (*Cicer arietinum* L.)—the effect of blanching pre-treatment on water intake and texture kinetics. *Journal of Food Engineering*, *78*(3), 810–819.

Goyeneche, R., Di Scala, K., & Roura, S. (2013). Biochemical characterization and thermal inactivation of polyphenol oxidase from radish (*Raphanus sativus* var. *sativus*). *LWT—Food Science and Technology*, *54*(1), 57–62.

Guiamba, I. R. F., & Svanberg, U. (2016). Effects of blanching, acidification, or addition of EDTA on vitamin C and β-carotene stability during mango purée preparation. *Food Science and Nutrition*, *4*(5), 706–715.

Guiamba, I. R. F., Svanberg, U., & Ahrné, L. (2015). Effect of infrared blanching on enzyme activity and retention of β-carotene and vitamin C in dried mango. *Journal of Food Science*, *80*(6), E1235–E1242.

Hamid, S. S., Wakayama, M., Ashino, Y., Kadowaki, R., Soga, T., & Tomita, M. (2020). Effect of blanching on the concentration of metabolites in two parts of *Undaria pinnatifida*, Wakame (leaf) and Mekabu (sporophyll). *Algal Research*, *47*, 101829.

Henley, J. P., & Sadana, A. (1985). Categorization of enzyme deactivations using a series-type mechanism. *Enzyme and Microbial Technology*, *7*(2), 50–60.

Hernández, H., Claramount, D., Kučerová, I., & Banout, J. (2017). The effects of modified blanching and oregano essential oil on drying kinetics and sensory attributes of dried meat. *Journal of Food Processing and Preservation*, *41*(5), e13161.

Icier, F., Cokgezme, O. F., & Sabanci, S. (2017). Alternative thawing methods for the blanched/non-blanched potato cubes: Microwave, ohmic, and carbon fiber plate assisted cabin thawing. *Journal of Food Process Engineering*, *40*(2), e12403.

Ismail, N., & Revathi, R. (2006). Studies on the effects of blanching time, evaporation time, temperature and hydrocolloid on physical properties of chili (*Capsicum annum* var *kulai*) puree. *LWT—Food Science and Technology*, *39*(3), 91–97.

Jabłońska-Ryś, E., Sławińska, A., Radzki, W., & Gustaw, W. (2016). Evaluation of the potential use of probiotic strain *lactobacillus plantarum* 299v in lactic fermentation of button mushroom fruiting bodies. *Acta Scientiarum Polonorum*, *15*(4), 399–407.

Jamali, S. N., Kashaninejad, M., Amirabadi, A. A., Aalami, M., & Khomeiri, M. (2018). Kinetics of peroxidase inactivation, color and temperature changes during pumpkin (*Cucurbita moschata*) blanching using infrared heating. *LWT—Food Science and Technology*, *93*, 456–462.

Kachhadiya, S., Kumar, N., & Seth, N. (2018). Process kinetics on physico-chemical and peroxidase activity for different blanching methods of sweet corn. *Journal of Food Science and Technology*, *55*(12), 4823–4832.

Kamal, M. M., Ali, M. R., Rahman, M. M., Shishir, M. R. I., Yasmin, S., & Sarker, M. S. H. (2019). Effects of processing techniques on drying characteristics, physicochemical properties and functional compounds of green and red chilli (*Capsicum annum* L.) powder. *Journal of Food Science and Technology*, *56*(7), 3185–3194.

Kar, J. R., Hallsworth, J. E., & Singhal, R. S. (2016). Glycine betaine-mediated protection of peas (*Pisum sativum* L.) during blanching and frozen storage. *International Journal of Food Properties*, *19*(11), 2510–2521.

Khani, M. R., Shokri, B., & Khajeh, K. (2017). Studying the performance of dielectric barrier discharge and gliding arc plasma reactors in tomato peroxidase inactivation. *Journal of Food Engineering*, *197*, 107–112.

Kim, A. N., Lee, K. Y., Rahman, M. S., Kim, H. J., Chun, J., Heo, J. H., et al. (2019). Effect of water blanching on phenolic compounds, antioxidant activities, enzyme inactivation, microbial reduction, and surface structure of samnamul (*Aruncus dioicus* var *kamtschaticus*). *International Journal of Food Science and Technology*, *55*(4), 1754–1762.

Knirsch, M. C., Santos, C. A., Vicente, A. A. M. O. S., & Penna, T. C. V. (2010). Ohmic heating—a review. *Trends in Food Science and Technology*, *21*(9), 436–441.

Koutidou, M., Grauwet, T., Loey, A. V., & Acharya, P. (2017). Impact of processing on odour-active compounds of a mixed tomato-onion puree. *Food Chemistry*, *228*, 14–25.

Krishnamurthy, K., Khurana, H. K., Jun, S., Irudayaraj, J., & Demirci, A. (2008). Infrared heating in food processing: An overview. *Comprehensive Reviews in Food Science and Food Safety*, *7*(1), 2–13.

Lara, L. M., Wilson, S. A., Chen, P., & Atungulu, G. G. (2019). The effects of infrared treatment on physicochemical characteristics of vegetable soybean. *Heliyon*, *5*(1), e01148.

Larrauri, M., Zunino, M. P., Zygadlo, J. A., Grosso, N. R., & Nepote, V. (2016). Chemical characterization and antioxidant properties of fractions separated from extract of peanut skin derived from different industrial processes. *Industrial Crops and Products*, *94*, 964–971.

Lehninger, A. L., Nelson, D. L., & Cox, M. M. (1995). *Princípios de Bioquímica*. São Paulo: Sarvier.

Leong, S. Y., Du, D., & Oey, I. (2018). Pulsed electric fields enhances calcium infusion for improving the hardness of blanched carrots. *Innovative Food Science and Emerging Technologies*, *47*, 46–55.

Levent, O., & Alpaslan, M. (2018). Effect of processing parameters on some physicochemical properties, sugar profile and rheological characterization of apricot sauce. *Journal of Food Measurement and Characterization*, *12*, 1072–1083.

Li, B., Kimatu, B. M., Li, C., Pei, F., Hu, Q., & Zhao, L. (2017). Analysis of volatile compounds in *L. edodes* blanched by hot water and microwave. *International Journal of Food Science and Technology*, *52*(7), 1680–1689.

Li, Y. H., Wu, Z. F., Tang, X., Yu, F., Wang, X. C., & Yang, M. (2019). Application of infrared-assisted steam blanching to improve enzymatic inactivation and quality retention of *Chrysanthemum indicum* L. flower. *Journal of Food Science and Technology*, *56*(9), 4274–4281.

Liburdi, K., Benucci, I., & Esti, M. (2019). Effect of microwave power and blanching time in relation to different geometric shapes of vegetables. *LWT—Food Science and Technology*, *99*, 497–504.

Ling, A. C., & Lund, D. B. (1978). Determining kinetic parameters for thermal inactivation of heat-resistant and heat-labile isozymes from thermal destruction curves. *Journal of Food Science*, *43*(4), 1307–1310.

Liu, F., Chang, X., Hu, X., Brennan, C. S., & Guo, X. (2017). Effect of thermal processing on phenolic profiles and antioxidant activities in *Castanea mollissima*. *International Journal of Food Science and Technology*, *52*(2), 439–447.

Liu, J., Wang, R., Wang, X., Yang, L., Zhang, Q., Shan, Y., et al. (2019). Effect of blanching and drying temperatures on the browning- related enzymes and physicochemical properties of lily bulb flours. *Journal of Food Processing and Preservation*, *43*(12), e14248.

Lopes, A. M., Toralles, R. P., & Rombaldi, C. V. (2014). Thermal inactivation of polyphenoloxidase and peroxidase in Jubileu clingstone peach and yeast isolated from its spoiled puree. *Food Science and Technology*, *34*(1), 150–156.

Lyco Manufacturing, Inc. (2020). *Cooker-cooler*. Available from: https://lycomfg.com/equipment/blanchers/cooker-cooler (19 March 2020).

Mahn, A., Saavedra, A., & Paz Rubio, M. (2018). Kinetic study of sulforaphane stability in blanched and unblanched broccoli (*Brassica oleracea* var. *italica*) florets during storage at low temperatures. *Journal of Food Science and Technology*, *55*(11), 4687–4693.

Manzocco, L., Anese, M., & Nicoli, M. C. (2008). Radiofrequency inactivation of oxidative food enzymes in model systems and apple derivatives. *Food Research International*, *41* (10), 1044–1049.

Megido, R. C., Desmedt, S., Blecker, C., Béra, F., Haubruge, E., Alabi, T., et al. (2017). Microbiological load of edible insects found in Belgium. *Insects*, *8*(1), 12.

Meher, J., Keshav, A., Mazumdar, B., & Keshav, S. (2018). Effect of blanching and thermal preservation on rheology of curry leaf puree. *Food Measure*, *12*, 105–117.

Moscetti, R., Raponi, F., Monarca, D., Bedini, G., Ferri, S., & Massantini, R. (2019). Effects of hot-water and steam blanching of sliced potato on polyphenol oxidase activity. *International Journal of Food Science and Technology*, *54*(2), 403–411.

Mozzoni, L. A., Chen, P., Morawicki, R. O., Hettiarachchy, N. S., Brye, K. R., & Mauromoustakos, A. (2009). Quality attributes of vegetable soybean as a function of boiling time and condition. *International Journal of Food Science and Technology*, *44*(11), 2089–2099.

Munyaka, A. W., Oey, I., Verlinde, P., Loey, A. V., & Hendrickx, M. (2009). Acidification, crushing and thermal treatments can influence the profile and stability of folate poly-γ-glutamates in broccoli (*Brassica oleracea* L. *var. italica*). *Food Chemistry*, *117*(3), 568–575.

Munyaka, A. W., Oey, I., Loey, A. V., & Hendrickx, M. (2010). Application of thermal inactivation of enzymes during vitamin C analysis to study the influence of acidification, crushing and blanching on vitamin C stability in broccoli (*Brassica oleracea* L *var. italica*). *Food Chemistry*, *120*(2), 591–598.

Nambi, V. E., Gupta, R. K., Kumar, S., & Sharma, P. C. (2016). Degradation kinetics of bioactive components, antioxidant activity, colour and textural properties of selected vegetables during blanching. *Journal of Food Science and Technology*, *53*(7), 3073–3082.

Nambi, V. E., Behera, G., Saharan, V. K., & Meena, V. S. (2016). Optimization of salt curing of bitter gourd (*Momordica charantia* L.) with blanching as pre-treatment using response surface methodology. *Nutrition and Food Science*, *46*(6), 742–752.

Ngobese, N. Z., & Workneh, T. S. (2018). 'Potato (*Solanum tuberosum* L.) nutritional changes associated with French fry processing: Comparison of low-temperature long-time and high-temperature short-time blanching and frying treatments. *LWT—Food Science and Technology*, *97*, 448–455.

Ngobese, N. Z., Workneh, T. S., & Siwela, M. (2017). Effect of low-temperature long-time and high-temperature short-time blanching and frying treatments on the French fry quality of six Irish potato cultivars. *Journal of Food Science and Technology*, *54*(2), 507–517.

Niu, L., Li, D., Liu, C., & Liu, F. (2015). Comparison of lipoxygenase activity characteristics in aqueous extracts from milk-stage sweet corn and waxy corn. *Food Science and Biotechnology*, *24*(3), 867–873.

Oner, M. E., & Walker, P. N. (2011). Shelf-life of near-aseptically packaged refrigerated potato strips. *LWT—Food Science and Technology*, *44*(7), 1616–1620.

Plazzotta, S., & Manzocco, L. (2019). High-pressure homogenisation combined with blanching to turn lettuce waste into a physically stable juice. *Innovative Food Science and Emerging Technologies*, *52*, 136–144.

Preciado-Iñiga, G. E., Amador-Espejo, G. G., & Bárcenas, M. E. (2018). Blanching and antimicrobial mixture (potassium sorbate–sodium benzoate) impact on the stability of a tamarillo (Cyphomandra betacea) sweet product preserved by hurdle technology. *Journal of Food Science and Technology*, *55*(2), 740–748.

Priecina, L., Karklina, D., & Kince, T. (2018). The impact of steam-blanching and dehydration on phenolic, organic acid composition, and total carotenoids in celery roots. *Innovative Food Science and Emerging Technologies*, *49*, 192–201.

Radošević, K., Srček, V. G., Bubalo, M. C., Brnčić, S. R., Takács, K., & Redovniković, I. R. (2017). Assessment of glucosinolates, antioxidative and antiproliferative activity of broccoli and collard extracts. *Journal of Food Compositions and Analysis*, *61*, 59–66.

Ranjan, S., Dasgupta, N., Walia, N., Thara Chand, C., & Ramalingam, C. (2017). Microwave blanching: An emerging trend in food engineering and its effects on *Capsicum annuum* L. *Journal of Food Process Engineering*, *40*(2), e12411.

Reis, F. R., Masson, M. L., & Waszczynskyj, N. (2008). Influence of a blanching pretreatment on color, oil uptake and water activity of potato sticks, and its optimization. *Journal of Food Process Engineering*, *31*(6), 833–852.

Ren, F., Perussello, C. A., Zhang, Z., Gaffney, M. T., Kerry, J. P., & Tiwari, B. K. (2018). Enhancement of phytochemical content and drying efficiency of onions (*Allium cepa* L.) through blanching. *Journal of the Science of Food and Agriculture, 98*(4), 1300–1309.

Richter Reis, F. (Ed.). (2017). *New perspectives on food blanching*. Cham: Springer.

Rizvi, A. F., & Tong, C. H. (1997). Fractional conversion for determining texture degradation kinetics of vegetables. *Journal of Food Science, 62*(1), 1–7.

Rudra Shalini, G., Shivare, U. S., & Basu, S. (2008). Thermal inactivation kinetics of peroxidase in mint leaves. *Journal of Food Engineering, 85*(1), 147–153.

Salas, J. J., Sánchez, C., García-González, D. L., & Aparicio, R. (2005). Impact of the suppression of lipoxygenase and hydroperoxide lyase on the quality of the green odor in green leaves. *Journal of Agricultural and Food Chemistry, 53*(5), 1648–1655.

Smith, P. G. (2011). *Introduction to food process engineering*. Boston: Springer.

Srimagal, A., Mishra, S., & Pradhan, R. C. (2017). Effects of ethyl oleate and microwave blanching on drying kinetics of bitter gourd. *Journal of Food Science and Technology, 54*(5), 1192–1198.

Tang, M., Sun, H., Zhang, Z., Zhao, J., Cao, J., Thakur, K., et al. (2019). Evaluation of hot water and microwave blanching on qualities and sensory characteristics of water dropwort (*Oenanthe javanica* DC.). *Journal of Food Processing and Preservation, 43*(10), e14104.

Tsai, C. L., Shiau, C. Y., & Sung, W. C. (2018). Effects of blanching and refrigerated storage on quality attributes of hybrid abalone (*Haliotidae discushannai* x *H. diversicolor diversicolor*). *Journal of Food Processing and Preservation, 42*(5), e13608.

Wahab, B. A., Adebowale, A. R. A., Sanni, S. A., Sobukola, O. P., Obadina, A. O., Kajihausa, O. E., et al. (2016). Effect of species, pretreatments, and drying methods on the functional and pasting properties of high- quality yam flour. *Food Science and Nutrition, 4*(1), 50–58.

Wampler, B., & Barringer, S. A. (2012). Volatile generation in bell peppers during frozen storage and thawing using selected ion flow tube mass spectrometry (SIFT-MS). *Journal of Food Science, 77*(6), C677–C683.

Wang, Y., Xing, J., Wang, R., & Guo, S. (2017). The analysis of the causes of protein precipitate formation in the blanched soymilk. *Food Chemistry, 218*, 341–347.

Wang, J., Fang, X. M., Mujumdar, A. S., Qian, J. Y., Zhang, Q., Yang, X. H., et al. (2017). Effect of high-humidity hot air impingement blanching (HHAIB) on drying and quality of red pepper (*Capsicum annuum* L.). *Food Chemistry, 220*, 145–152.

Wang, J., Yang, X. H., Mujumdar, A. S., Fang, X. M., Zhang, Q., Zheng, Z. A., et al. (2018). Effects of high-humidity hot air impingement blanching (HHAIB) pretreatment on the change of antioxidant capacity, the degradation kinetics of red pigment, ascorbic acid in dehydrated red peppers during storage. *Food Chemistry, 259*, 65–72.

Weil, M., Shum Cheong Sing, A., Méot, J. M., Boulanger, R., & Bohuon, P. (2017). Impact of blanching, sweating and drying operations on pungency, aroma and color of *Piper borbonense*. *Food Chemistry, 219*, 274–281.

Wickramasinghe, P. C. K., Murray, A. F., Sahoo, M. R., Dein, M., Luckett, C. R., Dia, V. P., et al. (2020). The effects of processing on *Garcinia xanthochymus* fruit beverage. *Journal of Food Measurement and Characterization, 14*, 55–68.

Wiktor, A., Nowacka, M., Anuszewska, A., Rybak, K., Dadan, M., & Witrowa-Rajchert, D. (2019). Drying kinetics and quality of dehydrated cranberries pretreated by traditional and innovative techniques. *Journal of Food Science, 84*(7), 1820–1828.

Xanthakis, E., Gogou, E., Taoukis, P., & Ahrné, L. (2018). Effect of microwave assisted blanching on the ascorbic acid oxidase inactivation and vitamin C degradation in frozen mangoes. *Innovative Food Science and Emerging Technologies, 48*, 248–257.

Xiao, H. W., Bai, J. W., Sun, D. W., & Gao, Z. J. (2014). The application of superheated steam impingement blanching (SSIB) in agricultural products processing – A review. *Journal of Food Engineering*, *132*, 39–47.

Xu, S., Sun, X., Lu, H., Yang, H., Ruan, Q., Huang, H., et al. (2018). Detecting and monitoring the flavor of tomato (*Solanum lycopersicum*) under the impact of postharvest handlings by physicochemical parameters and electronic nose. *Sensors*, *18*(6), 1847.

Xu, J., Wang, B., & Wang, Y. (2019). Electromagnetic fields assisted blanching-effect on the dielectric and physicochemical properties of cabbage. *Journal of Food Process Engineering*, *42*(8), e13294.

Yonny, M. E., Medina, A. V., Nazareno, M. A., & Chaillou, L. L. (2018). Enhancement in the oxidative stability of green peas by *Ilex paraguariensis* addition in a blanching process before their refrigerated and frozen storage. *LWT—Food Science and Technology*, *91*, 315–321.

Zhang, Z., Guo, C., Gao, T., Fu, H., Chen, Q., & Wang, Y. (2018). Pilot-scale radiofrequency blanching of potato cuboids: Heating uniformity. *Journal of the Science of Food and Agriculture*, *98*(1), 312–320.

Zhang, Z., Wang, J., Zhang, X., Shi, Q., Xin, L., Fu, H., et al. (2018). Effects of radio frequency assisted blanching on polyphenol oxidase, weight loss, texture, color and microstructure of potato. *Food Chemistry*, *248*, 173–182.

Zhang, Y., Kahl, D. H. W., Bizimungu, B., & Lu, Z. X. (2018). Effects of blanching treatments on acrylamide, asparagine, reducing sugars and colour in potato chips. *Journal of Food Science and Technology*, *55*(10), 4028–4041.

Zhao, W., Xie, W., Du, S., Yan, S., Li, J., & Wang, Q. (2016). Changes in physicochemical properties related to the texture of lotus rhizomes subjected to heat blanching and calcium immersion. *Food Chemistry*, *211*, 409–414.

Pasteurization in the food industry

9

Maryam Azizi-Lalabadi[a], *Najme Rezaee Moghaddam*[b], *and Seid Mahdi Jafari*[c]
[a]Research Center for Environmental Determinants of Health (RCEDH), Kermanshah University of Medical Sciences, Kermanshah, Iran, [b]Laboratory Instrumental Expert, IMEN KIA SANAT Engineering Company, Shiraz, Iran, [c]Department of Food Materials and Process Design Engineering, Gorgan University of Agricultural Sciences and Natural Resources, Gorgan, Iran

9.1 Introduction

The pasteurization process is a light thermal procedure that is applied to deactivate heat-resistant organisms including bacteria, yeasts, and molds. These microorganisms can create food spoilage if not eliminated. Pasteurization for dairy products is performed based on the International Dairy Federation (IDF, 1986) and known as the "thermal procedure that is used for substances with the least feasible health dangers arising from corruption microorganisms related to dairy products."

Pasteurization is also applied for most liquid and special foods such as fruit and vegetable juices, beer, wine, liquid eggs, soft drinks, syrups, soups, and ready-to-eat food. Cured and uncured meat products are often pasteurized at temperatures around 80°C for several minutes, which results in a limited shelf-life. However, they need to be kept in a refrigerator. Tomato and tomato paste are common food products that can be preserved by pasteurization (Deak, 2013).

It should be noted that in the pasteurization process, physicochemical and sensory alterations in the materials are at their least. Because pasteurization is a mild process, low acid materials and acidic products (pH > 4.5) are more suitable for pasteurization. In fact, the temperature which was used for these products is under 100°C. However, in acidic products, the major problem is the deactivation of yeasts, molds, and heat-resistant enzymes. As a general rule, the durability of dairy products in a refrigerator is between 7 and 10 days (Holdsworth, 2001).

Conduction of pasteurization and observing aseptic conditions increase the shelf life of products. However, the important issue in this subject is reduction or inhibition of the postpasteurization contamination (PPC). Therefore, the combination of the thermal process along with rapid cooling and storage at refrigeration temperatures is the favorable condition, as well as protection, maintenance, and public aspects of cleanliness that should be observed about materials and equipment (Martin, Boor, & Wiedmann, 2018). Pasteurization, in addition to protection of the product, is an integral part of the product processing and operation. For example, in producing cheese and dried milk powders, pasteurization can produce a salmonella-free product, which was regarded as a control point (CP) by Mettler (1994). Based on other research

Thermal Processing of Food Products by Steam and Hot Water. https://doi.org/10.1016/B978-0-12-818616-9.00009-2

studies (Cowell & Mermelstein, 1995; Thorne, 1986; Wilbey, 1993), the credit of this gentle process for deactivation of microorganisms such as bacteria, molds, and yeasts and increasing the shelf life of products has been given to Pasteur, who studied materials such as beer, wine, and vinegar. In fact, attention to this point that milk can spoil with pathogenic microorganisms and lead to illnesses such as tuberculosis has developed the mild temperature-long time process (LTLT pasteurization).

This process was first installed by Charles North in 1907 in New York (Wilbey, 1993). Satin has reviewed the early history of pasteurization (Satin, 2020). The idea of pasteurizing milk (as opposed to boiling it or drinking it raw) met with some fierce resistance when it was first mooted. Table 9.1 describes powerful reasons against the pasteurization process. The main objection is that pasteurization reduces the nutrition value and changes organoleptic properties of milk; therefore, some consumers prefer

Table 9.1 Reasons against pasteurization.

Object	Reasons
Hygiene procedures	➢ Pasteurization can be applied to cover milk with weak quality ➢ This thermal treatment can eliminate several of microorganisms in milk and hide the proof of insanitation ➢ Pasteurization encourages negligence and hinders the attempts to fabricate safe milk
Physical and bacteriological quality	➢ Pasteurization can affect on the combination of milk ➢ Pasteurization may eradicate the useful lactic acid bacteria as well as enzymes, antibodies, and hormones in milk ➢ Pasteurization may provoke the growth of bacteria in milk
Financial affairs	➢ Pasteurization legislates the right of selling low-quality milk ➢ Pasteurization is not needful in cases where the milk produced reaches the consumer immediately ➢ Pasteurization will enhance the price of milk because dealers have to purchase required equipment ➢ Maybe there are some consumers who request raw milk
Nutritional value	➢ Pasteurization can have an impact on the flavor of milk (enzymatic or nonenzymatic browning) ➢ Pasteurization considerably reduces the nutrition value of milk ➢ Raw milk has good effects on growth of infants, children, and patients
Safety and security	➢ Pasteurization cannot annihilate microorganism toxins in milk ➢ Pasteurization, by destroying tuberculosis of bovine origin in early life, would cause enhancement in pulmonary tuberculosis in adult life (raw milk does not expand tuberculosis) ➢Pasteurization leads to a wrong sense of security ➢Pasteurization would enhance the infant mortality

to use raw milk. It may also cause eradication of the useful lactic acid bacteria as well as enzymes, antibodies, and hormones in milk. Some opponents declared that the pasteurization technology can cover the weak quality of milk. It eliminates several microorganisms and hides the proof of insanitation.

In general, pasteurization as a thermal treatment in order to deactivate bacteria, molds, and yeasts has been one of the most widely applied processes of food maintenance for more than a century and has contributed considerably to the nutritional well-being of much of the world's population. However, pasteurization does not destroy all microorganisms and heat-resistant spore-forming bacteria. Hence, pasteurized products are not stable and must be protected by refrigeration or modified-atmosphere packaging to reduce the growth of microorganisms which stay viable in the product and will finally lead to spoilage. Depending on the kind of materials, the durability of pasteurized products could be different from several days (milk) to several weeks or more (fruit juices). Also, because of the low heat in pasteurization, the sensory quality and nutritive value of the products are often constant (Satin, 2020). The intensity of the applied thermal processing (time and temperature) and shelf life of products depends on the nature of materials, pH, heat resistance of the target bacteria, quality sensitivity of materials to heat damage, and the procedure of heating (Peng et al., 2017).

9.2 Basic pasteurization techniques for food

Conventional pasteurization processes usually employ continuous heat-transfer mechanisms to enhance the shelf life of food products. The low order of heat treatment, generally at a temperature below the boiling point of water, is used to inactivate all nonspore-forming pathogenic bacteria and the majority of spoilage microorganisms. Two types of processes can be applied in basic pasteurization: slow and rapid. Slow pasteurization uses pasteurization temperatures for several minutes, whereas rapid or flash pasteurization uses pasteurization at higher temperatures just for a few seconds (Deak, 2013).

9.2.1 Low temperature long time

High temperatures in these processes can destroy or decrease nutritional values of food products such as proteins or polysaccharides and cause considerable alteration in organoleptic properties (Patil, Ananthan, Nanavati, Nataraj, & Prasad, 2019). The use of low temperatures, along with receiving enough microorganism reduction, would be a suitable process which leads to a decrease in energy and cost. Therefore, low temperature long time (LTLT) pasteurization or holder pasteurization has been created. The LTLT process would decrease costs and energy input as well as maintain the quality of food products with an acceptable reduction in microbial count. The only problem with this process is the short shelf life of food products because of the use of low temperatures. For example, researchers evaluated the effect of different methods of pasteurization on bactericidal action of human milk (Patil et al., 2019). Human milk can prepare enough nutrients, immune and growth factors, digestive enzymes, and bioactive compounds for growth and development of the body, especially in infants (Lönnerdal, 2013; Martin, Ling, & Blackburn, 2016). Also, milk lactoferrin shows great antimicrobial properties against microorganisms (Van'T Land, Boehm, & Garssen, 2010). Therefore, LTLT pasteurization of human milk deactivates microorganisms and preserves the milk nutrition value (Daniels et al., 2017). The optimum temperature for LTLT pasteurization of milk is 62.5°C for 30min. For human milk, LTLT pasteurization is more common than high temperature short time (HTST) pasteurization or the flash method that needs up-to-date instruments and a large amount of milk (Giribaldi, Arslanoglu, Tonetto, & Moro, 2011). With all the advantages of LTLT pasteurization, because of some problems such as mild and slow heating and cooling, its suitability for lower volumes, the fact that it has not been prepared for automation compared to HTST, the lower shelf life of food products, and so on, HTST pasteurization also became common.

9.2.2 High temperature short time

Pasteurization uses thermal treatment for food compounds via subjecting them to a low temperature for a short time; the higher the temperature, the shorter the time needed. The goal of this procedure is to decline a number of bacteria, molds, and yeasts or other microorganism that can cause disease; therefore it leads to a reduced health hazards to people. Various pasteurization processes are admitted via public health agencies; however, the HTST procedure is the industrial standard method because of its plain adaptation to processing (Jo, Benoist, Barbano, & Drake, 2018; Ranieri, Huck, Sonnen, Barbano, & Boor, 2009). It should be noted that the combination of time and temperature which will adequately decline a number of bacteria alters based on the nature and conditions of the food compounds such as viscosity, fat, solid contents, and so on. Jo et al. (2018) evaluated the flavor of milks processed by HTST (Jo et al., 2018). First, they announced that the taste and odor of milk and dairy crops are essential factors which have an effect on people's acceptance and the durability of products. For example, in raw milk, flavor is affected by the quality of milk, microbial counts, and heat-resistant enzymes (Cadwallader, 2020). By using thermal processing, microbial growth and enzymatic activities can be declined or stopped in order to warrant safety and security of milk as well as enhance the shelf life of it. It should be noted that under these conditions, the taste and odor of milk would be different from those of raw milk (Broersen, 2020). In the United States, HTST processing is conducted as a typical milk processing (72°C for 15s) and increases the durability of products for about 3weeks (Andriani, Mashito, & Handayani, 2021). HTST processing causes different reactions such as Maillard, lipid degradation, and thermal denaturation of whey protein and other proteins in the milk during preservation and storage of products, which can affect their color (Kokkinidou & Peterson, 2014; Miyaji et al., 2021). In the Maillard process, the reaction between sugar (lactose) and protein (amino group) leads to produce a broad range of flavored ingredients including aldehydes, sulfur and nitrogen substances, maltol, and di-acetyl (Munekata et al., 2020). Also, fat decomposition causes β-oxidation of free fatty acids and creates methyl ketones. β-Lactoglobulin protein generates volatile sulfur substances, which involve unfolding and exposure of reactive sulfides; also, these volatile compounds are the sources of the aroma portion of food (Guo, He, Wu, Zeng, & Chen, 2020; Zhang, Kang, Zhang, & Lorenzo, 2021).

9.3 Ultra-high-temperature pasteurization

Ultra-high-temperature (UHT) pasteurization includes thermal processing of compounds such as milk, cream, liquid whole egg, and so on at 135–154°C for 3 s. These products can be packed in sterile, hermetically sealed packages (aseptic packaging) and can be maintained without refrigeration for at least 6 months (USFDA, 2018). Aseptic packaging involves the use of sterile compounds in a sterile package under sterile conditions. Durability of these compounds is not confined via microorganisms' spoilage. It can be confined by precipitation of particulate substances and also the existence of off-flavors (cooked, bitter, sweet, and stale flavor), which are related to free sulfhydryls, aldehydes, and ketones (Lee, Barbano, & Drake, 2017). In the UHT method, by applying high temperature, not only the product gets sterile quickly but also the incoming fluid is heated. This process is economically viable in both heating and cooling costs and saves energy and increases the energy regeneration efficiency to >90%. The UHT method can be classified in three steps, including a heating part, a holding part, and a cooling part (Fig. 9.1). Because of the use of liquid compounds in this method, the food products with significant contents of fat will require to be homogenized between the heating and cooling sections. The optimum temperature

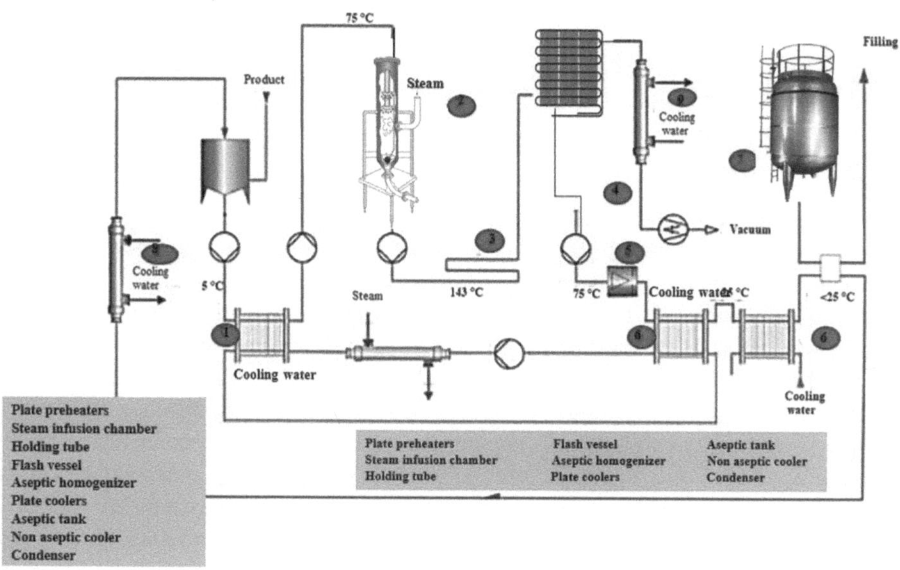

Fig. 9.1 Ultra-high-temperature (UHT) pasteurization process.

for fat homogenization is between 70 and 80°C, which confirms the fat changed state and improves the size decline procedure. For example, in UHT milk, the fat globules should be decreased from their normal size range of 1–10 μm (Liang, Qi, Wang, Jin, & Mcclements, 2017). Liu et al. evaluated the impact of UHT pasteurization on the properties of liquid whole eggs (Liu et al., 2020). UHT pasteurization is frequently applied to sterilize liquid whole eggs, especially in China. Liu et al. (2020) announced that the emulsifying characteristics of sterilized liquid whole eggs are low and decreased during the storage time. However, using HUT pasteurization can enhance the emulsification stability index to 21.9%. Also, the foaming ability of UHT pasteurized eggs increases to 33.38% compared to 15.48% for unsterilized liquid eggs. It should be noted that other properties of liquid eggs such as viscosity, gel strength, and soluble protein content are better than those of unsterilized liquid eggs, and they improve during the storage time.

9.4 Novel pasteurization techniques

As mentioned above, the classical pasteurization process not only eliminates pathogenic bacteria in food products but also can cause negative effects on sensory, nutritional, and physiochemical properties of compounds (Salinas-Roca, Elez-Martínez, Welti-Chanes, & Martín-Belloso, 2017). These days, the quality and safety of food products with minimal processing are important for consumers. Therefore, changing people's attitude toward healthy food consumption can be a motivating factor to develop novel thermal and nonthermal processes as alternatives to conventional heat treatment (Sattar et al., 2020). The new pasteurization processes can be efficient in destruction or deactivation of foodborne microorganisms from a wide range of foods, including from solid to liquid (Gomez-Gomez, Brito-De La Fuente, Gallegos, Garcia-Perez, & Benedito, 2020). There are several new methods for food pasteurization, such as microwave, pulsed electric field, irradiation, Ohmic heating, and solar energy.

9.4.1 Microwave

Microwaves are specified as an electromagnetic spectrum in the combination of infrared and radio frequencies, with a wavelength of 1 mm to 1 m and functioning at frequencies of 300 MHz to 3 GHz (Sierra-Lopera, Ciro-Velásquez, Torres-Trujillo, & Montoya-Campuzano, 2017). Microwave pasteurization can be applied as an alternative method for conventional thermal processing in liquid and viscous compounds. Indeed, this technique has an efficient role in enhancing the color, taste, and texture and maintaining nutritional values and consequently enhances the speed of the come-up-time stage of pasteurization, in comparison to conventional pasteurization techniques (Bagade & Patil, 2021). Several researchers conducted studies about the high effect of microwave deactivation against microorganisms and enzymes (Kubo et al., 2020; Soni, Smith, Thompson, & Brightwell, 2020). In the microwave technique, electromagnetic energy inactivated bacteria, molds, yeasts, and so on because of dielectric disruption.

There are two mechanisms for deactivation of these organisms. The first one is the thermal effect via heat produced by microwaves interacting with water and food products, and the second one is related to the nonthermal deactivation with microwaves such as electroporation, cell membrane rupture, and magnetic field coupling. Cavalcante et al. stated that nonthermal effects of microwaves on the organisms would be more efficient compared to the surrounding medium and hence can annihilate microbial cells significantly (Cavalcante, Dos Santos Funcia, & Gut, 2021). The main reason in this process is electroporation, which creates membrane disruption and consequently leakage of materials from the membrane (Cavalcante et al., 2021).

9.4.2 Pulsed electric field

Pulsed electric field (PEF) is one of the nonthermal pasteurization methods that have been taken into large consideration. This process has a high speed and can produce compounds with minimal processes, minimal bacterial contamination, and maintenance of freshness. PEF systems are composed of a pulse generator, treatment chambers, a fluid-handing system, and monitoring systems (Yildiz, Pokhrel, Unluturk, & Barbosa-Cánovas, 2021). PEF pasteurization treatment chambers play an important role because there are houses for electrodes and transfer a high voltage to food products (Salehi, 2020). Accordingly, the scheme of the treatment chamber is one of the fundamental agent expansions of the PEF treatment as a nonthermal processing (Gómez et al., 2019). Indeed, treatment chambers can apply a constant electric field to compounds with the least increment of the temperature, and the electrodes should be designed to reduce the impact of electrolysis (Maniglia, Castanha, Rojas, & Augusto, 2021). Based on the purpose, various PEF treatment chambers for microbial deactivation of liquid foods are designed. The PEF process may be either static or continuous. In the static processing, separate sections of fluid food products are treated as a unit by placing the entire fluid to a PEF treatment chamber. This procedure exerts monotonous field strength to all elements of food products to be processed. In the continuous method, the processed food products are flowing into and given up from the PEF treatment chamber in a stable stream by a pump (Sridhar, Ponnuchamy, Kumar, & Kapoor, 2021), and the creation of the treatment chamber is a step-by-step conduct from static treatment chambers to continuous treatment chambers (Sridhar et al., 2021).

The popular static treatment chambers are U-shaped static treatment chambers, parallel plate static treatment chambers, disk-shaped static treatment chambers, wire-cylinder static treatment chambers, rod-rod static treatment chambers, and sealed static treatment chambers. This trend for continuous treatment chambers is as follows: cooling system and coaxial treatment chambers (Huang & Wang, 2009). Generally, static treatment chambers are mostly proper for laboratory applications, and for commercial actions, continuous treatment chambers are more effective. For the well-organized treatment, a several number of the important design factors must be remarked, including electric-field strength, treatment time, treatment temperature, electric field distribution, and so on. Despite the well-equipped and fully designed treatment chamber, there are still shortcomings in the PEF. For instance, the PEF process is not suitable for any type of products and are mainly used for liquid foods. Also, the PEF process showed antimicrobial properties against *Escherichia coli* and *Saccharomyces cerevisiae*. Moreover, the high expense for PEF equipment and the small market for this process provided PEF technology only a limited success (Arshad et al., 2020).

9.4.3 Irradiation

Manufacture of safe food has a premier significance to maintain the health of the community. Regardless of continuous attempts, foodborne disease stays a critical challenge. In most countries, the increase in the number of foodborne diseases is related to consumption of fresh products, nuts, and dry foodstuffs. The most disease outbreaks are connected to *E. coli* and *Listeria* spp., which are seen in fresh-cut spinach, lettuce, and cantaloupe. Therefore, irradiation is getting a great deal of notice as a potential nonthermal technology for reducing the microbial count in order to warrant the safety of fresh fruits and vegetables (Zhang, Ha, Seck, & Zhou, 2020). Thermal pasteurization is an effective process for liquid foods such as fruit juices and milk, but this method is not suitable for solid foods, especially seafood, poultry, and meat. Irradiation as a cold pasteurization method is a more efficient process without significant changes in quality and taste of products (Thayer, 1996). The electromagnetic wavelengths are different from 102 m (AM radio signals) to 10–10 m (X-rays) and based on the wavelength electromagnetic waves will yield various impacts. Radiation is determined as a procedure in which electromagnetic waves or particles pass through a medium. Ionizing radiation includes both direct (electrons, protons, alpha particles, heavy ions) and indirect radiation (X-ray, gamma rays) (Podgoršak, 2006).

Radiation pasteurization of food products can be conducted with different rays such as α-, β-, γ-, X-rays and electrons that can effectively inactivate foodborne microorganisms in beef, pork, lamb, and fish (Fig. 9.2). In this process, demolition of pathogenic non-spore-forming bacteria and parasitic-like trichina takes place. Radiation pasteurization can protect people against diseases such as salmonellosis; hemorrhagic diarrhea, which is caused by *E. coli* O157:H7; and gastroenteritis by *Vibrio vulnificus*. Also, irradiation can increase the durability of fruits and vegetables and can destruct insects and mites in a more environmentally friendly manner. In addition, long-time animal nourishment research studies have exhibited that radiation-pasteurized or radiation-sterilized foods are safe and nutritious for humans (Hassanpour, Rezaie, & Baghizadeh, 2019; Lima, Vieira, Santos, & De Souza, 2018). Radiation pasteurization has been approved by the US Food and Drug Administration (FDA),

Fig. 9.2 Different irradiation for preserving food products.

the US Department of Agriculture (USDA), the US Army, the World Health Organization (WHO), the Codex Alimentarius Commission, the American Medical Association, the American Dietetic Association, the Institute of Food Technologists, and the health authorities of approximately 40 countries (Delorme et al., 2020). Irradiated food does not expose radioactive materials, and used rays (gamma, X-rays, or electrons) cannot make them radioactive. In fact, appropriate exposure to radiation can be applied to destroy quickly growing cells such as those in cancers. These quickly growing cells of insects or pathogenic bacteria are killed when food is irradiated but do not have effects on the food structure because the cells are not multiplying (Choudhary & Bandla, 2012). Moreover, irradiation has little impact on some vitamins such as B1 in pork and C in fruits, but based on the results, it has been computed that even if all of the pork in the United States were to be irradiated, only 2.3% of vitamin B1 in the diet of Americans would be lost and a small quantity of vitamin C in fruits is changed to another equally usable form of the vitamin (Barkaoui et al., 2020; Dionísio, Gomes, & Oetterer, 2009).

9.4.4 Ohmic heating

Ordinary heating procedures are essentially composed of heat-transfer systems of conduction, convection, and radiation in both steady and unsteady states (Goullieux & Pain, 2014; Sun, 2014). The interior resistance by conduction causes heterogeneous treatment and the remarkable loss of compound quality. To address this issue, further processing employing electrical energy directly in the food technology has obtained interest in recent decades. Ohmic heating is an electro-heating process based on the transmission of electrical current through foodstuffs having electrical resistance (Marra, Zell, Lyng, Morgan, & Cronin, 2009). In this process, heat is produced right away into the foodstuffs; its value is straightly related to the voltage gradient and the electrical conductivity, and the constant heat production causes uniform temperature distribution (Varghese, Pandey, Radhakrishna, & Bawa, 2014). The explicit benefit of the Ohmic process toward the ordinary process is the absence of high-wall temperatures and restricting heat-transfer coefficient commitments. Another superiority of the Ohmic process is keeping the color and nutritional value of compounds, a short processing time, and a higher efficiency (Darvishi, Saba, Behroozi-Khazaei, & Nourbakhsh, 2020; Kumar, 2018). The Ohmic process is applied in a different range of utilizations including preheating, blanching, pasteurization, sterilization, and extraction of foodstuffs (Leizerson & Shimoni, 2005; Shao, Zhao, Zou, Li, & Dai, 2021). With regard to the approval of USDA and FDA for the Ohmic process in the pumpable food products, it is currently being applied commercially all over the world (USA, Japan, UK, and other various European countries) to pasteurize pumpable food products (syrups such as whole fruits, fruit juices, egg, milk, etc.) (Icier & Bozkurt, 2011). The Ohmic process can be applied as an ongoing in-line heater for cooking and sterilization of viscous and liquid food compounds. Hence, the process control is dependent on electrical system parameters, flow properties in the pumping system, food properties, and other design criteria. Also, in industrial-scale production, other parameters such as the voltage gradient, frequency, temperature range, electrical conductivity differences, concentration and type of food, electrode type used, and flow properties as significant design factors of the liquid foodstuffs are considered. Cho et al. evaluated pasteurization of fermented red pepper paste by Ohmic heating (Cho, Yi, & Chung, 2016). They worked on Gochujang, a Korean fermented food with red pepper paste. This product has a high viscosity, making the application of a plate heat exchanger difficult. In addition, the low heat-conduction coefficient of it increases the time of thermal processing in the sterilization procedure. Consequently, overheating of the heating surface causes reduction in food quality along with development of off-flavors and discoloring reactions (Ryu et al., 2021). They stated that HTST and UHT sterilization are not good approaches for these high viscous products and do not show a high efficiency in declining microbial counts (Cho & Song, 2021). They suggested Ohmic technology for this pasteurization. Ohmic heating has no restrictions about the penetration depth as long as the inherent electrical resistance of the products is not too high (Cho et al., 2016).

9.4.5 Solar energy

Based on some research studies, waterborne illnesses have an influence on the large population of people mostly in developing countries (Panchal et al., 2018). Pichel et al. announced almost 2.5 billion cases of disease and about 5 million deaths each year, which can be related to consumption of contaminated water (Pichel, Vivar, & Fuentes, 2019). Louis Pasteur (1822–1895), by introducing pasteurization theory, stated that heating water to an enough degree of temperature for a specified time period annihilates pathogen microorganisms (table 11.2) (Hartzell, 2019; Rossi et al., 2019). Generally, it is considered that solar pasteurization to 65°C will make it secure and safe to drink. The time needed to pasteurize reduces exponentially with temperature (Manfrida, Petela, & Rossi, 2017). Indeed, a relevant but different water treatment concept is solar pasteurization, where microorganisms are eliminated with straight action of determined wavelengths of the solar spectrum, independent of temperature (Adler, Gavilan, McKay, & Campos, 2018).

Solar pasteurization as a simple, renewable, and cheap method for pasteurization of drinking water in distant regions as well as cooking is used. This pasteurization is dependent on solar energy to warm the water. Regarding this, consumption of healthy water relies on the destruction of its microorganisms by heat, but it should be noted that pasteurization temperatures do not eliminate all kinds of microorganisms in the water. In fact, in solar pasteurization, the source of energy is the sun, which is cheap in comparison to the milk pasteurization method. Applying solar pasteurization is highlighted in distant, high-altitude sunny regions, where power is not accessible. In solar pasteurization, the ultraviolet spectrum in the range of 200–400 nm eliminates microorganisms efficiently (Adler et al., 2018; Panchal, Patel, & Sathyamurthy, 2021). As a result, it is favorable to use clear containers which have great transfer in this wavelength. Solar pasteurization has a lot of privileges, for example, environmental protection, warrant of superior human health in distant regions, addressing the difficulty of energy and decline expense, a new utilization of solar energy applications, and the low cost of such energy which only needs essential equipment to establish it (Panchal, Patel, & Parmar, 2020) (Table 9.2).

Table 9.2 Evaluation of solar pasteurization systems.

Sample	Results	References
Milk pasteurization	They prepared the innovative design of a solar pasteurizer made by flat-plate collectors and tested for 3 months, September, October, and November 2009. They found that the maximum and minimum pasteurized amounts are 73.9 and 37.3 L at 63 and 72°C, respectively	Atia, Mostafa, Abdel-Salam, and El-Nono (2010)
Milk pasteurization	They prepared a solar milk pasteurizer consisting of a flat-plate collector and stainless-steel vat of an 80 L capacity. They found that the low-cost solar milk pasteurizer is used for arid and remote areas where electricity or gas are not available	Wayua, Okoth, and Wangoh (2013)
Milk pasteurization	Compared the existing layout of the pasteurization system with a modified plant and obtained improvement in energy payback time	Modi and Prajapat (2014)
Milk pasteurization	They prepared a solar pasteurization system consisting of the tank and evacuated tubes. They found that 72°C is the best temperature for the killing of bacteria and the purpose of pasteurization	Dobrowsky, Carstens, De Villiers, Cloete, and Khan (2015)
Milk pasteurization	They have carried out the study on a solar pasteurization system and found the energy payback time and saving of electricity	Pandagale, Madhuri, and Khambaklar (2016)
Disinfection of water	The system uses commercial solar thermal collectors arranged in a self-compensating natural circulation system using no electricity and sized for an adequate residence time for disinfection by pasteurization. This system has the ability to produce typically 40–80 kg/day of the collector surface depending on the location, representing a high-tech alternative for the purpose of water purification	Manfrida et al. (2017)
Disinfection of water	The chemical and microbial quality of water produced by two solar pasteurization systems were assessed in order to determine whether these decentralized systems may be	Reyneke, Cloete, Khan, and Khan (2018)

Continued

Table 9.2 Continued

Sample	Results	References
	viable solutions in providing inhabitants of periurban informal settlements with an alternative water source. The results for the chemical analyses indicated that the roofing material utilized for the catchment system could influence the chemical quality of tank water samples. A reduction in the level of the chemical contaminants detected may then be achieved by combining the current solar pasteurization technology with filtration systems	
	In this work, a complete life cycle assessment and exergo-environmental analysis are performed in order to calculate and compare the eco-profiles of two solar pasteurization technologies: a natural circulation and a thermostatic valve system. The results show that natural circulation systems are generally more environmentally sustainable (0.30 mPt/L) than the thermostatic valve system (0.83 mPt/L) thanks to the higher productivity of treated water	Rossi et al. (2019)
Waterborne pathogens	Solar pasteurization is a cost-effective and environmentally sustainable technique. Moreover, typical UV disinfection doses are also designed to disinfect drinking water by inactivating the complete range of pathogens (mostly bacteria and viruses). However, the DNA repairing photoreactivation mechanism reduces the overall disinfection capacity	Rajkhowa (2020)
	Using various coatings' materials and components in the manufacturing of solar pasteurization systems with emphasis on increasing the efficiency. The results reveal that use of PV/T support and certain design modifications leads to the reduction in the total pasteurization time of raw milk	Taj et al. (2021)

9.5 Factors affecting pasteurization

Stability of quality is one of the most considerable factors which have an effect on food product commercial quality. With regard to that, in the pasteurization process, only vegetative spores are inactivated; hence, maintaining the quality of food can be affected by several factors. Some of these factors are time/temperature conditions, declining post-pasteurization processing, and storage temperatures. In fact, maintaining quality can be increased by understanding and evaluating the pasteurization procedure (Colonna, Durham, & Meunier-Goddik, 2011).

9.5.1 Time and temperature

Many societies consider standard principles for product pasteurization such as thermal processing as essential. Therefore, attention to the minimum temperature and time required for food products is very important. Pasteurization is a thermal moderate process for preserving food products. The time and temperature which are applied in the pasteurization process can destroy and decrease vegetative pathogens and spoilage bacteria. It should be noted that spores of microorganisms can survive in the thermal pasteurization process; hence, their growth should be restricted during storage. The time and temperature needed to eliminate thermoduric bacteria and bacteria spores are 63–64°C for 30 min and 80°C for 10 min, respectively. These kinds of microorganisms are great concerns in the pasteurization process because they are able to survive and grow at refrigerated temperatures, mostly *Bacillus cereus*. Heat-resistant enzymes are another considerable factor as well. The enzymes such as proteases and lipases are resistant to heat or are not totally inactivated (Escuder-Vieco et al., 2018; Ranieri et al., 2009).

9.5.2 Process characterization

Evaluation of pasteurization methods is dependent on pasteurization units (PUs) and "P." Based on the results, PU is defined as a 60°C (140°F) for 1 min. Also, with regard to that, the standard *Z* value is equal to 10°C (18°F); thus, the number of PUs is obtained by

$$\mathrm{PU} = 10^{(\mathrm{T}-60)/10)} \times t\,(\,\mathrm{min}\,).$$

As shown, a temperature of 63°C for 30 min would have a value of nearly 60 (Ağçam, Akyıldız, & Dündar, 2018), whereas for HTST treatment (72°C/15 s), it would give only 3.96, which is related to the large *Z*-value. Rachon, Rice, Pawlowsky, and Raleigh (2018) announced that the *Z*-value for beer spoilage microorganisms was 7°C, and based on the mentioned equation, the PU for this process is 5.6. The second factor is "P," which was defined by Kessler (1981). This is based on a reference temperature of 72°C and a *Z*-value of 8°C (or Kelvin). Generally, treatment conditions of 72°C for 15 s are defined as providing safe pasteurization procedures for milk and are given an arbitrary "P" value of 1 (Rachon et al., 2018).

9.5.3 Postpasteurization contamination

The postpasteurization process can be used as an effective factor to assess the level of microorganisms that remain alive after pasteurization. Thus, it should be one of the principal goals in all pasteurization procedures to decline the quantity of post-pasteurization contamination. Indeed, most of the contaminations in the pasteurization process are related to the postpasteurization contamination, especially the end of the holding tube. Martin et al. (2018) reported that postpasteurization contamination has an important role in maintaining quality of food products, chiefly milk and cream. Based on their results, a remarkable decrease in food quality over a long period, with attention to postpasteurization contamination, can improve significantly (Martin et al., 2018). Prasantha and Wimalasiri (2019) conducted some research about HTST milk pasteurization. They announced that eliminating postpasteurization contamination can improve the quality of milk for 13–17 days at 10°C and 25–40 days at 4°C (Prasantha & Wimalasiri, 2019). Generally, evaluating postpasteurization contamination is dependent on great attention being paid to scouring, disinfecting, and sterilizing.

9.5.4 Storage conditions

Under storage conditions, storage temperature has a significant role in the food chain. The storage conditions of the product after the product leaves the store and reaches the consumer's refrigerator are considered; for example, the storage temperature in the transport vehicle should be considered as an important factor in the quality of the product. In fact, consumers have a considerable role to ensure and maintain the safety of heat-treated products (Martini et al., 2018). Gogliettino et al. (2020) stated that decreasing the storage temperature will enhance the maintaining quality. For instance, increasing the temperature to 3°C decreases the shelf life by 50% and an increase to 10°C will decline the maintaining quality by more than half (Gogliettino et al., 2020). Bezie (2019) conducted some research about the effect of storage temperature on the milk storage time. They stated that at 8°C, the shelf life of milk ranged between 6 and 11 days, whereas at 5°C, it ranged between 10 and 16 days. Another factor is the storage temperature that affects the microbial flora of food (Bezie, 2019). The prevailing flora in cream without postpasteurization treatment are *B. cereus* types. The critical temperature for growth of *Bacillus* spp. is 6°C. Using temperatures below 5°C, it would be sufficient to eliminate significant growth of *B. cereus* (Bezie, 2019).

9.6 Influence of pasteurization on food quality properties

Heat treatment during the process of food products can affect food qualities such as color, texture, and flavor. In the following, there are brief explanations about these parameters.

9.6.1 Texture

The texture of the food product as well as fruits and vegetables is frequently affected by thermal processes such as pasteurization or blanching. In many food treatments such as cooking, nursing homes, hospitals, and restaurants, it is suggested to use enough heating under thermal pasteurization conditions for food products. Moreover, blanching is applied at a temperature more than 80°C to deactivate enzymes, which leads to a decrease in product quality during processes and storage. Therefore, using a high temperature during the process results in the deterioration of the tissue structure, which causes an unacceptable food texture (Peng et al., 2017). Among food products, fruits and especially vegetables mostly need to maintain their structure and texture even if exposed to the thermal treatments for a long time. In fact, texture is a significant qualitative property which is evaluated by consumers, and the deterrence or restriction of such degradation is important to ameliorate the food value. Imaizumi et al. reported that to prevent or decrease tissue degradation and texture deterioration, moderate heat treatments are used (Imaizumi, Tanaka, & Uchino, 2019). For instance, 50–80°C is applied for some kinds of vegetables such as potato, sweet potato, pumpkin, and carrot (Chhe, Imaizumi, Tanaka, & Uchino, 2018; Gallego-Castillo & Ayala-Aponte, 2018; Gong et al., 2019; Jansrimanee & Lertworasirikul, 2017). At these temperatures, the pectin methyl esterase enzyme causes the demethylation of pectin molecules. Regarding this, the role of pectin molecules is to make connection between cells, and demethylated pectin molecules are not comfortably broken down by β-elimination and lead to a crosslinking by divalent ions such as Ca^{2+}, hence contributing to strengthening tissue structures (Imaizumi et al., 2017). In conclusion, tissue decomposition of vegetables is prevented by moderate thermal processing during cooking.

9.6.2 Color

Fruits and fruit juices are often affected with heat treatments in pasteurization for a long time in order to develop microbial deactivation. Indeed, the main objective of this process is increasing the shelf life of products, but the heat process during pasteurization can affect the product quality and (color and flavor) cause the production of unfavorable sensory properties and nutritional losses. Color is one of the most considerable characteristics of food products and drinks. It is the initial factor which consumers focus on in stores and encourages or discourages consumers from buying the products (Jacucci, Vignolini, & Schertel, 2020). Color factors in red fruit juices such as red orange, grape, berry, and pomegranate juices are important (Becerril, Nerín, & Silva, 2021). In these beverages, anthocyanin pigments exist, which are unfortunately sensitive to heat treatments and decompose to a brown color. In other words, alteration of color in red beverages is related to the decomposition of anthocyanins and the production of brown pigments (Lamce, Kycyk, Gozhdari, Zejnelhoxha, & Kukali, 2020). These color alterations intensely affect consumer behavior and lead to a loss of marketability of processed products.

9.6.3 Flavor

The flavor of food products is an important factor which influences the consumer's total acceptance, durability, and other properties of the products (Guo et al., 2020). For example, in dairy products such as milk, the aroma and taste are affected by raw milk quality at first (microbial counts and heat-resistant enzymes) (Júnior et al., 2019). Thermal processing can decrease or eliminate microbial growth and enzyme activity to improve safety and enhance shelf life, but during these processes, the aroma and taste may alter and differ from that of raw milk (Zhu & Guo, 2021). The aroma and taste of pasteurized milk are caused by different reactions during thermal processing or storage, which can also affect the color (Jansson et al., 2019). Significant alterations in aroma during thermal processing of milk are dependent to Maillard reactions (Kokkinidou & Peterson, 2014; Troise, Colantuono, & Fiore, 2020; Zhu, Poojary, Andersen, & Lund, 2020), lipid decomposition (Tiwari & Asgar, 2017), and thermal denaturation of proteins (Mehta, 2015). In fact, lactose and amino acids participate in Maillard reactions during heat treatment, and this procedure produces a various range of aroma compounds, including Strecker aldehydes, sulfur- and nitrogen-containing compounds, maltol, and diacetyl (Huan, Zhou, Zhao, Xu, & Peng, 2005; Su & Wiley, 1998; Tiwari & Asgar, 2017).

In addition, lipid decomposition during thermal processing of milk generates β-oxidation of free fatty acids creating methyl ketones (Tiwari & Asgar, 2017). Also, β-lactoglobulin in serum protein releases volatile sulfur compounds with heat denaturation, which contains folding and exposure of reactive sulfides (Kühn, Considine, & Singh, 2006; Tiwari & Asgar, 2017). The aroma and taste of ultrapasteurization milk are different from those of HTST milk, and they are related to the specific sulfurous, cooked, cabbage, and caramelized compound (Lee et al., 2017; Zhu et al., 2020). These different aroma are the disadvantages of ultrapasteurization milk and lead to reduce the consumer acceptance (Panchal et al., 2020; Zhang, Jin, Zhang, & Yu, 2020). Lee et al. reported that most of the people including adults and children like the flavor of HTST milk more than direct steam injection-ultrapasteurization or indirect heat-ultrapasteurization milks (Lee et al., 2017).

9.7 Pasteurization of different food products

9.7.1 Milk and milk-based drinks

Most studies about pasteurization are related to the dairy products, especially milk. For example, comprehensive International Dairy Federation (IDF) reviewed this process in 1986 (IDF, 1986). In fact, pasteurized milk as a favorable type of milk under thermal processing in different societies is largely consumed. Raw milk quality is an important factor, and using suitable thermal processing leads to a reduced microbial growth. For example, in the United Kingdom, the bacteria count in raw milk was 104–105 cfu/mL. In addition, by using high hygienic procedures in the field, these numbers just decrease one unit (103 cfu/mL). It should be noted that keeping milk in the refrigerator causes the increase of Gram-negative psychotropic bacteria in the milk and turns them to prevailing flora; moreover, when the microbial count in milk is more than 106cfu/mL, heat-resistant enzymes also create a problem (Cheng, Barbano, & Drake, 2019b). Generally, all these microorganisms are sensitive to heat and quickly eliminated by the pasteurization process, and their existence in freshly pasteurized milk represent postprocessing contamination. Rysstad and Kolstad stated that the best quality of raw milk would have a low quantity of psychotropic and thermoduric bacteria (Rysstad & Kolstad, 2006) and there would be no *Bacillus cereus* in it. In the process of pasteurization, there are some key important factors. With regard to that, pasteurization is moderate thermal processing which denaturized 5%–15% of whey protein, but this is not enough to enhance significantly the contents of free sulfhydryl group activity or to induce production of hydrogen sulfide and creation of any cooked flavor (Zulueta, Barba, Esteve, & Frígola, 2013).

Moreover, the level of whey protein denaturation is higher in skim milk concentrates by the ultrafiltration method, which enhances with the increment in the concentration factor (Bonczar, Walczycka, & Duda, 2016). The tendency to use milk-based drinks is getting enhanced. In these drinks, milk is the basic component, generally 80%–90% of the volume; the milk may be either fresh or reconstituted. Among milk-based drinks, flavored milks are becoming more popular, for example, strawberry, banana, and chocolate, likely being the most popular. In addition, in these compounds, the total solids content is more than in that in normal milk, which leads to a decreased water activity; however, the pH is not much changed. Heat treatment procedures are similar to those for milk. It should be noted that additional materials have negative effects on microbial load, enzyme levels, or stability of milk to heat. Raw material quality becomes an additional quality control point. Blending and stirring of raw milk may be needed within their production, which cause enhanced lipolytic activity and soapy off-flavors (Cheng, Barbano, & Drake, 2019a).

9.7.2 Cream

Many of the public subjects that have been considered for milk will also be linked to cream, such as the four principal control points of raw material quality, time/temperature conditions, eliminating postpasteurization contamination, and low-temperature storage (Lianou, Moschonas, Nychas, & Panagou, 2018). Post-pasteurization contamination is an important factor which affects the weak quality of compounds. The main difference of milk and cream is related to the fat. Indeed, creams are defined by their fat contents which have a range of 9%–55% (Sakkas, Spiliopoulos, & Moatsou, 2020). These classifications are as follows: half cream or coffee cream, 9%–12%; single cream, 18%; whipping cream, 30%–35%; double cream, 48%; and clotted cream at greater than 50%. Generally, cream may be more sensitive to mechanical detriment; consequently, positive pumps may be chosen instead of centrifugal pumps. In spite of the fact that there is some evidence which shows that fat can approximately protect microorganisms against thermal deactivation, in the United Kingdom, thermal processing regulations for cream are similar to those for milk (a minimum of 72°C/15 s). The important point is that in the industry, higher temperatures (more than 72°C) are usually used to pasteurize cream in order to compensate the alteration in the residence time distribution which is caused by higher viscosity of cream in comparison to milk (Uysal, Boyaci, & Sumnu, 2019).

9.7.3 Ice cream

Ice cream is a soft, sweetened frozen snack or dessert along with milk solids and cream which is produced by freezing the ice cream mix and combination air within the freezing procedures. This dessert consists of skim milk powder, fat (dairy or vegetable origin), sugar, emulsifiers, stabilizers, flavorings, colors, and water. In different countries, there are several rules about these ingredients. For instance, in the United Kingdom, ice cream consists of more than 5% fat and 7.5% milk solids and not fat (Schmidt, 2004). Based on the regulations, the fat content of dairy ice cream must be more than that of milk, and in the case of lack of fat, hydrogenated vegetable oils are applied. The standard total solids of the ice cream mixture are between 30% and 40%, and incorporated air to the mixture within the refrigeration process is measured by the overrun. In the United Kingdom and other countries, pasteurization of ice cream after mixing is a legal requirement conduct and also HTST pasteurization of ice cream is more severe than that of milk.

Finally, it is essential to keep cool the ice cream mixture below 7.2°C during 1.5 h (Hammer & Sanders, 2017). It should be noted that the ice cream mixture should not be maintained above 7.2°C for more than 1 h before pasteurization. After pasteurization and cooling, the ice cream mixture is ordinarily aged for several hours at low temperatures before being frozen. Therefore, there is no necessity for the ice cream mixture to have a long shelf life and there are no microbiological standards. The last but not the least is that the high microbial growth (coliforms) in the ice cream would most likely have arisen from postpasteurization contamination as a result of weak hygiene, cleaning, and disinfection (Homayouni et al., 2018).

9.7.4 Eggs

Recently, many industries tend to use commercial egg forms in the liquid form. Liquid egg is extensively applied in the cooking industry and is very susceptible to microbial spoilage with a short shelf life. Liquid egg may have vast microbiological flora; the most important pathogenic microorganism which is related to eggs is *Salmonella*. Moreover, *Enterococcus faecalis* is also a concern. *Salmonellae* are approximately susceptible to thermal processing (63–65°C for 30–60 s) (Cunningham, 2017). Indeed, most of the thermal processing rules arrange conditions well in excess of these in order to achieve more safety. In most societies, regulation is based on public health necessities for salmonella-free eggs (Lechevalier et al., 2017). For instance, in the United States, the thermal processing is conducted at 64°C for 4 min. Also, in the United Kingdom, the requirement rule for thermal processing was 64.4°C for 2.5 min with negative oc-amylase activity. It should be noted that albumen in egg white cannot tolerate such high temperatures as whole egg. Pasteurized liquid egg is maintained refrigerated and supplied as such, but its durability is short (for about 6 days) (Lechevalier et al., 2017).

9.7.5 Fruit juices and fruit drinks

All the fruits nearly have acidic conditions (pH < 4.5). In general, fruits are the primary substances for producing juices, purees, nectars, and other fruit-based drinks. In this process, fruits are exposed to thermal processing based on the pH levels, initial microbial count, and the freshness or reconstitution of juice. The important reasons for thermal processing are the deactivation or elimination of the spoilage microorganisms including yeasts, molds, and lactic acid and acetic acid bacteria as well as some of the enzymes which are responsible for browning within procedures and storage. It should be noted that browning within the production process will be enzymatic, whereas after thermal processing and storage, it will be nonenzymatic (Ağçam et al., 2018). The pasteurization process conditions for acidic juices are 85°C for 15 s or 95°C for 2 s. However, there are no specific thermal processing regulations, unlike those for dairy products, and the conditions which are applied commercially are not often recorded. Bhalerao and Chakraborty reported the decimal reduction times at 75°C for some microorganisms including 0.004 s for *S. cerevisiae*, 0.053 s for *Lactobacillus fermentum*, and 0.02 s for *Aspergillus niger* spores (Bhalerao & Chakraborty, 2021). These microorganisms are easily inactivated by the applied pasteurization conditions. A temperature of 74°C will eliminate the main spoilage flora in fruit juices. For example, in the United Kingdom, there is an increasing demand for apple and black currant fruit juice consumption (Müller, Pasin, Sarkis, & Marczak, 2021). The pasteurization process for fruits with Mediterranean and tropical origin is as follows: these fruits are often achieved in the United Kingdom as frozen concentrates, next defrosted, diluted back to near its original solids content, pasteurized, and finally packaged. Generally, thermal processing is conducted in various phases, for instance, straightaway after juice or puree is produced (apple juice: 85–90°C) or as pretreatment to hot filling or after cooling for the production of frozen concentrates or as a pretreatment for evaporation or drying (Benjamin & Gamrasni, 2020).

9.7.6 Tomatoes

Tomatoes with a moisture content of 94%–95% are processed in different forms such as juices (up to 12% Brix), purees (12%–22% Brix), and pastes (between 21% and 31% Brix). With regard to that, tomatoes have pH 4.3–4.5, and thermal processing has a significant effect on their texture; it not only controls the microbial contamination but also affects the tomato viscosity, which is derived from the pectin of various enzymes (Jeż, Błaszczak, Penkacik, & Amarowicz, 2020). Suitable properties of tomatoes include a perfect body, low physical loss during transportation, high soluble solids, and an acceptable color, and aroma and taste which are important in juices, but less so for sauces. For thermal processing, tomatoes are washed and cut and next subjected to either a hot-break or cold-break heating process. Thermal processing for hot break is used at 95°C for 5–8 min and will inactivate the pectin enzymes, which produce a product with a higher viscosity with a lower flavor, and that for cold break is conducted at 65°C for 10 min, producing a fresher product. After thermal processing, the skin and seeds of tomatoes are removed. Hot-break juice would now be at 70° to

80°C, with a pH of about 4–4.2 and a Brix of 4.5. Then titratable acidity and serum viscosity will be evaluated and citric acid and salt may be incorporated, and a high amount of juice will also evaporate. The evaporation process occurs steadily all over the growing season. Next, tomato concentrates are subjected to a final thermal process at 95°C for 2–3 min prior packaging. Moreover, the direct-steam injection method or tubular heat exchangers can be used (Iniesta, Perez-Conesa, Garcia-Alonso, Ros, & Periago, 2009).

The processes can be extended to other food products such as vinegar, pickled products, beverages and soft drinks, jams, and so on.

9.8 Conclusion

This section has discussed main developments in the pasteurization process, which are important factors in preparing high quality of products as well as high safety. Traditional thermal processing was accepted by the industry by emerging of deactivation procedures. They are well developed both scientifically and technically and widely adjusted by the industry with a main aim to receive the necessary level of safety based on elimination of pathogenic microorganisms of concern and organisms and enzymes which lead to economical spoilage. Discovery of new techniques which apply modern thermal and nonthermal physical and chemical high-tech processes and their confirmation with regulatory agencies internationally requires reviews of the definitions of traditional pasteurization methods and introduction of novel requirements for establishment and design of the preservation process. It is also obvious that new high technologies which are applied for pasteurization can amend to manufacture high-quality, safe, and nutritious food.

References

Adler, I., Gavilan, H., McKay, P., & Campos, L. (2018). An innovative approach to solar thermal water pasteurisation. *Journal of Environmental Research*, *1*, 1.

Ağçam, E., Akyıldız, A., & Dündar, B. (2018). Thermal pasteurization and microbial inactivation of fruit juices. In *Fruit juices* Elsevier.

Andriani, D., Mashito, R., & Handayani, M. (2021). Offline quality control for optimization of pasteurized milk production parameters. In *IOP conference series: Earth and environmental science* (p. 012074). IOP Publishing.

Arshad, R. N., Abdul-Malek, Z., Ahmad, M. H., Buntat, Z., Nawawi, Z., Kumara, C. P., et al. (2020). Coaxial treatment chamber for liquid food treatment through pulsed electric field. *Indonesian Journal of Electrical Engineering and Computer Science*, *19*, 1169–1176.

Atia, M., Mostafa, M., Abdel-Salam, M., & El-Nono, M. (2010). Solar energy utilization for milk pasteurization. *Renewable and Sustainable Energy Reviews*, *92*, 1–8.

Bagade, S. B., & Patil, M. (2021). Recent advances in microwave assisted extraction of bioactive compounds from complex herbal samples: A review. *Critical Reviews in Analytical Chemistry*, *51*, 138–149.

Barkaoui, S., Madureira, J., Santos, P. M., Margaça, F. M., Miloud, N. B., Mankai, M., et al. (2020). Effect of ionizing radiation and refrigeration on the antioxidants of strawberries. *Food and Bioprocess Technology*, *13*, 1516–1527.

Becerril, R., Nerín, C., & Silva, F. (2021). Bring some colour to your package: Freshness indicators based on anthocyanin extracts. *Trends in Food Science & Technology*, *111*, 495–505.

Benjamin, O., & Gamrasni, D. (2020). Microbial, nutritional, and organoleptic quality of pomegranate juice following high-pressure homogenization and low-temperature pasteurization. *Journal of Food Science*, *85*, 592–599.

Bezie, A. (2019). The effect of different heat treatment on the nutritional value of milk and milk products and shelf-life of milk products. A review. *Journal of Dairy and Veterinary Sciences*, *11*, 1–8.

Bhalerao, P. P., & Chakraborty, S. (2021). Integrated calculation of pasteurization time: A case study for thermal inactivation kinetics of a mixed fruit beverage. *Journal of Food Process Engineering*, e13761.

Bonczar, G., Walczycka, M., & Duda, I. (2016). The changes of proteins fractions shares in milk and fermented milk drinks. *Acta Scientiarum Polonorum. Technologia Alimentaria*, *15*.

Broersen, K. (2020). Milk processing affects structure, bioavailability and immunogenicity of β-lactoglobulin. *Foods*, *9*, 874.

Cadwallader, D. C. (2020). *Sensory and instrumental analysis of flavors due to scalping and migration from packaging in fluid milk*. North Carolina State University.

Cavalcante, T. A. B. B., Dos Santos Funcia, E., & Gut, J. A. W. (2021). Inactivation of polyphenol oxidase by microwave and conventional heating: Investigation of thermal and non-thermal effects of focused microwaves. *Food Chemistry*, *340*, 127911.

Cheng, N., Barbano, D. M., & Drake, M. (2019a). Effects of milk fat, casein, and serum protein concentrations on sensory properties of milk-based beverages. *Journal of Dairy Science*, *102*, 8670–8690.

Cheng, N., Barbano, D. M., & Drake, M. A. (2019b). Effect of pasteurization and fat, protein, casein to serum protein ratio, and milk temperature on milk beverage color and viscosity. *Journal of Dairy Science*, *102*, 2022–2043.

Chhe, C., Imaizumi, T., Tanaka, F., & Uchino, T. (2018). Effects of hot-water blanching on the biological and physicochemical properties of sweet potato slices. *Engineering in Agriculture, Environment and Food*, *11*, 19–24.

Cho, W.-I., & Song, S.-H. (2021). Novel inactivation methods of Doenjang (fermented soybean paste) by high pressure and ohmic heating. *Food Science and Biotechnology*, *30*, 513–520.

Cho, W.-I., Yi, J. Y., & Chung, M.-S. (2016). Pasteurization of fermented red pepper paste by ohmic heating. *Innovative Food Science & Emerging Technologies*, *34*, 180–186.

Choudhary, R., & Bandla, S. (2012). Ultraviolet pasteurization for food industry. *International Journal of Food Science and Nutrition Engineering*, *2*, 12–15.

Colonna, A., Durham, C., & Meunier-Goddik, L. (2011). Factors affecting consumers' preferences for and purchasing decisions regarding pasteurized and raw milk specialty cheeses. *Journal of Dairy Science*, *94*, 5217–5226.

Cowell, N. D., & Mermelstein, N. (1995). Who introduced the tin can? A new candidate. *Food Technology*, *49*, 61–64.

Cunningham, R. (2017). Egg-product pasteurization. In *Egg science and technology* CRC Press.

Daniels, B., Schmidt, S., King, T., Israel-Ballard, K., Amundson Mansen, K., & Coutsoudis, A. (2017). The effect of simulated flash-heat pasteurization on immune components of human milk. *Nutrients*, *9*, 178.

Darvishi, H., Saba, M. K., Behroozi-Khazaei, N., & Nourbakhsh, H. (2020). Improving quality and quantity attributes of grape juice concentrate (molasses) using ohmic heating. *Journal of Food Science and Technology*, *57*, 1362–1370.

Deak, T. (2013). *Food safety management: Chapter 17. Thermal treatment*. Elsevier Science.

Delorme, M. M., Guimarães, J. T., Coutinho, N. M., Balthazar, C. F., Rocha, R. S., Silva, R., et al. (2020). Ultraviolet radiation: An interesting technology to preserve quality and safety of milk and dairy foods. *Trends in Food Science & Technology*, *102*, 146–154.

Dionísio, A. P., Gomes, R. T., & Oetterer, M. (2009). Ionizing radiation effects on food vitamins: A review. *Brazilian Archives of Biology and Technology*, *52*, 1267–1278.

Dobrowsky, P., Carstens, M., De Villiers, J., Cloete, T., & Khan, W. (2015). Efficiency of a closed-coupled solar pasteurization system in treating roof harvested rainwater. *Science of the Total Environment*, *536*, 206–214.

Escuder-Vieco, D., Espinosa-Martos, I., Rodríguez, J. M., Corzo, N., Montilla, A., Siegfried, P., et al. (2018). High-temperature short-time pasteurization system for donor milk in a human milk bank setting. *Frontiers in Microbiology*, *9*, 926.

Gallego-Castillo, S., & Ayala-Aponte, A. A. (2018). Changes in physical properties of sweet potato due to effects of thermal pre-treatments for puree production. *Dyna*, *85*, 135–142.

Giribaldi, M., Arslanoglu, S., Tonetto, P., & Moro, G. E. (2011). Effect of two pasteurization methods on the protein content of human milk. *Frontiers in Bioscience*, *3*, 818–829.

Gogliettino, M., Balestrieri, M., Ambrosio, R. L., Anastasio, A., Smaldone, G., Proroga, Y. T., et al. (2020). Extending the shelf-life of meat and dairy products via PET-modified packaging activated with the antimicrobial peptide MTP1. *Frontiers in Microbiology*, *10*, 2963.

Gómez, B., Munekata, P. E., Gavahian, M., Barba, F. J., Martí-Quijal, F. J., Bolumar, T., et al. (2019). Application of pulsed electric fields in meat and fish processing industries: An overview. *Food Research International*, *123*, 95–105.

Gomez-Gomez, A., Brito-De La Fuente, E., Gallegos, C., Garcia-Perez, J. V., & Benedito, J. (2020). Non-thermal pasteurization of lipid emulsions by combined supercritical carbon dioxide and high-power ultrasound treatment. *Ultrasonics Sonochemistry*, *67*, 105138.

Gong, C., Zhao, Y., Zhang, H., Yue, J., Miao, Y., & Jiao, S. (2019). Investigation of radio frequency heating as a dry-blanching method for carrot cubes. *Journal of Food Engineering*, *245*, 53–56.

Goullieux, A., & Pain, J.-P. (2014). Ohmic heating. In *Emerging technologies for food processing* Elsevier.

Guo, J., He, Z., Wu, S., Zeng, M., & Chen, J. (2020). Effects of concentration of flavor compounds on interaction between soy protein isolate and flavor compounds. *Food Hydrocolloids*, *100*, 105388.

Hammer, B., & Sanders, L. (2017). A bacteriological study of the method of pasteurizing and homogenizing the ice cream mix. *Bulletin*, *16*, 1.

Hartzell, W. (2019). Solar water pasteurizers. In *Providing safe drinking water in small systems* Routledge.

Hassanpour, M., Rezaie, M. R., & Baghizadeh, A. (2019). Practical analysis of aflatoxin M1 reduction in pasteurized milk using low dose gamma irradiation. *Journal of Environmental Health Science and Engineering*, *17*, 863–872.

Holdsworth, D. (2001). *Continuous thermal processing of foods—pasteurization and UHT sterilization*. Wiley Online Library.

Homayouni, A., Javadi, M., Ansari, F., Pourjafar, H., Jafarzadeh, M., & Barzegar, A. (2018). Advanced methods in ice cream analysis: A review. *Food Analytical Methods*, *11*, 3224–3234.

Huan, Y., Zhou, G., Zhao, G., Xu, X., & Peng, Z. (2005). Changes in flavor compounds of dry-cured Chinese Jinhua ham during processing. *Meat Science*, *71*, 291–299.

Huang, K., & Wang, J. (2009). Designs of pulsed electric fields treatment chambers for liquid foods pasteurization process: A review. *Journal of Food Engineering*, *95*, 227–239.

Icier, F., & Bozkurt, H. (2011). Ohmic heating of liquid whole egg: Rheological behaviour and fluid dynamics. *Food and Bioprocess Technology*, *4*, 1253–1263.

IDF. (1986). Production and utilization of ewe's and goat's milk. In IDF (Ed.), *Proceedings of the international seminar, Athens, Greece, September 23–25, 1985*. Brussels, Belgium: International Dairy Federation Publisher. Bulletin no. 202/1986, 222 pp.

Imaizumi, T., Szymańska-Chargot, M., Pieczywek, P. M., Chylińska, M., Kozioł, A., Ganczarenko, D., et al. (2017). Evaluation of pectin nanostructure by atomic force microscopy in blanched carrot. *LWT*, *84*, 658–667.

Imaizumi, T., Tanaka, F., & Uchino, T. (2019). Effects of mild heating treatment on texture degradation and peroxidase inactivation of carrot under pasteurization conditions. *Journal of Food Engineering*, *257*, 19–25.

Iniesta, M. D., Perez-Conesa, D., Garcia-Alonso, J., Ros, G., & Periago, M. J. S. (2009). Folate content in tomato (Lycopersicon esculentum). Influence of cultivar, ripeness, year of harvest, and pasteurization and storage temperatures. *Journal of Agricultural and Food Chemistry*, *57*, 4739–4745.

Jacucci, G., Vignolini, S., & Schertel, L. (2020). The limitations of extending nature's color palette in correlated, disordered systems. *Proceedings of the National Academy of Sciences*, *117*, 23345–23349.

Jansrimanee, S., & Lertworasirikul, S. (2017). Effect of sodium alginate coating on osmotic dehydration of pumpkin. *International Food Research Journal*, *24*.

Jansson, T., Waehrens, S. S., Rauh, V., Danielsen, B. P., Sørensen, J., Bredie, W. L., et al. (2019). Effect of green tea catechins on physical stability and sensory quality of lactose-reduced UHT milk during storage for one year. *International Dairy Journal*, *95*, 25–34.

Jeż, M., Błaszczak, W., Penkacik, K., & Amarowicz, R. (2020). Quality parameters of juice obtained from hydroponically grown tomato processed with high hydrostatic pressure or heat pasteurization. *International Journal of Food Science*, *2020*.

Jo, Y., Benoist, D., Barbano, D., & Drake, M. (2018). Flavor and flavor chemistry differences among milks processed by high-temperature, short-time pasteurization or ultra-pasteurization. *Journal of Dairy Science*, *101*, 3812–3828.

Júnior, J. C. R., De Oliveira, A. M., Silva, F. G., Garcia, L. N. H., De Oliveira Lobo, C. M., Alexandrino, B., et al. (2019). Influence of the microbiological quality of raw milk on the shelf life of pasteurized milk. *Semina: Ciências Agrárias*, *40*, 1469–1476.

Kessler, H.-G. (1981). *Food engineering and dairy technology*. Verlag A. Kessler.

Kokkinidou, S., & Peterson, D. G. (2014). Control of Maillard-type off-flavor development in ultrahigh-temperature-processed bovine milk by phenolic chemistry. *Journal of Agricultural and Food Chemistry*, *62*, 8023–8033.

Kubo, M. T., Siguemoto, E. S., Funcia, E. S., Augusto, P. E., Curet, S., Boillereaux, L., et al. (2020). Non-thermal effects of microwave and ohmic processing on microbial and enzyme inactivation: A critical review. *Current Opinion in Food Science*, *35*, 36–48.

Kühn, J., Considine, T., & Singh, H. (2006). Interactions of milk proteins and volatile flavor compounds: Implications in the development of protein foods. *Journal of Food Science*, *71*, R72–R82.

Kumar, T. (2018). A review on ohmic heating technology: Principle, applications and scope. *International Journal of Agriculture, Environment and Biotechnology*, *11*, 679–687.

Lamce, F., Kycyk, O., Gozhdari, K., Zejnelhoxha, S., & Kukali, E. (2020). The influence of stems and oak chips on the fermentation and color stabilization during maturation of red wines produced by cv. Shesh i Zi. *Journal of Central European Agriculture*, *21*, 129–134.

Lechevalier, V., Guérin-Dubiard, C., Anton, M., Beaumal, V., Briand, E. D., Gillard, A., et al. (2017). Pasteurisation of liquid whole egg: Optimal heat treatments in relation to its functional, nutritional and allergenic properties. *Journal of Food Engineering, 195*, 137–149.

Lee, A., Barbano, D., & Drake, M. (2017). The influence of ultra-pasteurization by indirect heating versus direct steam injection on skim and 2% fat milks. *Journal of Dairy Science, 100*, 1688–1701.

Leizerson, S., & Shimoni, E. (2005). Effect of ultrahigh-temperature continuous ohmic heating treatment on fresh orange juice. *Journal of Agricultural and Food Chemistry, 53*, 3519–3524.

Liang, L., Qi, C., Wang, X., Jin, Q., & Mcclements, D. J. (2017). Influence of homogenization and thermal processing on the gastrointestinal fate of bovine milk fat: In vitro digestion study. *Journal of Agricultural and Food Chemistry, 65*, 11109–11117.

Lianou, A., Moschonas, G., Nychas, G.-J. E., & Panagou, E. Z. (2018). Growth of Listeria monocytogenes in pasteurized vanilla cream pudding as affected by storage temperature and the presence of cinnamon extract. *Food Research International, 106*, 1114–1122.

Lima, F., Vieira, K., Santos, M., & De Souza, P. M. (2018). Effects of radiation technologies on food nutritional quality. In A. V. Díaz, & R. M. García-Gimeno (Eds.), *Descriptive food science* (p. 137). IntechOpen.

Liu, T., Lv, B., Zhao, W., Wang, Y., Piao, C., Dai, W., et al. (2020). Effects of ultrahigh temperature pasteurization on the liquid components and functional properties of stored liquid whole eggs. *BioMed Research International, 2020*.

Lönnerdal, B. (2013). Bioactive proteins in breast milk. *Journal of Paediatrics and Child Health, 49*, 1–7.

Manfrida, G., Petela, K., & Rossi, F. (2017). Natural circulation solar thermal system for water disinfection. *Energy, 141*, 1204–1214.

Maniglia, B. C., Castanha, N., Rojas, M. L., & Augusto, P. E. (2021). Emerging technologies to enhance starch performance. *Current Opinion in Food Science, 37*, 26–36.

Marra, F., Zell, M., Lyng, J., Morgan, D., & Cronin, D. (2009). Analysis of heat transfer during ohmic processing of a solid food. *Journal of Food Engineering, 91*, 56–63.

Martin, N. H., Boor, K. J., & Wiedmann, M. (2018). Symposium review: Effect of post-pasteurization contamination on fluid milk quality. *Journal of Dairy Science, 101*, 861–870.

Martin, C. R., Ling, P.-R., & Blackburn, G. L. (2016). Review of infant feeding: Key features of breast milk and infant formula. *Nutrients, 8*, 279.

Martini, M., Salari, F., Altomonte, I., Ragona, G., Piazza, A., Gori, R., et al. (2018). Effects of pasteurization and storage conditions on donkey milk nutritional and hygienic characteristics. *Journal of Dairy Research, 85*, 445–448.

Mehta, B. M. (2015). Chemical composition of milk and milk products. In *Handbook of food chemistry* (pp. 511–553). Springer.

Mettler, A. (1994). Present day requirements for effective pathogen control in spray dried milk powder production. *International Journal of Dairy Technology, 47*, 95–107.

Miyaji, K., Kuwano, Y., Murakami, Y., Hirata, S., Imayoshi, Y., Maruyama, H., et al. (2021). Off-flavors generated during long-term ambient storage of pasteurized drinking yogurt from skim milk. *Bioscience, Biotechnology, and Biochemistry, 85*, 391–400.

Modi, A., & Prajapat, R. (2014). Pasteurization process energy optimization for a milk dairy plant by energy audit approach. *International Journal of Scientific and Technology Research, 3*, 181–188.

Müller, W. A., Pasin, M. V. A., Sarkis, J. R., & Marczak, L. D. F. (2021). Effect of pasteurization on Aspergillus fumigatus in apple juice: Analysis of the thermal and electric effects. *International Journal of Food Microbiology*, *338*, 108993.

Munekata, P. E., Domínguez, R., Budaraju, S., Roselló-Soto, E., Barba, F. J., Mallikarjunan, K., et al. (2020). Effect of innovative food processing technologies on the physicochemical and nutritional properties and quality of non-dairy plant-based beverages. *Foods*, *9*, 288.

Panchal, H., Patel, R., Chaudhary, S., Patel, D., Sathyamurthy, R., & Arunkumar, T. (2018). Solar energy utilisation for milk pasteurisation: A comprehensive review. *Renewable and Sustainable Energy Reviews*, *92*, 1–8.

Panchal, H., Patel, R., & Parmar, K. (2020). Application of solar energy for milk pasteurisation: A comprehensive review for sustainable development. *International Journal of Ambient Energy*, *41*, 117–120.

Panchal, H., Patel, R., & Sathyamurthy, R. (2021). Investigation and performance analysis of solar milk pasteurisation system. *International Journal of Ambient Energy*, *42*, 522–529.

Pandagale, V., Madhuri, G., & Khambaklar, V. (2016). Integration of solar thermal energy in milk processing plant. *International Journal of Engineering Science Technology*, *1*, 1–5.

Patil, S., Ananthan, A., Nanavati, R. N., Nataraj, G., & Prasad, P. (2019). Effect of different methods of pasteurization on bactericidal action of human milk: A prospective observational study. *The Indian Journal of Medical Research*, *150*, 504.

Peng, J., Tang, J., Barrett, D. M., Sablani, S. S., Anderson, N., & Powers, J. R. (2017). Thermal pasteurization of ready-to-eat foods and vegetables: Critical factors for process design and effects on quality. *Critical Reviews in Food Science and Nutrition*, *57*, 2970–2995.

Pichel, N., Vivar, M., & Fuentes, M. (2019). The problem of drinking water access: A review of disinfection technologies with an emphasis on solar treatment methods. *Chemosphere*, *218*, 1014–1030.

Podgoršak, E. B. (2006). *Radiation physics for medical physicists*. Springer.

Prasantha, B., & Wimalasiri, K. (2019). Effect of HTST thermal treatments on end-use quality characteristics of goat milk. *International Journal of Food Science*, *2019*.

Rachon, G., Rice, C. J., Pawlowsky, K., & Raleigh, C. P. (2018). Challenging the assumptions around the pasteurisation requirements of beer spoilage bacteria. *Journal of the Institute of Brewing*, *124*, 443–449.

Rajkhowa, S. (2020). Heat, solar pasteurization, and ultraviolet radiation treatment for removal of waterborne pathogens. In *Waterborne pathogens* Elsevier.

Ranieri, M., Huck, J., Sonnen, M., Barbano, D., & Boor, K. (2009). High temperature, short time pasteurization temperatures inversely affect bacterial numbers during refrigerated storage of pasteurized fluid milk. *Journal of Dairy Science*, *92*, 4823–4832.

Reyneke, B., Cloete, T. E., Khan, S., & Khan, W. (2018). Rainwater harvesting solar pasteurization treatment systems for the provision of an alternative water source in peri-urban informal settlements. *Environmental Science: Water Research & Technology*, *4*, 291–302.

Rossi, F., Parisi, M. L., Maranghi, S., Manfrida, G., Basosi, R., & Sinicropi, A. (2019). Environmental impact analysis applied to solar pasteurization systems. *Journal of Cleaner Production*, *212*, 1368–1380.

Rysstad, G., & Kolstad, J. (2006). Extended shelf life milk—Advances in technology. *International Journal of Dairy Technology*, *59*, 85–96.

Ryu, J.-A., Kim, E., Kim, M.-J., Lee, S., Yoon, S.-R., Ryu, J.-G., et al. (2021). Physicochemical characteristics and microbial communities in gochujang, a traditional Korean fermented hot pepper paste. *Frontiers in Microbiology*, *11*, 3543.

Sakkas, L., Spiliopoulos, M., & Moatsou, G. (2020). Partial substitution of sheep and goat milks of various fat contents by the respective sweet buttermilks: Effect of cream heat treatment. *LWT*, *133*, 109926.

Salehi, F. (2020). Physico-chemical properties of fruit and vegetable juices as affected by pulsed electric field: A review. *International Journal of Food Properties*, *23*, 1036–1050.

Salinas-Roca, B., Elez-Martínez, P., Welti-Chanes, J., & Martín-Belloso, O. (2017). Quality changes in mango juice treated by high-intensity pulsed electric fields throughout the storage. *Food and Bioprocess Technology*, *10*, 1970–1983.

Satin, M. (2020). *Food irradiation: A guidebook*. CRC Press.

Sattar, S., Imran, M., Mushtaq, Z., Ahmad, M. H., Arshad, M. S., Holmes, M., et al. (2020). Retention and stability of bioactive compounds in functional peach beverage using pasteurization, microwave and ultrasound technologies. *Food Science and Biotechnology*, *29*, 1381–1388.

Schmidt, K. A. (2004). Dairy: Ice cream. In *Food processing—Principles and applications* (pp. 287–296). Wiley.

Shao, L., Zhao, Y., Zou, B., Li, X., & Dai, R. (2021). Ohmic heating in fruit and vegetable processing: Quality characteristics, enzyme inactivation, challenges and prospective. *Trends in Food Science & Technology*, *118*, 601–616.

Sierra-Lopera, L. M., Ciro-Velásquez, H. J., Torres-Trujillo, P. I., & Montoya-Campuzano, O. I. (2017). Inactivation of Bacillus licheniformis in liquid coffee extract by microwave heating in domestic oven and precise temperature measurements using an optical fiber sensor. *Dyna*, *84*, 302–308.

Soni, A., Smith, J., Thompson, A., & Brightwell, G. (2020). Microwave-induced thermal sterilization—A review on history, technical progress, advantages and challenges as compared to the conventional methods. *Trends in Food Science & Technology*, *97*, 433–442.

Sridhar, A., Ponnuchamy, M., Kumar, P. S., & Kapoor, A. (2021). Food preservation techniques and nanotechnology for increased shelf life of fruits, vegetables, beverages and spices: A review. *Environmental Chemistry Letters*, *19*, 1715–1735.

Su, S., & Wiley, R. (1998). Changes in apple juice flavor compounds during processing. *Journal of Food Science*, *63*, 688–691.

Sun, D.-W. (2014). *Emerging technologies for food processing*. Elsevier.

Taj, F., Dar, A., Dwivedi, M., Naik, H. R., Farzana, W., & Khan, S. A. (2021). Design strategies and recent advances in utilisation of solar energy for pasteurisation. *MRS Energy & Sustainability*, 1–15.

Thayer, D. (1996). *Radiation pasteurization of food*. Council for Agricultural Science and Technology. Issue paper number 7, April 1996.

Thorne, S. (1986). *The history of food preservation*. Barnes and Noble Books.

Tiwari, P. K., & Asgar, S. (2017). Diversification in flavoured milk: A review. *International Journal of Clinical and Biomedical Research*, 15–20.

Troise, A. D., Colantuono, A., & Fiore, A. (2020). Spray-dried olive mill wastewater reduces Maillard reaction in cookies model system. *Food Chemistry*, *323*, 126793.

USFDA. (2018). *CFR-code of federal regulations title 21*. Washington, DC, USA: US Food and Drug Administration.

Uysal, R. S., Boyaci, I. H., & Sumnu, G. (2019). Determination of pasteurization treatment of liquid whole egg by measuring physical and rheological properties of cake cream. *Journal of Food Process Engineering*, *42*, e13167.

Van'T Land, B., Boehm, G., & Garssen, J. (2010). Breast milk: Components with immune modulating potential and their possible role in immune mediated disease resistance. In *Dietary components and immune function* Springer.

Varghese, K. S., Pandey, M., Radhakrishna, K., & Bawa, A. (2014). Technology, applications and modelling of ohmic heating: A review. *Journal of Food Science and Technology*, *51*, 2304–2317.

Wayua, F. O., Okoth, M. W., & Wangoh, J. (2013). Design and performance assessment of a flat-plate solar milk pasteurizer for arid pastoral areas of Kenya. *Journal of Food Processing and Preservation*, *37*, 120–125.

Wilbey, R. (1993). Pasteurization of foods: Principles of pasteurization. In *Encyclopedia of food science, food technology and nutrition* (pp. 3437–3441). Academic Press.

Yildiz, S., Pokhrel, P. R., Unluturk, S., & Barbosa-Cánovas, G. V. (2021). Shelf life extension of strawberry juice by equivalent ultrasound, high pressure, and pulsed electric fields processes. *Food Research International*, *140*, 110040.

Zhang, H., Ha, T. M. H., Seck, H. L., & Zhou, W. (2020). Inactivation of Escherichia coli O157: H7 and Salmonella typhimurium in edible bird's nest by low-energy X-ray irradiation. *Food Control*, *110*, 107031.

Zhang, Y., Jin, S., Zhang, Y. Y., & Yu, X. (2020). How country of origin influences Chinese consumers' evaluation of imported milk? *China Agricultural Economic Review*, *13*(1).

Zhang, J., Kang, D., Zhang, W., & Lorenzo, J. M. (2021). Recent advantage of interactions of protein-flavor in foods: Perspective of theoretical models, protein properties and extrinsic factors. *Trends in Food Science & Technology*, *111*, 405–425.

Zhu, H., Poojary, M. M., Andersen, M. L., & Lund, M. N. (2020). Trapping of carbonyl compounds by epicatechin: Reaction kinetics and identification of epicatechin adducts in stored UHT milk. *Journal of Agricultural and Food Chemistry*, *68*, 7718–7726.

Zhu, Z., & Guo, W. (2021). Recent developments on rapid detection of main constituents in milk: A review. *Critical Reviews in Food Science and Nutrition*, *61*, 312–324.

Zulueta, A., Barba, F. J., Esteve, M. J., & Frígola, A. (2013). Changes in quality and nutritional parameters during refrigerated storage of an orange juice–milk beverage treated by equivalent thermal and non-thermal processes for mild pasteurization. *Food and Bioprocess Technology*, *6*, 2018–2030.

Thermal sterilization of food in cans and pouches

10

*Ghani Albaali**
King Talal Faculty of Business and Technology (KTFBT), Princess Sumaya University for Technology, Amman, Jordan

Nomenclature

$\boldsymbol{A}$	reaction frequency factor (s^{-1})
$\boldsymbol{a}$	surface area (m^2)
$\boldsymbol{Bi}$	Biot number, $Bi = \frac{h\delta}{k}$
$\boldsymbol{C_o}$	initial concentration ($kg\,m^{-3}$)
$\boldsymbol{C_P}$	specific heat of liquid food ($J\,kg^{-1}\,K^{-1}$)
$\boldsymbol{D}$	diffusion coefficient ($m^2\,s^{-1}$)
$\boldsymbol{D_T}$	decimal reduction time at temperature T (min)
$\boldsymbol{D_{Tref}}$	decimal reduction time at T_{ref}
$\boldsymbol{d_v}$	vertical dimension (m)
$\boldsymbol{E}$	activation energy (kJ $(kg\,mol)^{-1}$)
$\boldsymbol{F}$	thermal death time (min)
$\boldsymbol{g}$	acceleration due to gravity (ms^{-2})
$\boldsymbol{G_r}$	Grashof number, $G_r = \frac{g\beta\Delta T L^3 \rho^2}{\mu^2}$
$\boldsymbol{h}$	heat transfer coefficient ($W\,m^{-2}\,K^{-1}$)
$\boldsymbol{h_s}$	surface heat transfer coefficient ($W\,m^{-2}\,K^{-1}$)
$\boldsymbol{k_T}$	reaction rate constant at temperature T (s^{-1})
$\boldsymbol{k}$	thermal conductivity ($W\,m^{-1}\,K$)
$\boldsymbol{Nu}$	Nusselt number, $Nu = \frac{hd_v}{k}$
$\boldsymbol{n_1}$	initial concentration of microorganisms at the starting of the thermal process
$\boldsymbol{n_2}$	final concentration of microorganisms at the end of the thermal process
$\boldsymbol{Pr}$	Prandtl number, $Pr = \frac{C_p\mu}{k}$
$\boldsymbol{Q}$	rate of heat transfer (W)
$\boldsymbol{R}$	gas constant (kJ $(kg\,mol)^{-1}\,K^{-1}$)
$\boldsymbol{t}$	heating time (min)
$\boldsymbol{t_i}$	time at the beginning of the thermal process (min)
$\boldsymbol{t_f}$	time at the end of cooling (min)
$\boldsymbol{T}$	temperature (°C)
$\boldsymbol{T_{bu}}$	bulk fluid temperature (°C)
$\boldsymbol{T_s}$	surface temperature (°C)

* Previously, Food Science & Process Engineering Group, Department of Chemical & Materials Engineering, The University of Auckland, New Zealand.

Thermal Processing of Food Products by Steam and Hot Water. https://doi.org/10.1016/B978-0-12-818616-9.00002-X

T_w wall temperature (°C)
T_{ref} reference temperature (°C)
v_x velocity in x direction ($m\,s^{-1}$)
v_y velocity in y direction ($m\,s^{-1}$)
v_z velocity in z direction ($m\,s^{-1}$)
Z temperature-dependent characteristic (i.e., temperature increase required to cause a reduction of the one-log cycle in DT)
α thermal diffusivity of bacteria or vitamins ($m^2\,s^{-1}$)
β thermal expansion coefficient (K^{-1})
μ apparent viscosity (Pa s)
ρ density ($kg\,m^{-3}$)
δ diameter (m)

10.1 Introduction

Preservation of food has occupied and continued to occupy a large part of man's time and effort. Early man preserved his food supply by applying the methods of preservation available in his environment, such as sun drying, salting, and fermentation, which were used to provide food in periods when fresh foods were not available. As civilization developed, the demand for large quantities of better quality processed food also increased. This leads to the development of a large food-preservation industry, aimed to supply food that is satisfying, nutritious, and economical. Thermal sterilization of foods is the most significant part of this industry (Karel, Fennema, & Lund, 1975). Other methods of sterilization such as pulsed electric field (Barbosa-Canovas & Zhang, 2000; Barbosa-Canovas, Gongora-Nieto, & Swanson, 1998; Martin, Qin, Chang, Barbosa-Canovas, & Swanson, 1997; Qin, Barbosa-Canovas, Swanson, Pedrow, & Olsen, 1998; Qin, Zhang, Barbosa-Canovas, Swanson, & Pedrow, 1994; Sepulveda-Ahumada, Ortega-Revas, & Barbosa-Canovas, 2000; Vega-Mercado et al., 1997; Vega-Mercado, Gongora-Nieto, Barbosa-Canovas, & Swanson, 1999; Zhang, Barbosa-Canovas, & Swanson, 1995), ultrahigh hydrostatic pressure (Barbosa-Canovas, Pothakamury, Palou, & Swanson, 1997; Furukawa & Hayakawa, 2000; Palou, Lopez-Malo, Barbosa-Canovas, & Swanson, 1999; Sancho et al., 1999), and ultraviolet treatment (Farid, Chen, & Dost, 2000) have been widely studied. However, it will take some time before any of these methods may be applied to an industrial scale.

10.2 Thermal sterilization principles

Two different methods of conventional thermal processing are known, aseptic processing in which the food product is sterilized prior to packaging, and canning in which the product is packed and then sterilized (Barbosa-Canovas, Ma, & Barletta, 1997). Food after being canned has to undergo thermal treatment to deactivate most organisms (i.e., sterilization). In 1981, the food industry in the United States of America alone processed more than 16.3 billion kg of food products in approximately 37 billion containers (Kumar, Bhattacharya, & Blaylock, 1990).

The objective of thermal sterilization is to make the food safe for human consumption by destroying the pathogenic microorganisms. In reality, the death of microorganisms is expected to begin at an early stage of heating at locations near the wall, where the temperature approaches the retort temperature very quickly. A sterilizer (retort) is the unit operation in which food is heated at a high temperature and then held at that temperature for a period sufficient to kill the microorganisms of concern from the food product. A sterile product is one in which no viable microorganisms are present. A viable organism is one that can reproduce when exposed to conditions that are optimum for its growth. Temperature slightly above the maximum for bacterial growth results in the death of vegetative bacterial cells, whereas bacterial spores can survive at much higher temperatures. Since bacterial spores are far more heat resistant than vegetative cells, they are of primary concern in most sterilization processes. Saturated steam is the most common and highly desirable heating medium used for commercial sterilization of canned foods. The conventional canning operation consists of the following operations:

1. Preparing the food (cleaning, cutting, grading, blanching, etc.).
2. Filling the container.
3. Sealing the container.
4. Placing the container in a batch or continuous retort where it is heated for a time sufficient for commercial sterility.
5. Cooling of the container, which is usually done using a cold shower.

The still retort is the oldest type of equipment used for thermal processing. It is still used in large canning plants for metal- and glass-packed products. The sterilization method using still retort consists of loading the containers, which are placed into baskets immediately after seaming and then placed into an appropriately designed iron vessel (retort), closing the vessel, and heating the containers with steam. A controller regulates the temperature, and the duration of heating is determined by the rate of heat transfer into the containers. The introduction of the steam into the retort should be done with care since it is necessary to displace all of the air in the retort. The presence of air during thermal sterilization processing can result in underprocessing since steam–air mixtures result in lower heat transfer rates than steam or water alone (Karel et al., 1975).

Still, retorts are usually arranged either vertically or horizontally. The metal shell pressure vessel is fitted with steam inlet, water inlet, outlet ports for venting air during retort come-up and for draining, outlet ports for venting the retort at the end of the cycle, and safety and pressure relief valve. A pocket for instruments, thermometer, temperature-recording probe, and a pressure gauge is located on the side of the vessel (Rahman, 1999). The operating cycle for this type of retort involves bringing the retort up to a temperature of around 100°C. The steam is then allowed to pass through the vessel so that all air in the retort and between the cans is removed (venting) before the retort is finally brought up to the operating pressure and processing temperature (Rahman, 1999). At the end of the processing time, the steam is turned off and a mixture of cooling water and air is introduced into the retort to cool the cans. The purpose of the air is to maintain the pressure in the retort following the condensation of the residual steam after the initial introduction of cooling water. The containers may deform if this pressure is not maintained, which is due to the pressure difference between inside and outside the container.

The recent concern in thermal sterilization of foods is to improve the rates of heating, to increase production rates, and minimize damage to product quality.

10.3 Heat transfer

The heat transfer mechanisms through the liquid food in the can during the sterilization process of liquid canned food are classified as conduction-heated, convection-heated, or combined convection and conduction heated (Herson & Hulland, 1980). The conduction is defined as the movement of heat by direct transfer of molecular energy within solids, while the convection is the transfer of heat in the fluid by groups of molecules or fluid bulk that moves as a result of differences in the density of the fluid (buoyancy force). Usually, these two types of heat transfer occur simultaneously but one type may be more important than the other. Steady-state heat transfer happens when there is no alteration in temperature with time. However, in most applications in food processing, the temperature of the food heating or cooling medium is constantly changing, and unsteady-state heat transfer is more usually found. Under these conditions, the calculations of heat transfer are extremely difficult but are simplified by making several assumptions (Fellows, 1996).

The thermal conductivities of a variety of food materials are available in the literature. The most notable feature of food products is their low values of thermal conductivity compared to metals. In metals, the electrons transmit most of the heat energy, whereas in foods, where water is the main constituent, the free electron concentration is low and the transfer mechanism involves primarily vibration of atoms and molecules (Karel et al., 1975). However, the thermal conductivity of liquid food is close to that of water.

10.3.1 Unsteady-state heat conduction

In food processing, there are many situations where the temperature at some point in a product is a function of time. Heating or cooling of particulate materials is the most notable example, such as heating and cooling products in containers (canning process). The calculations of unsteady-state heat transfer are commonly complicated and included to the use of the Fourier equation, which is written in terms of partial differential equations in three dimensions (Karel et al., 1975).

In unsteady-state heat transfer, the temperature within a food during processing depends on the time and position. These temperature changes are influenced by

1. Initial temperatures of the heated body.
2. The temperature of the heating medium.
3. The surface heat transfer coefficient.
4. The thermal conductivity, specific heat, and density of the food.
5. The thickness of the heated body.

The basic equation for unsteady-state heat transfer in one direction (x) is

$$\frac{\partial T}{\partial t} = \frac{k}{\rho C_p} \frac{\partial^2 T}{\partial x^2} \tag{10.1}$$

where $\partial T/\partial t$ is the change of temperature with time, k is the thermal conductivity ($W m^{-1} K^{-1}$), ρ is the density ($kg m^{-3}$), and C_p is the specific heat capacity ($J kg^{-1} K^{-1}$).

10.3.2 Convection

The resulting changes in the density due to the variation of temperature lead to establishing the convection current, which occurs as a result of the bulk movement of the fluid, such as the liquid movement inside a can during the sterilization process. Heat transfer by conduction occurs simultaneously but is generally small compared to convection heat transfer (Karel et al., 1975). The rate of convection heat transfer is governed by Newton's law of cooling. This law states that the rate of heat transfer by convection is directly proportional to the heat transfer area and the temperature difference between the hot and cold fluid. The rate of heat transfer from a hot fluid to the surface of food is written as follows:

$$Q = h_s a (T_{bu} - T_s) \tag{10.2}$$

where Q is the rate of heat transfer (W), a is the surface area (m^2), T_s is the surface temperature, T_{bu} is the bulk fluid temperature, and h_s is the surface or film heat transfer coefficient ($W m^{-2} K^{-1}$). The surface heat transfer coefficient is a measure of the resistance to heat flow, caused by the boundary film and is therefore equivalent to the term $k/\Delta x$ in the conduction equation. The heat transfer coefficient is dependent on the physical properties of the fluid food used in thermal sterilization such as the density, viscosity, and specific heat. It is also related to gravity, which causes circulation due to the changes in density, the temperature difference, and the length or diameter of the container under investigation. The formulae, which relate these factors, are expressed as dimensionless numbers as shown below:

$$\textit{Nusselt number}, Nu = \frac{h d_v}{k} \tag{10.3}$$

$$\textit{Prandtl number}, Pr = \frac{C_p \mu}{k} \tag{10.4}$$

$$\textit{Grashof number}, Gr = \frac{{d_v}^3 \rho^2 g \beta \Delta T}{\mu^2} \tag{10.5}$$

where h is the heat transfer coefficient at the solid–liquid interface ($W m^{-2} K^{-1}$), d_v is the vertical dimension (m), k is the thermal conductivity of the fluid ($W m^{-1} K^{-1}$),

C_p is the specific heat capacity ($J\,kg^{-1}\,K^{-1}$), ρ is the density ($kg\,m^{-3}$), μ is the viscosity ($N\,s\,m^{-2}$), g is the acceleration due to gravity ($m\,s^{-2}$), β is the thermal expansion coefficient (K^{-1}), and ΔT is the temperature difference.

10.3.3 Unsteady-state heat transfer by conduction and convection

When a solid piece of food is heated or cooled by a fluid, the resistance to heat transfer is controlled by the surface heat transfer coefficient (h) and the thermal conductivity of the food (k). These two factors are related by the Biot number (Bi):

$$\text{Biot number, } Bi = \frac{h\delta}{k} \tag{10.6}$$

where h ($W\,m^{-2}\,K^{-1}$) is the heat transfer coefficient, δ is the diameter (m), and k is the thermal conductivity ($W\,m^{-1}\,K^{-1}$). At small Bi values (<0.2), the surface film is the main resistance to heat flow; in this case, the assumption of a negligible internal (conductive) heat transfer resistance of a solid food exposed to a convective heat transfer condition is valid (Barbosa-Canovas, Ma, & Barletta, 1997). However, in most applications, the thermal conductivity of the food limits the rate of heat transfer ($Bi > 1$). The calculations are complicated, and a series of charts are available to solve the unsteady-state equations for simple shaped foods. These charts relate the dimensionless temperature to Fourier number Fo (a dimensionless number which relates the thermal diffusivity, the size of the piece, and the time of heating or cooling) and the Biot number Bi (Fellows, 1996).

10.4 Effects of heat treatment during sterilization

The rate of heat transfers from the heating medium (steam) through the can wall into the outer layer of food is high. From this point, penetration of heat into the slowest heating zone (SHZ) (i.e., the location of the lowest heating zone) of the food is controlled by thermal properties of the food itself (Bhowmik & Tandon, 1987). The severe heat treatment during the thermal sterilization process produces substantial variations in nutritional and sensory qualities of foods. It is very important to have information on both the rate of heat penetration and its influence on the deactivation of microorganisms, to determine the processing time for a given food.

10.4.1 Heat penetration

The study of heat penetration into foods is of quite importance in thermal processing, which represents the most common technique used in food preservation. In thermal sterilization, heat is transferred from pressurized water or steam through the container wall into the food. Usually, the steam condensation heat transfer coefficient is very high and is not a limiting factor in heat transfer. The factors that influence the rate of heat penetration are:

1. Product specification.
2. Container size.
3. Retort's temperature.
4. Container shape.
5. Container type.

The heat penetration rate is measured by placing a thermocouple at the thermal center of a container to record temperatures in the food during sterilization, by assuming that all other points in the container receive more heat and are therefore adequately processed. Fellows (1996) stated that the thermal center is at the geometric center of the cylindrical containers for conductive heating foods and approximately one-third up from the base of the container for convective heating foods (Fig. 10.1). This is in agreement with the finding of Barbosa-Canovas, Ma, and Barletta (1997), who also stated that the lowest temperature (coldest) point for a solid product will be at the center of the can, whereas in liquid-type products, it will be on the lower part of the can axis. However, the exact position of the SHZ may change progressively in convection heating.

10.4.2 Heat resistance of microorganisms

Clostridium botulinum is the most dangerous heat-resistant, spore-forming pathogen that is likely to be present in low acid foods (pH > 4.5).

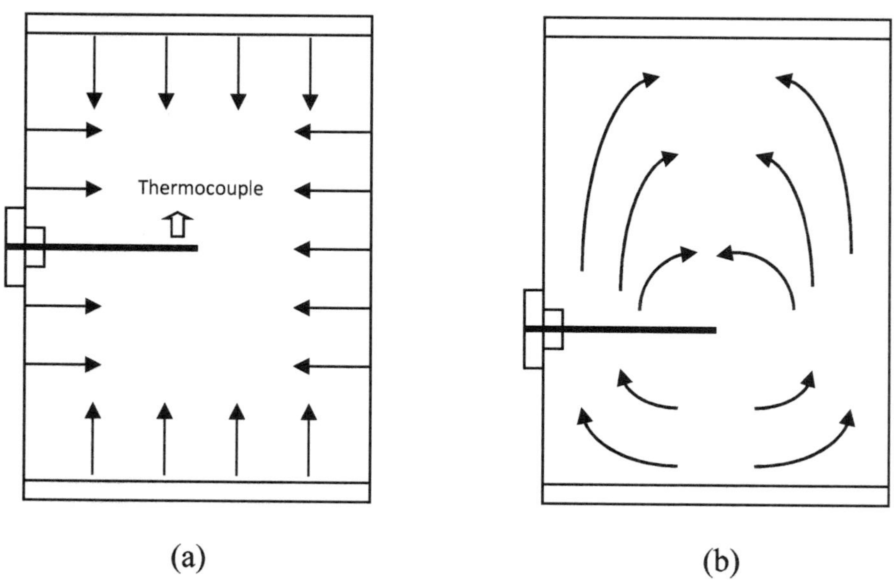

Fig. 10.1 Heat transfer in container by (A) conduction and (B) convection (Fellows, 1996).

Table 10.1 Heat resistance of some spore-forming bacteria used as a basis for heat sterilization processes for low-acid foods (Fellows, 1996).

Microorganism	Z (°C)	D_{121} (min)	Typical foods
Thermophilic (35–55 °C)			
Bacillus stearothermophilus	10	4.0	Vegetables, milk
Clostridium Thermosaccharolyticum	7.2–10	3.0–4.0	Vegetables
Mesophilic (10–40 °C)			
Clostredium sporogenes	8.8–11.1	0.8–1.5	Meats
Bacillus subtilis	4.1–7.2	0.5–0.76	Milk products
C. botulinum toxins A and B	5.5	0.1–0.3	Low-acid foods
Psychrophilic (−5 to 1.5°C)			
C. botulinum toxin E	10	3.0 (60°C)	Low-acid foods

Rigorous procedures and regulations are established by government agencies for the thermal sterilization processing of low-acid canned foods because of the wide-spread of public health concern about the anaerobic *C. botulinum*, a spore-forming microorganism that produces a toxin deadly to humans, even in very small amounts (Barbosa-Canovas, Ma, & Barletta, 1997). It can grow under anaerobic conditions inside a closed container to produce a potent exotoxin; therefore, its deactivation is a minimum demand in the thermal sterilization of foods. Generally, foods receive more than this minimum treatment, because of the presence of more heat-resistance spoilage bacteria (Table 10.1). In low-acid foods, processing could be based on the deactivation of spores, which are more heat resistant than those of *C. botulinum* since some spores do exist, such as *Bacillus stearothermophilus*. This organism's spores are extremely heat resistant (up to 20 times more resistant than *C. botulinum* spores). Growth of these spores results in "flat sour" spoilage because the acid is produced with little or no gas. Optimum growth temperatures for these flat sour thermophiles vary from about 35 to 55 °C, as shown in Table 10.1.

10.5 Effect of heat on microbial population

The deactivation of microorganisms in thermal processing takes place logarithmically and a sterile product cannot, therefore, be produced with certainty no matter how long the processing time. However, the probability of survival of a single microorganism can be predicted using details of heat resistance of the microorganisms and the temperature of time heating. This gives rise to a concept known as commercial sterility. In the canning industries, the trials of accelerated storage on a randomly selected can of food secure that these levels of commercial sterility are maintained before foods are released for retail sale.

The effect of heat on the microbial population is due to the denaturation of proteins, which destroys enzyme activity and enzyme-controlled metabolism in microorganisms. The destruction rate is generally taken as a first-order reaction. Differences between populations of the vegetative microbial as compared to the populations of microbial spores are evident within short periods of time after initial exposure to high temperature. The reduction pattern for both of them is the same, following the logarithmic reduction. When the microbial populations as a function of time are presented on semi-logarithmic coordinates, with the microbial population on the vertical logarithmic scale and time on standard horizontal coordinates, a linear reduction in microbial population with time is observed at a given temperature. This is known as the logarithmic order of death and is described by a death rate curve or survivor curve. A typical relationship for a microbial spore population with time on semi-logarithmic coordinates (i.e., death rate curve) is shown in Fig. 10.2 (Fellows, 1996).

The time required to devastate 90% of the microorganisms and to decrease their numbers by a factor of 10 is defined as the decimal reduction time (D_T). It is defined as the time required for a reduction of the one-log cycle in the microbial population. D_T value has been applied throughout thermal processing literature to quantify the influence of temperature on microbial populations. D_T values differ for different microbial species (Table 10.1). The larger D_T value at a given temperature indicates higher thermal resistance of the microbial population. A plot of decimal reduction time as a function of temperature on a semilog coordinates results in a linear relationship, which is referred to as the thermal death time (TDT) curve for a given microbial population (Fig. 10.3).

The slope of the TDT curve is termed the Z-value, which is defined as the temperature increase required to cause a reduction of the one-log cycle in D_T. It is also defined as the temperature difference that causes a tenfold change in the rate of microbial deactivation. The value of Z can be used to estimate the effect of changing temperature on the processing time, using the equation (Fryer, Pyle, & Rielly, 1997):

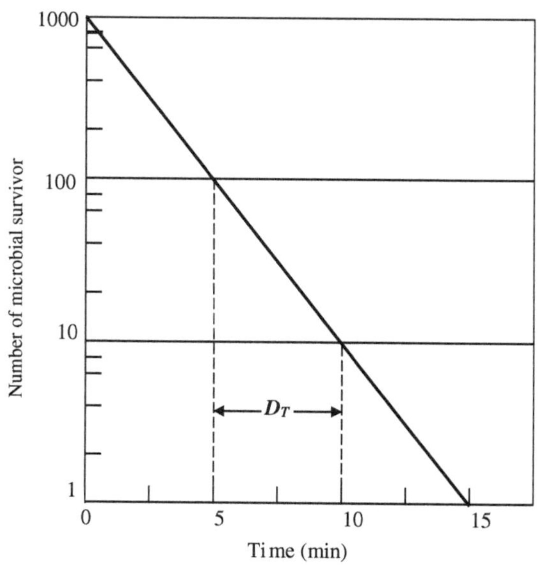

Fig. 10.2 Death rate curve of microbial population (Fellows, 1996).

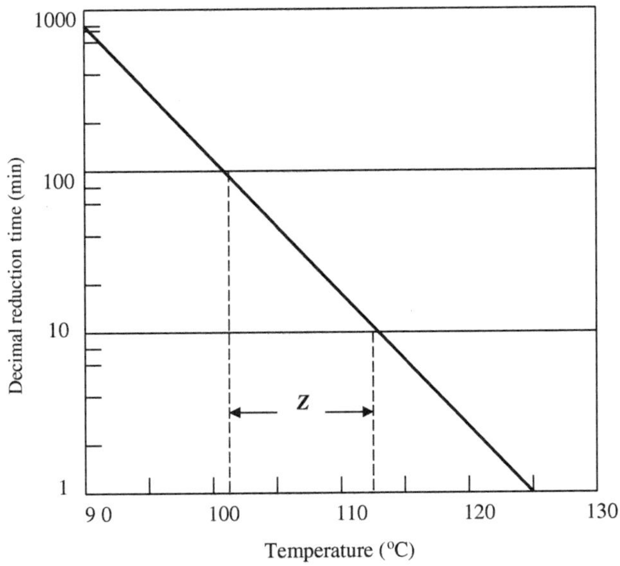

Fig. 10.3 Thermal death time (TDT) curve of microbial population (Fellows, 1996).

$$\text{Time required at temperature } T = \frac{\text{Time required at temperature } T_{\text{ref}}}{10^{\frac{(T-T_{\text{ref}})}{Z}}} \tag{10.7}$$

The deactivation of microorganisms is temperature defendant, and it follows that the value of decimal reduction time will decrease with increasing temperature. The D_T is a characteristic of the type of spore and is found experimentally changing with temperature according to:

$$D_T = D_{\text{Tref}} \cdot 10^{\frac{(T_{\text{ref}}-T)}{Z}} \tag{10.8}$$

where D_T is the decimal reduction time at temperature T, D_{Tref} is the decimal reduction time at T_{ref}, T_{ref} is an arbitrary reference temperature, and Z is a temperature-dependent characteristic microorganism assumed to be constant over normal processing conditions (David, Graves, & Carlson, 1996).

D_T and Z values are important parameters usually used to identify the heat resistance of an enzyme, a microorganism, or a chemical component of a food. It is also used to quantify the thermal process needed for a given microbial population. A large Z-value proposed that a given rise in exposure temperature for the microbial population results in a small change in D_T. This is due to the microbial population containing vegetative cells or microbial spores that have greater heat resistance to high temperatures than populations recognized by a lower Z-value (Heldman & Hartel, 1997).

The third important quantitative parameter in thermal processing is the TDT (F) or F-value. TDT defined as the time required to destroy a given percentage of microorganisms at a given Z-value and temperature. This parameter stated the decrease in the population of microbial pathogens required to establish product safety. The F-value represents the total time–temperature combination received by a food and is quoted with suffixes indicating the retort temperature and the Z-value of the target microorganism. A reference F-value F_{ref} is used to describe processes that operate at 121°C which are based on a microorganism with a Z-value of 10°C. F-value is defined by Eq. (10.9), which gives the relationship between F and the change in the concentration of spores in containerized food, as a result of heating or cooling (David et al., 1996; Karel et al., 1975).

$$F^{Z}_{T_{\text{ref}}} = D_{\text{Tref}}(\log n_1 - \log n_2) \tag{10.9}$$

where n_1 is the initial concentration of microorganisms at the starting of the thermal process and n_2 is the final concentration of microorganisms at the end of the thermal process. For the spores of *C. botulinum*, researchers have experimentally used n_1 less than or equal to 10^{12} spores mL^{-1} with an endpoint of n_2 less than 1 spore mL^{-1}, at D_{121} value of 0.1–0.2 min and Z-value of 10°C. Thus Eq. (10.9) can be written as

$$F^{Z}_{T_{\text{ref}}} = D_{\text{T}}\left(\log 10^{12} - \log 1\right) = 12D \tag{10.10}$$

which present the most common relationships ($F = 12D$) for *C. botulinum* in commercial sterilization. A $12D$ process is a process that minimizes cell numbers by 12 decimal reductions. The type of microorganism that is expected to contaminate the raw material determines the level of survival. In foods that contain more heat-resistant spoilage microorganisms than *C. botulinum*, the application of $12D$ would result in overprocessing and excessive loss of quality. In practice, a $5D$ or $8D$ process is therefore used to give the most economical level of food spoilage consistent with sufficient food quality and safety (David et al., 1996; Karel et al., 1975).

Eq. (10.9) forms the basis for the definition of a process F-value (F_{process}), which is defined as the time–temperature effect of a thermal process expressed as time in minutes at a single-reference temperature T_{ref} for a given value of microorganisms. It is defined by Eq. (10.11) (David et al., 1996):

$$\left(F^{Z}_{T_{\text{ref}}}\right)_{\text{Process}} = \int_{t_i}^{t_f} \frac{1}{10^{(T_{\text{ref}} - T_t)/Z}} \cdot \text{d}t \tag{10.11}$$

where t_i and t_f are the time at the beginning and end of the thermal process (min), respectively, that is, at end of cooling (min). For a successful sterilization process, this time must be greater than or equal to the required F-value at temperature T_{ref}. This approach is widely used in the food industry. An F-value of 3 min is taken as acceptable industrially, although process design values of at least 6 min are used in practice (Fryer et al., 1997).

10.6 Effect of heat on nutritional properties

The food contains a great number of nutrients. These nutrients are used for growth maintenance, tissue repair, and cell reproduction in the human body. Vitamins, fats, carbohydrates proteins, and minerals comprise the general classes of nutrients. Many food-processing operations, especially those do not involve heat, have a little or no effect on the nutritional quality of foods. Examples include mixing, cleaning, sorting, freeze-drying, and pasteurization. Heat processing is a major cause of change to the nutritional properties of food. It destroys some types of heat-labile vitamin (Fellows, 1996). The destruction of many vitamins by heat follows a similar first-order reaction to microbial destruction. In general, both D_{T} and Z values in nutrients are higher than those of microorganisms and enzymes. As a result, nutritional properties are better to conserve by the use of higher temperatures and shorter times during the sterilization process. It is, therefore, possible to select a particular time-temperature combination from a TDT curve, to optimize a process for nutrient retention. This concept forms the basis of individual quick blanching, ultrahigh-temperature sterilization, high-temperature short-time (HTST) pasteurization, and HTST extrusion.

10.7 Reaction kinetics of quality change

Thermal sterilization of food results in a change in the quality of food (biochemical changes). These changes include positive impacts such as the decrease in the microbial population and negative impact such as the loss of nutrients. Most reactions occurring in food follow well-established kinetics. The thermal deactivation of microorganisms, most nutrients, quality factors (color, texture, and flavor), and enzymes generally follow first-order reaction kinetics. That is the destruction rate of each of these components that is linearly dependent on the concentration of the component. This relationship is frequently referred to as a "logarithmic order of inactivation or destruction" since this is regardless of the initial concentration, a constant fraction of the remaining substrate reacts per unit time. The first-order response describes that the changes occurring during processing of food may be expressed mathematically as:

$$-\frac{\mathrm{d}C}{\mathrm{d}t} = k_{\mathrm{T}}C \tag{10.12}$$

where C is the quality attribute concentration, t is the exposure time, and, k_{T} is the reaction rate constant. Most changes occurring during thermal processing of food are more complicated (Heldman & Hartel, 1997). It has been accepted in practice to use relatively simple expressions to describe the change of quality in food, as well as to use parameters similar to the rate constant (k_{T}). By using the form of Eq. (10.12),

$$\ln\frac{C}{C_o} = -k_{\mathrm{T}}t \tag{10.13}$$

where C_o is the quality attribute initial concentration. From the values of the reaction rate constant (k_T) and the quality attribute initial concentration, the remaining amount of the attribute can be estimated for any time of exposure to a process.

The relationships between the microbial populations and time, as well as the impact of temperature, are quite similar to those used to describe kinetic parameters in first-order chemical reactions explained above. The reaction rate constant (k_T) was used to describe the change in concentration of a reactant as a function of time. In microbial populations, D_T value is used to describe the same relationship. It follows that the relationship between k_T and the D_T is given by the following equation:

$$k_T = \frac{2.303}{D_T} \tag{10.14}$$

Two principal methods are used to describe the dependence of the reaction rate constant on temperature: (1) the Arrhenius equation and (2) TDT curves, which are explained above in Section 10.4. The dependence of the reaction rate constant on temperature as described by the Arrhenius equation is:

$$k_T = A \exp\left(\frac{-E}{R_g T}\right) \tag{10.15}$$

where A is the pre-exponential factor (s^{-1}); E is the activation energy (kJ (kg mol)$^{-1}$); R_g is the universal gas constant (kJ (kg mol)$^{-1}$ K^{-1}); T is the temperature (°C); k_T is the reaction rate constant at temperature T (s^{-1}).

The Arrhenius equation is very useful and important in representing the influence of temperature on the rate of change in product quality during processing and storage (Fellows, 1996). It has a sounder theoretical base compared to that of F-value (Fryer et al., 1997).

10.8 Studies on thermal sterilization

Thermal sterilization of canned foods has been one of the most widespread methods used for food preservation during the 20th century and has significantly contributed to the nutritional well-being of much of the world's population (Teixeira & Tucker, 1997). The objective of thermal sterilization is to produce safe and high-quality food at a price that the consumer is able and willing to pay. As explained earlier, it is a function of several factors such as the product heating rate, surface heat transfer coefficient, initial food temperature, heating medium come-up time, and Z-value for the quality factor and target F_{ref} value (Silva, Hendrickx, Oliveira, & Tobback, 1992).

In thermal sterilization of food, the heating medium (steam or hot water) temperatures can deviate significantly from their design values during the heating phases. Such deviations may seriously endanger public safety due to underprocessing (under sterilization) of food, or waste energy, effort and quality due to overprocessing of food (Datta, Teixeira, & Manson, 1986). For these reasons, online retort control in thermal sterilization has been well studied by Teixeira and Tucker (1997), Teixeira and Manson (1982), Datta et al. (1986), and Giannoni-Succar and Hayakawa (1982), to assure safety, quality, and process efficiency of thermally processed canned foods.

10.8.1 Thermal sterilization of food in cans

In the design of thermal food process operations, the temperature in the SHZ and the thermal center of the food during the process must be known. Conduction, natural convection, and forced convection are important means of heat transfer in the thermal processing of food. Foods like canned tuna, thick syrups, purees, and concentrates are usually assumed to be heated by pure conduction. For these foods, the required processing time is generally determined by analytical or numerical solution of the heat conduction equation (Datta et al., 1986). Dincer, Varlik, and Gun (1993) have studied the transient heat transfer during the sterilization of canned foods. Their model was based on solving the conduction equation with constant wall temperature as a boundary condition. It was established experimentally that during heating the convective circulating stream rises along the can wall and falls in the area center of the can (Hiddink, 1975). It was also established both experimentally (Nickerson & Sinskey, 1972) and theoretically (Datta & Teixeira, 1988) that the SHZ in a convection-heated food in a cylindrical can was a torrid continuously altering its location.

While traditionally the variation in the temperature and the SHZ is measured using thermocouples, there is a growing interest toward the use of mathematical models to predict the food temperature during the thermal sterilization process (Datta & Teixeira, 1987, 1988; Naveh, Kopelman, & Pflug, 1983; Nicolai, Verboven, Scheerlinck, & De Baerdemaeker, 1998; Teixeira, Dixon, Zahradnik, & Zinsmeister, 1969).

Most of the numerical studies have been carried out for water as a model liquid food product having a constant viscosity (Kumar & Bhattacharya, 1991). The influence of natural convection heating during the sterilization process of sodium carboxylmethyl cellulose (CMC) has been studied by Kumar et al. (1990) and Kumar and Bhattacharya (1991) using CFD. Lanoiselle, Candau, and Debray (1986) developed a linear recursive model to represent the heat transfer inside a can during sterilization in a retort and predicted the temperature of canned foods during thermal processing. Akterian (1994) developed a numerical model for the determination of the unsteady-state temperature field in conduction-heated canned foods of various shapes under convective boundary conditions. The numerical predictions of the transient temperature and velocity profiles in a still retort during natural convection heating of canned liquid foods have been carried out by Datta and Teixeira (1988). They presented a considerable internal circulation at the bottom of the can and showed that the SHZ has a doughnut shape and is close to the bottom of the can at about one-tenth of the can height. Thermal sterilization process of a viscous liquid food in a metal can, sit in an upright position and heated from the sidewall in a still retort, was simulated by Kumar et al. (1990). The equations of continuity, momentum, and energy conservation for an axisymmetric case were solved to provide plots of temperature, velocity, and streamlines for natural convection heating. They also presented a simulation for the same can when its bottom and top surfaces were insulated (Kumar et al., 1990). The results indicated that natural convection again tends to push the slowest heating region to the bottom of the can. The sterilization of a can filled with a viscous liquid sodium CMC, heated by condensing steam on all sides and the case where the top is insulated, was simulated by Ghani, Farid, and Chen (1998). The temperature difference from top to bottom of the can at the end of heating was 12°C for the can with top insulated and 10°C for the case of can heated from all sides. The results show also that the locations of the SHZ for both cases are similar, which is due to the strong effect of the natural convection current, even with the use of highly viscous liquid food.

In the study of Ghani, Farid, Chen, and Richards (1999a), natural convection heating within a can of liquid food during sterilization is simulated by solving the governing equations for continuity, momentum, and energy conservation for an axisymmetric case using a commercial computational fluid dynamics (CFD) package (PHOENICS). Transient flow patterns and temperature profiles within model liquids (CMC and water) have been predicted. The model liquids, CMC and water, were assumed to have constant properties except for the viscosity (temperature-dependent) and density (Boussinesq approximation). It has been shown that the action of natural convection forces the SHZ to migrate toward the bottom of the can as expected. The magnitude of the axial velocity was found to be in the range of 10^5–$10^4\,\mathrm{m\,s^{-1}}$ for CMC (Fig. 10.4A) and of 10^2–$10^1\,\mathrm{m\,s^{-1}}$ for water; these magnitudes vary with time and position in the can. The time required for the SHZ to reach the sterilization temperature of 100°C was 1800 s for CMC (Fig. 10.4B) and only 150 s for water.

Thermal sterilization of canned liquid food (carrot-orange soup) in a metal can lie horizontally and heated at 121°C from all sides is simulated for a three-dimensional geometry (Ghani, Farid, & Chen, 2002a, 2002b). Temperature and velocity profiles and shapes of the SHZ during natural convection heating of canned liquid foods are presented and studied. The simulation shows clearly the action of natural convection, which forces the SHZ to migrate toward the bottom of the can as expected (Fig. 10.5).

The secondary flow formation and its effect on the shape of the SHZ are evident. The results of this work are compared with those for vertical can. It shows faster heating in the vertical can, which is expected due to the enhancement of natural convection caused by its longer height.

When a microbial population in food is exposed to an elevated temperature, changes in individual microbial cells will result in a reduction of the population over time. Any high temperatures above 60°C would cause such reductions in the microbial population, such as those for *Salmonella* spp., *Listeria monocytogenesis*, or *Escherichia*. To have a similar impact on microbial spores, such as *C. botulinum*, this processing temperature must exceed 100°C. In both situations, the effect of temperature results in a similar pattern of microbial population reduction. In reality, the death of microorganisms is expected to start at an early stage of heating, especially near the wall, where the temperature approaches 121°C very fast. Therefore, it is necessary to

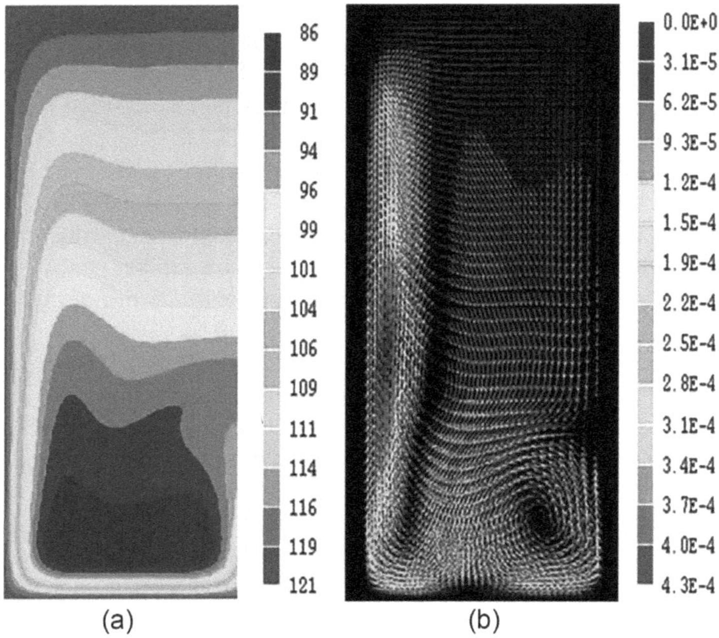

Fig. 10.4 Temperature and velocity profile for CMC in a cylindrical can heated by condensing steam after 1157 s (Ghani et al., 1999a).

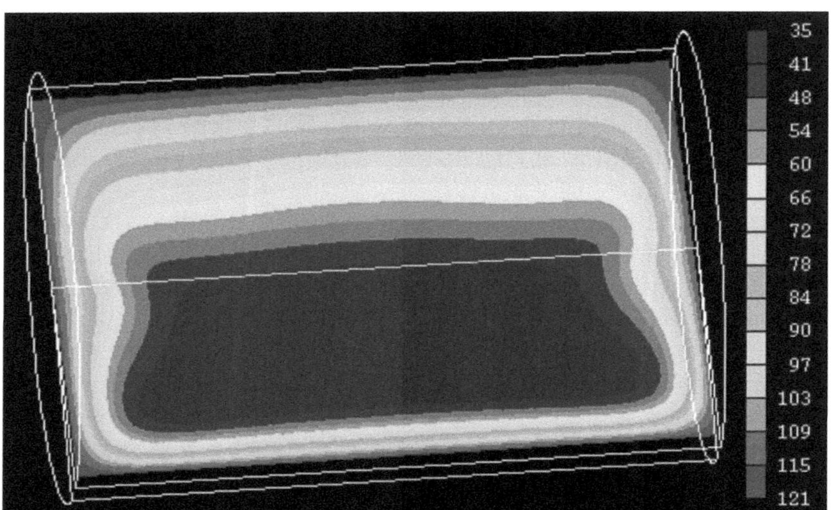

Fig. 10.5 Temperature profiles of carrot-orange soup in a 3D cylindrical can lying horizontally and heated by condensing steam after 600 s at a radial-vertical plane (Ghani et al., 2002a, 2002b).

solve the partial differential equation, governing the bacteria concentration (Eq. 10.16), coupled with the equations of continuity, momentum, and energy.

$$\frac{\partial C}{\partial t} + v_x \frac{\partial C}{\partial x} + v_y \frac{\partial C}{\partial y} + v_z \frac{\partial C}{\partial z} = D\left[\frac{\partial^2 C}{\partial x^2} + \frac{\partial^2 C}{\partial y^2} + \frac{\partial^2 C}{\partial z^2}\right] - k_T C \tag{10.16}$$

where C is the concentration of the bacteria in food liquid (number of bacteria m^{-3}), t is the heating time (s), v_x, v_y, and v_z are the velocity components in the x, y, and z directions, respectively ($m\,s^{-1}$), D is the diffusion coefficient of bacteria ($m^2\,s^{-1}$), and k_T is the reaction rate constant of bacteria at temperature T (s^{-1}).

Datta (1991) has followed this approach to study the sterilization of liquid food in a nonagitated cylindrical heated from all sides and described the computational procedure for obtaining the full range of biochemical changes during processing without explicitly following the liquid elements. The computational results by Datta (1991) are presented in the form of transient distribution of a sterilization parameter and showed that the lowest sterilization achieved by any portion of the fluid in the system was considerably more than the sterilization level normally calculated by using the temperature of the SHZ. In the analysis presented by Datta (1991), the diffusion was neglected and the bacteria were assumed to undergo convective motion only.

A natural convection heating of a viscous liquid in a cylindrical can, heated from all sides as well as a computational procedure for describing the changes in live bacteria concentration and its transient spatial distributions during sterilization processing, is presented and studied by Ghani, Farid, Chen, and Richards (1999b). Saturated steam at 121°C was the heating media and the high viscous liquid (sodium CMC) was used as the model liquid food, having constant properties except for the viscosity (temperature-dependent) and density (Boussinesq approximation). The governing equations of continuity, momentum, and energy were solved together with that of the concentration mentioned above, using the finite volume method (FVM) and a CFD code PHOENICS. Arrhenius equation was used to describe the kinetics of bacteria death and the influence of temperature on the reaction rate constant. The bacteria were assumed to diffuse in the liquid food following the modified Brownian diffusion equation. Transient flow pattern, live bacteria concentration, and temperature profiles are presented in this work. The results show that the action of natural convection forces the SHZ to migrate toward the bottom of the can. The simulations also show how the concentration of the live bacteria depends on both temperature distribution and flow pattern (Fig. 10.6). The effect of diffusion on the rate of sterilization is negligible in the cases simulated in this study.

Profiles of temperature, bacteria concentration, and the concentrations of vitamins C (ascorbic acid), B_1 (thiamine), and B_2 (riboflavin) in a can filled with cherry juice during thermal sterilization have been investigated and studied through a numerical simulation (Ghani, Farid, Chen, & Richards, 2001). Different heating medium temperatures of 121, 130, and 140°C were tested. In this simulation, the continuity, momentum, and energy equations are solved numerically, together with that of bacteria and vitamins concentrations, combined with reaction kinetics models. In this

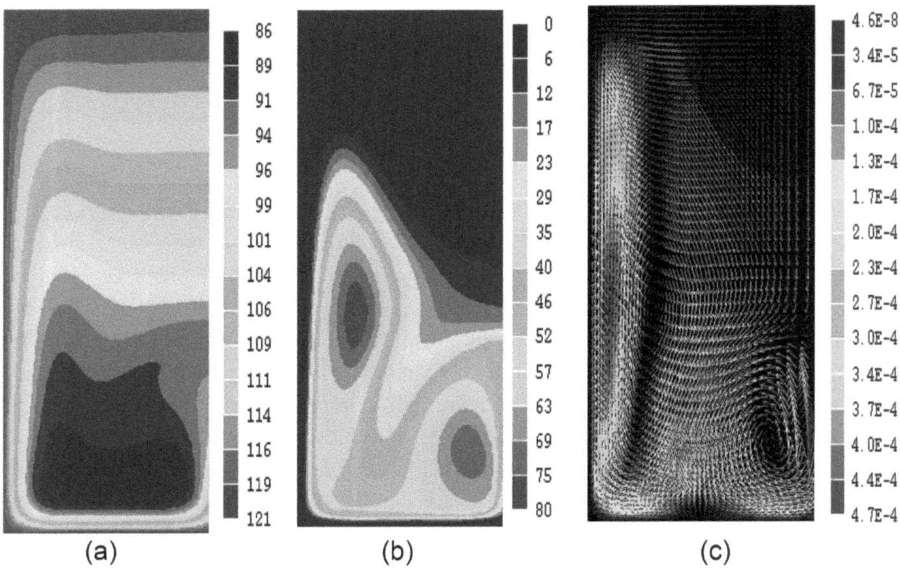

Fig. 10.6 Temperature (A), bacterial deactivation (B), and flow pattern (C) profiles in a can filled with CMC and heated by condensing steam after 19 min. The right-hand side of each figure is the center line (Ghani et al., 1999b).

work, natural convection that occurs during thermal sterilization of concentrated cherry juice −74 degree Brix in a two-dimensional cylindrical can be heated from all sides has been studied. The simulations show clearly the dependency of the concentration of live bacteria and different vitamins on both the temperature distribution and flow pattern as sterilization progresses. The results also show that the best sterilization temperature may not be always 121°C, depending on the quality requirements imposed on individual food material of concern. The effect of can lying horizontally and rotated axially in a still retort during the process of thermal sterilization of liquid food is also simulated and presented (Ghani, Farid, & Zarrouk, 2003).

10.8.2 Thermal sterilization of food in pouches

Retortable flexible containers are laminate structures that are thermally processed like a can. Pouch materials provide superior barrier properties for long shelf-life, toughness, and puncture resistance, seal integrity, and withstand the rigors of thermal sterilization processing. Retortable flexible containers may be retort pouches or semi-rigid containers. It is perhaps the most considerable advance in food packaging since the development of the metal can (Mermelstein, 1978). The structure of the retortable pouch used today is made from a laminate of three materials: an outer layer of 12 μm polyester film for strength, adhesive laminated to a middle layer of 9–18 μm aluminum foil as a moisture, light, and gas barrier, which is laminated to the inner layer of 76 μm polypropylene film as the heat seal and food-contact material. The use of polyester film is to provide high-temperature resistance, toughness, and

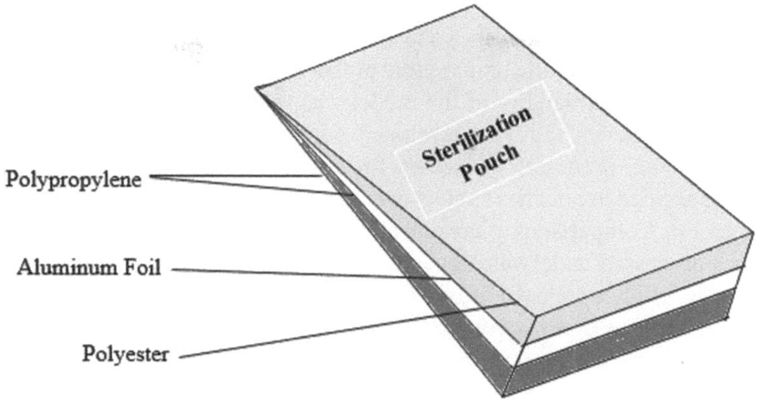

Fig. 10.7 Sketch of retort pouch and its layers.

printability (Rahman, 1999). The retort pouch with its multilayer polymer foil is shown in Fig. 10.7.

Several articles were published on the benefit of using pouches in food processing (Lampi, 1980; Mermelstein, 1978). Sterilization of food in cans has been well studied both experimentally and theoretically since the 1920s. The theory and practice of establishing and evaluating thermal processes for cans can be found in several sources, and the mathematical modeling of thermal processing has also been thoroughly reviewed (Hawakawa, 1977a, 1977b, 1978; Holdsworth, 1985). The flexible retort pouches were introduced to the market in the early 1960s; little information is available on the temperature distribution within the pouch during the sterilization process. The results of sterilization of food in cans cannot be extended to pouches due to the complicated geometry of the later. Pouch analysis will require further computer modeling in a three-dimensional domain.

A numerical solution of the heat conduction equation 1D to obtain an optimal temperature profile for the pouches and to achieve a minimum loss in the sensory and nutritional quality of the processed food is presented by Ohlsson (1980). The geometrical configuration of the pouch used in this study was assumed to be an infinite slab to conveniently apply the finite difference method (FDM). The calculations were performed at different conditions and sterilization temperatures to choose the optimal sterilization temperature with minimal quality changes. Manson, Stumbo, and Zahradnik (1970) developed a model to evaluate the thermal processing of conduction-heated foods in a rigid rectangular container by solving the differential heat conduction equation in 3D using the FDM. The history of the time–temperature predicted at different locations in the container was used to assess lethality and nutrient retention during sterilization.

The effect of the heat sterilization process for canned foods on the quality and nutrient retention has been a major concern in thermal processes of food since the beginning of the canning industry. Castillo, Barreiro, and Salas (1980) developed a model to predict the retention of nutrients with first-order kinetics of thermal degradation in foods packaged in retortable pouches and sterilized by thermal conduction. The model was effective in predicting the temperature at the center of the container at the end of the heating period. The validity of the model was verified experimentally on a simulated food. Hawakawa (1977b) developed computerized models to estimate the appropriate thermal processes of canned foods based on the Ball formula method, which can be applied to pouches subject to constant retort temperature during the sterilization process. Comparisons of general and Ball formula methods were also studied, for pouches processed underwater in a still vertical retort (Spinak & Wiley, 1982).

All previous studies are used for rectangular containers. The assumption of a rectangular container was necessary to apply easy numerical techniques for thermal sterilization evaluation such as the FDM used in most of these studies. This assumption led to an overestimation of process time resulting in less retention of nutrients and greater loss of sensory quality in the food (Tandon & Bhowmik, 1986).

A computer model to evaluate the thermal processing of a 2D pouch containing a conduction-heated food was developed by Bhowmik and Tandon (1987). In this model, a transient two-dimensional heat conduction equation was solved using a modified FDM for a pouch filled with food. Temperature distributions and sterilizing values predicted in this work were compared with similar values obtained by applying a finite element method, and the results were found close between the two methods. The study of Terajima (1975) presented experimentally and theoretically the overall heat transfer coefficient from the heating medium to a pouch containing liquid food products. The overall heat transfer coefficient from the heating medium to the food in the pouch is found mainly governed by the heat transfer coefficient from the inner surface of the pouch to the contents. The circulation rate of heated water has been identified as a critical processing factor, which is important for the maintenance of uniform retort temperature during processing.

In the study of Ghani, Farid, and Chen (2001), transient temperature, velocity profiles, and the migration of the SHZ during natural convection heating of food in a three-dimensional pouch (Fig. 10.8) are simulated and presented for the first time.

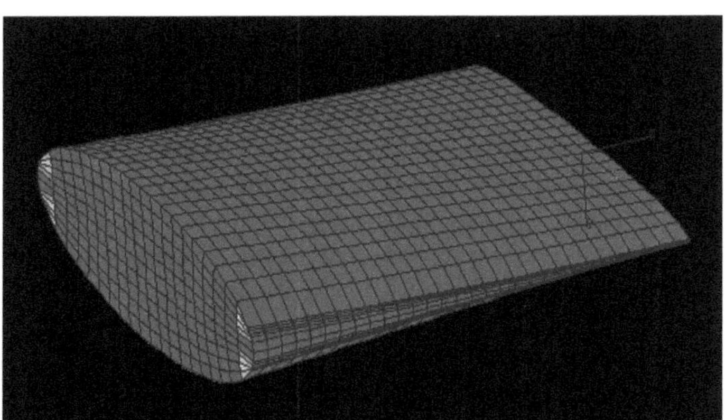

Fig. 10.8 Pouch geometry and grid mesh (Ghani, Farid, & Chen, 2001).

The simulation result showed that the maximum axial velocity of the soup used was 10^{-2}–10^{-4} mm s^{-1}, which was due to the small height of the pouch and high viscosity of the soup. The SHZ was found to migrate into a region within 30%–40% of the pouch height above the bottom and at a distance of approximately 20%–30% of the pouch length from its widest end. The concentration of the viable bacteria (*B. stearothermophilus* spores), profile temperature, and velocity vector in a 3D pouch filled with beef-vegetable soup during thermal sterilization is presented and analyzed by Ghani et al. (2002a, 2002b). In this simulation, the partial differential equations of the continuity, momentum, and energy equations are solved numerically, together with the concentration equation of viable bacteria, based on FVM of analysis. The model liquid food used is assumed to have constant properties except for the density (Boussinesq approximation) and viscosity (temperature dependent). The simulation results show clearly the dependency of the concentration of live bacteria on both the temperature distribution and the flow pattern as sterilization proceeds (Fig. 10.9). Experimental measurements are done to validate the bacteria concentration profiles. The experimentally measured concentrations of *B. stearothermophilus* spores during different periods of heating are compared in the study with those obtained from simulations, showing good agreement.

A theoretical analysis of heating and cooling cycle during sterilization of the 3D pouch filled with the carrot-orange soup was presented and analyzed (Ghani, Farid, & Chen, 2003). Both the shape of the SHZ and the slowest cooling zone (SCZ) with the transient temperature were presented and studied. The whole heating and cooling cycles of 3600 and 1200 s, respectively, were covered in this simulation. Saturated steam at 121°C and water at 20°C were assumed to be the heating and cooling media, respectively. The liquid food model used in the simulation has a temperature-dependent viscosity and density. At the end of heating, the SHZ was found settled into a region within 30%–40% of the pouch height above the bottom and at a distance of approximately 20%–30% of the pouch length from its deepest

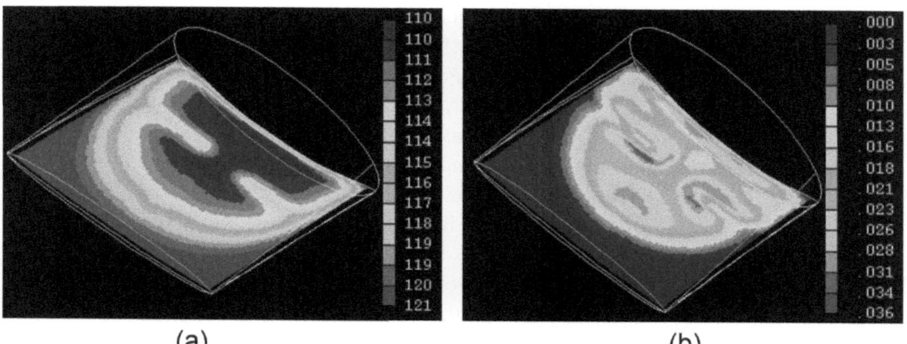

Fig. 10.9 Temperature (A) and relative concentration (B) profiles of *Bacillus stearothermophilus* at 30% height from the bottom of pouch filled with beef-vegetable soup and heated by condensing steam after a period of 1500 s (Ghani et al., 2002a, 2002b).

end. In the cooling cycle, the SCZ was found to develop in the core of the pouch and gradually migrated toward the widest end and was found settled in a region about 60%–70% of the pouch height.

Other methods of thermal sterilization of food using retort food pouches are presented and studied, such as assessment of the long-term stability of retort pouch foods (Catauro & Perchonok, 2012), modeling the retort process for Indian traditional foods (Gokhale & Lele, 2012), assessment of the long-term stability of retort pouch foods to support extended duration spaceflight (), development of retort process (Abhishek et al., 2014; Shahab, Kumar, Tamilselvan, & Nadanasabapathi, 2013), development and evaluation of egg-based ready-to-eat products (Kumar et al., 2015), and energy efficiency measures for batch retort sterilization (Peesel, Philippa, Schumma, Hesselbacha, & Walmsley, 2016). The processing of Chettinad style goat meat curry, ready-to-drink traditional Thari Kanchi payasam, chopped mussel meat, ready-to-serve crab sandwich in retort pouches, and the product development and process optimization is also studied (Rajkumar, Dushyanthan, & Das, 2010; Shahab et al., 2013; Sreelakshmi, Manjusha, Nagalakshmi, Chouksey, & Venkateshwarlu, 2015; Tribuzi, Falcão de Aragao, & Laurindo, 2015).

References

Abhishek, V., Kumar, R., George, J., Nataraju, S., Lakshmana, J. H., Kathiravan, T., et al. (2014). Development of retort process for ready-to-eat (RTE) soy-peas curry as a meat alternative in multilayer flexible retort pouches. *International Food Research Journal, 21*(4), 1553–1558.

Akterian, S. G. (1994). Numerical simulation of unsteady heat conduction in arbitrary shaped canned foods during sterilization processes. *Journal of Food Engineering, 21*, 343–354.

Barbosa-Canovas, G. V., Pothakamury, U. R., Palou, E., & Swanson, B. G. (1997). *Nonthermal preservation of foods*. New York, USA: Marcel Dekker Inc.

Barbosa-Canovas, G. V., Ma, L., & Barletta, B. (1997). *Food engineering laboratory manual*. Lancaster, PA: Technomic Publishing Company, Inc.

Barbosa-Canovas, G. V., & Zhang, Q. H. (2000). *Pulsed electric field in food processing: Fundamentals aspects and applications*. Lancaster, PA: Technomic Publishing Company, Inc.

Barbosa-Canovas, G. V., Gongora-Nieto, M. M., & Swanson, B. G. (1998). Non thermal electrical methods in food preservation. *Food Science and Technology International, 4* (5), 363–370.

Bhowmik, S. R., & Tandon, S. (1987). A method of thermal process evaluation of conduction heated foods in retortable pouches. *Journal of Food Science, 52*(1), 202–209.

Castillo, P. F., Barreiro, J. A., & Salas, G. R. (1980). Prediction of nutrient retention in thermally processed heat conduction food packaged in retortable pouches. *Journal of Food Science, 45*, 1513.

Datta, A. K., Teixeira, A. A., & Manson, J. E. (1986). Computer based retort control logic for on-line correction process deviations. *Journal of Food Science, 51*(2), 480–483.

Datta, A. K., & Teixeira, A. A. (1987). *Numerical modeling of natural convection heating in canned liquid foods*. St. Joseph, MI: Transaction of American Society of Agricultural Engineers. No. 86-6516.

Datta, A. K., & Teixeira, A. A. (1988). Numerically predicted transient temperature and velocity profiles during natural convection heating of canned liquid foods. *Journal of Food Science*, *53*(1), 191–195.

Catauro, P. M., & Perchonok, M. H. (2012). Assessment of the long-term stability of retort pouch foods to support extended duration spaceflight. *Journal of Food Science*, *77*(1), S29–S39.

Datta, A. K. (1991). Mathematical modeling of biochemical changes during processing of liquid foods and solutions. *Biotechnology Progress*, *7*, 397–402.

David, J. R. D., Graves, R. H., & Carlson, V. R. (1996). *Aseptic processing and packaging of food*. Boca Raton, FL: CRC Press Inc.

Dincer, I., Varlik, C., & Gun, H. (1993). Heat transfer rate variation in a canned food during sterilization. *International Communications in Heat and Mass Transfer*, *20*, 301–309.

Farid, M. M., Chen, X. D., & Dost, Z. (2000). Ultraviolet sterilization of orange juice. In *8th International Congress on Engineering and Food, April 2000, Pueblo, Mexico*.

Fellows, P. J. (1996). *Food processing technology (principles and practise)*. UK: Woodhead Publishing Series in Food Science and Technology.

Fryer, J. P., Pyle, D. L., & Rielly, C. D. (1997). *Chemical engineering for the food industry*. Blackie Academic & Professional.

Furukawa, S., & Hayakawa, I. (2000). Investigation of desirable hydrostatic pressure required to sterilize Bacillus stearothermophilus IFO 12550 spores and its sterilization properties in glucose, sodium chloride and ethanol solutions. *Journal of Food Research International*, *33*, 901–905.

Ghani, A. G., Farid, M. M., & Chen, X. D. (1998). A CFD simulation of the coldest point during sterilization. In *The 26th Australian Chemical Engineering Conference*. 28–30 September 1998, Port Douglas, Queensland, No. 358.

Ghani, A. G., Farid, M. M., Chen, X. D., & Richards, P. G. (1999a). Numerical simulation of natural convection heating of canned food by computational fluid dynamics. *Journal of Food Engineering*, *41*(1), 55–64.

Ghani, A. G., Farid, M. M., Chen, X. D., & Richards, P. G. (1999b). An investigation of deactivation of bacteria in canned liquid food during sterilization using computational fluid dynamics (CFD). *Journal of Food Engineering*, *42*(4), 207–214.

Ghani, A. G., Farid, M. M., & Chen, X. D. (2001). Thermal sterilization of canned food in a 3-D pouch using computational fluid dynamics. *Journal of Food Engineering*, *48*, 147–156.

Ghani, A. G., Farid, M. M., Chen, X. D., & Richards, P. G. (2001). A computation fluid dynamics study on the effect of sterilization temperature on bacteria deactivation and vitamin destruction. *Journal of Process Mechanical Engineering*, *15*(E), 9–17.

Ghani, A. G., Farid, M. M., & Chen, X. D. (2002a). Numerical simulation of transient temperature and velocity profiles in a horizontal can during sterilization using computational fluid dynamics. *Journal of Food Engineering*, *51*(2002), 77–83.

Ghani, A. G., Farid, M. M., & Chen, X. D. (2002b). Theoretical and experimental investigation on the thermal inactivation of *Bacillus stearothermophilus* during thermal sterilization in food pouches. *Journal of Food Engineering*, *51*(2002), 221–228.

Ghani, A. G., Farid, M. M., & Zarrouk, S. J. (2003). The effect of can rotation on sterilization of liquid food using computational fluid dynamics. *Journal of Food Engineering*, *57*, 9–16.

Ghani, A. G., Farid, M. M., & Chen, X. D. (2003). A computational and experimental study of heating and cooling cycles during thermal sterilization of liquid in pouches using CFD. *Journal of Process Mechanical Engineering*, *217*, 1–9.

Giannoni-Succar, E. B., & Hayakawa, K. A. (1982). Correction factor of deviant thermal processes applied to packaged heat conduction food. *Journal of Food Science*, *47*(2), 642–646.

Gokhale, S. V., & Lele, S. S. (2012). Retort process modelling for Indian traditional foods. *Journal of Food Science and Technology*, *51*(11).

Hawakawa, K. (1977a). Review on computerized prediction of nutrients in thermally processed canned food. *Journal of the AOAC*, *60*, 1243.

Hawakawa, K. (1977b). Mathematical models for estimating proper thermal processes and their computer implementation. *Advances in Food Research*, *23*, 75.

Hawakawa, K. (1978). A critical review of the mathematical procedures for determining proper heat sterilization processes. *Food Technology*, *32*(3), 59.

Heldman, D. R., & Hartel, R. W. (1997). *Principles of food processing*. New York: Chapman & Hall.

Herson, A. C., & Hulland, E. D. (1980). *Canned foods. Thermal processing and microbiology*. Edinburgh: Churchill Livingstone.

Hiddink, J. (1975). *Natural convection heating of liquids, with reference to sterilization of canned food*. Agricultural Research Report No. 839 Wageningen, The Netherlands: Center for Agricultural Publishing and Documentation.

Holdsworth, S. D. (1985). Optimization of thermal processing—A review. *Journal of Food Engineering*, *4*, 89.

Karel, M., Fennema, O. R., & Lund, D. B. (1975). *Principles of food science, Part II*. New York: Marcel Dekker, Inc.

Kumar, A., Bhattacharya, M., & Blaylock, J. (1990). Numerical simulation of natural convection heating of canned thick viscous liquid food products. *Journal of Food Science*, *55*(5), 1403–1411.

Kumar, A., & Bhattacharya, M. (1991). Transient temperature and velocity profiles in a canned non-Newtonian liquid food during sterilization in a still—Cook retort. *International Journal of Heat and Mass Transfer*, *34*(4/5), 1083–1096.

Kumar, R., George, J., Kumar, D., Jayaprahash, C., Nataraju, S., Lakshmana, J. H., et al. (2015). Development and evaluation of egg based ready-to-eat (RTE) products in flexible retort pouches. *African Journal of Food Science*, *9*(4), 243–251.

Lampi, R. A. (1980). Retort pouch: The development of a basic packaging concept in today's high technology era. *Journal of Food Process Engineering*, *4*, 1–18.

Lanoiselle, J. L., Candau, Y., & Debray, E. (1986). Predicting internal temperature of canned food during thermal processing using a linear recursive model. *Journal of Food Science*, *60* (3), 715–719.

Manson, J. E., Stumbo, C. R., & Zahradnik, J. W. (1970). Evaluation of lethality and nutrient retentions of conduction heating foods in rectangular containers. *Food Technology*, *24*(11), 109–113.

Martin, O., Qin, B. L., Chang, F. J., Barbosa-Canovas, G. V., & Swanson, B. G. (1997). Inactivation of Escherichia coli in skim milk by high intensity pulsed electric fields. *Journal of Food Process Engineering*, *20*, 317–336.

Mermelstein, N. H. (1978). Retort pouch earns 1978 IFT food technology industrial achievement award. *Food Technology*, *32*(6).

Naveh, D., Kopelman, I. J., & Pflug, I. J. (1983). The finite element method in thermal processing of foods. *Journal of Food Science*, *48*(4), 1086–1093.

Nicolai, B. M., Verboven, B., Scheerlinck, N., & De Baerdemaeker, J. (1998). Numerical analysis of the propagation of random parameter fluctuations in time and space during thermal food processes. *Journal of Food Engineering*, *38*, 259–278.

Nickerson, J. T., & Sinskey, A. J. (1972). *Microbiology of foods and food processing*. NY: Elsevier Publishing Co.

Ohlsson, T. (1980). Optimal sterilization temperature for flat containers. *Journal of Food Science*, *45*, 848–852.

Palou, E., Lopez-Malo, A., Barbosa-Canovas, G. V., & Swanson, B. G. (1999). High pressure treatment in food preservation. In M. S. Rahman (Ed.), *Handbook of food preservation* (pp. 815–854). New York: Marcel Dekker.

Peesel, R. H., Philippa, M., Schumma, G., Hesselbacha, J., & Walmsley, T. G. (2016). Energy efficiency measures for batch retort sterilization in the food processing industry. *Chemical Engineering Transaction*, *52*. ISBN 978-88-95608-42-6; ISSN 2283-9216.

Qin, B.-L., Zhang, Q., Barbosa-Canovas, G. V., Swanson, B. G., & Pedrow, P. D. (1994). Inactivation of microorganisms by pulsed electric fields with different waveforms. *IEEE Transactions on Dielectrics and Electrical Insulation*, *1*(6), 1047–1057.

Qin, B.-L., Barbosa-Canovas, G. V., Swanson, B. G., Pedrow, P. D., & Olsen, R. G. (1998). Inactivating microorganisms using a pulsed electric field continuos treatment system. *IEEE Transactions on Industry Applications*, *34*(1), 43–50.

Rahman, M. S. (1999). Handbook of food preservation. *Food science and technology. A series of monographs, textbooks, and reference books*. New York: Marcel Dekker, Inc.

Rajkumar, V., Dushyanthan, K., & Das, A. K. (2010). Retort pouch processing of Chettinad style goat meat curry—A heritage meat product. *Journal of Food Science and Technology*, *47*(4), 372–379.

Sancho, F., Lambert, Y., Demazeau, G., Largeteau, A., Bouvier, M., & Narbonne, J. F. (1999). Effect of ultra-high hydrostatic pressure on hydrosoluble vitamins. *Journal of Food Engineering*, *39*, 247–253.

Sepulveda-Ahumada, D. R., Ortega-Revas, E., & Barbosa-Canovas, G. V. (2000). Quality aspects of cheddar cheese obtained with milk pasteurized by pulsed electric fields. *Journal of Institution of Chemical Engineers*, *78*, 65–71.

Shahab, M., Kumar, H. R., Tamilselvan, K., & Nadanasabapathi, S. (2013). Development and evaluation of shelf stable retort processed ready-to-drink (RTD) traditional Thari Kanchi payasam in flexible retort pouches. *International Food Research Journal*, *20*(4), 1765–1770.

Silva, C., Hendrickx, M., Oliveira, F., & Tobback, P. (1992). Optimal sterilization temperatures for conduction heating foods considering finite surface heat transfer coefficients. *Journal of Food Science*, *57*(3), 743–748.

Spinak, S. H., & Wiley, R. C. (1982). Comparisons of the general and Ball formula methods for retort pouch process calculations. *Journal of Food Science*, *47*, 880–884.

Sreelakshmi, K. R., Manjusha, L., Nagalakshmi, K., Chouksey, M. K., & Venkateshwarlu, G. (2015). Ready-to-serve crab sandwich spread in retort pouch: Product development and process optimization. *Journal of Aquatic Food Product Technology*, *24*(4), 315–329.

Tandon, S., & Bhowmik, S. R. (1986). Evaluation of thermal processing of retortable pouches filled with conduction-heated foods considering their actual shapes. *Journal of Food Science*, *51*(3), 709–714.

Teixeira, A. A., Dixon, J. R., Zahradnik, J. W., & Zinsmeister, G. E. (1969). Computer optimization of nutrient retention in thermal processing of conduction-heated foods. *Food Technology*, *23*(6), 134–140.

Teixeira, A. A., & Manson, J. E. (1982). Computer control of batch retort operations with online correction of processes deviations. *Food Technology*, *36*, 85–90.

Teixeira, A. A., & Tucker, G. S. (1997). On-line retort control in thermal sterilization of canned foods. *Food Control*, *8*(1), 13–20.

Terajima, Y. (1975). Over-all heat transmission from the heating medium (steam and water) to the content of the retortable pouch. *Canners Journal*, *54*(1), 73–79.

Tribuzi, G., Falcão de Aragao, J. M., & Laurindo, J. B. (2015). Processing of chopped mussel meat in retort pouch. *Food Science and Technology*, *35*(4), 612–619.

Vega-Mercado, H., Martin, O., Qin, B.-L., Chang, F. J., Gongora-Nieto, M. M., & Barbosa-Canovas, G. V. (1997). Non-thermal food preservation: Pulsed electric fields. *Trends in Food Science and Technology*, *5*(8), 151–157.

Vega-Mercado, H., Gongora-Nieto, M. M., Barbosa-Canovas, G. V., & Swanson, B. G. (1999). Non-thermal preservation of liquid foods using pulsed electric fields. In M. S. Rahman (Ed.), *Handbook of food preservation* (pp. 783–814). New York: Marcel Dekker.

Zhang, Q., Barbosa-Canovas, G. V., & Swanson, B. G. (1995). Engineering aspects of pulsed electric field pasteurization. *Journal of Food Engineering*, *25*(2), 261–281.

Section D

Application of heat exchangers in the food industry

Heat exchangers in the dairy industry

11

Abdur Rehman[a], Tong Qunyi[a], Aiman Karim[a], Asia Noreen[b], Shabbir Ahmad[c], Muhammad Usman[d], and Seid Mahdi Jafari[e,f]

[a]State Key Laboratory of Food Science and Technology, Jiangnan University, Wuxi, Jiangsu, China, [b]Enzyme Biotechnology Lab, Department of biochemistry, University of Agriculture, Faisalabad, Pakistan, [c]Department of Food Science and Technology, MNS-University of Agriculture, Multan, Pakistan, [d]Beijing Advance Innovation Center for Food Nutrition and Human Health, School of Food and Chemical Technology, Beijing Technology and Business University, Beijing, China, [e]Department of Food Materials and Process Design Engineering, Gorgan University of Agricultural Sciences and Natural Resources, Gorgan, Iran, [f]Nutrition and Bromatology Group, Department of Analytical Chemistry and Food Science, Faculty of Science, Universidade de Vigo, Ourense, Spain

Abbreviations

DSSHE dynamic scraped surface heat exchangers
HTST high temperature short time
LTE low temperature extruder
LTLT low temperature long time pasteurization
OHTC overall heat-transfer coefficient
PHE plate heat exchanger
SSHE scraped surface heat exchangers
UHT ultrahigh temperature

11.1 Introduction

Heat exchangers are a fundamental aspect of dairy processing, and dairy producers are heavily relied on by these equipment producers to ensure that their milk products satisfy industrial standards and are safe for consumption. Liquid milk is a relatively stable fluid that presents few challenges in terms of handling and processing (Bansal & Chen, 2006). Because of milk's ability to be easily temperature regulated, most dairy companies use simple plate-type heat exchangers to perform tasks such as pasteurization, sterilization, and other food processing operations. Despite the simplicity of handling liquid milk, other dairy products, such as yogurt, butter, curds, and cheese, all require different processing and storing temperatures (Karoui & De Baerdemaeker, 2007). If handled incorrectly, they may have their key parameters damaged by routine procedures. Therefore, it is important for companies processing dairy products to select a heat exchanger that is best suited to the type of product they are producing.

Thermal Processing of Food Products by Steam and Hot Water. https://doi.org/10.1016/B978-0-12-818616-9.00003-1

Some important parameters that need to be considered when selecting a heat exchanger are product texture, viscosity, particle size, and how it behaves during routine production and processing.

Before heat treatment was introduced as a common practice for processing dairy products, most milk products were considered a potential source of infection because they are an ideal growth medium for many microorganisms that can cause harmful diseases such as tuberculosis and typus (Ahmedsham, Amza, & Tamiru, 2018). A substantial amount of research went underway to determine a treatment that could be used to make milk safer for consumption. It was not until the mid-19th century that a solution to this problem was finally discovered. Pasteurization is a term coined by Louis Pasteur, who studied the effects of heat on microorganisms found in dairy products. His findings led him to propose the use of heat as a preservative and treatment technique. All common pathogenic organisms that likely occur in milk are killed by relatively minor heat treatment, making the product aseptic, which will only slightly affect the milk's physical and chemical properties.

The ability to control the milk product's temperature throughout all processing stages is one of the most important aspects of assuring product quality in the dairy industry. Hot water and low-pressure steam are the two most commonly used heating mediums (Palaniandy, Adam, Hung, & Naning, 2020). A specific amount of heat is transferred from the heating medium to the milk products so that milk's temperature increases and the heating medium's temperature decreases correspondingly. These two substances must display a significant temperature difference to transfer heat effectively. The heat will flow from the warmer substance to the cooler substance, and the larger the temperature difference, the more rapid the heat flow. The rate of heat transfer begins to decrease as the two substances start to approach an equilibrium value. Heat can be transferred in three ways: conduction, convection, or radiation. Conduction and convection are the two most common methods used by a heat exchanger to regulate a substance's temperature (Agyenim, Hewitt, Eames, & Smyth, 2010). In a plate heat exchanger (PHE), a partition is placed between the product (milk) and the heating or cooling medium (water). The partition is heated on the medium side and cooled on the product side. There is a boundary layer located on either side of the partition where heat is transferred to by conduction and convection. Once the liquids begin to approach the boundary layer, their velocity decreases to almost zero, so they can effectively transfer heat through the partition. This process continues until the heat-transfer rate starts to decrease as the two substances approach an equilibrium temperature.

11.2 An overview on different heat exchangers applied in the dairy industry

Heat exchangers are used for multiple treatment processes, including pasteurization, sterilization, and other food processing needs. Most dairy producing companies use a PHE to perform these tasks because of their versatility in design and size (Agarwal, Sikand, & Shanthi, 2014).

11.2.1 Plate heat exchangers

PHEs are another type of heat exchangers that are widely used in a variety of chemical processes and other industrial applications, with a focus on the food industry for a variety of reasons, including a compact design, a large surface area in a small volume that can be modified as needed by increasing or decreasing the number of plates, and advances in material technology. It is made up of numerous thin plates (Fig. 11.1) that are slightly spaced from one another and have a path for fluid flow for heat transfer (Abu-Khader, 2012; Agarwal et al., 2014). Pasteurization and sterilization operations both rely on PHEs. They have a higher overall heat-transfer coefficient (OHTC) than tubular ones and are often more compact. However, maintenance costs may be high as gaskets may have to be replaced at regular intervals, and the integrity of the plates must be inspected on a regular basis, particularly in the regeneration section, where a broken or leakage plate could allow raw products, such as raw milk, to contaminate already pasteurized or sterilized milk (Lewis & Jun, 2011).

The PHE, which was designed over 70 years ago, has a wide range of applications in the dairy and food beverage industries. PHEs are ideal for liquids with a low viscosity (0.5 $Pa\,s^{-1}$). The corresponding diameter of the particles should be less than $<$0.3 cm if suspended solids are present. Larger particles can bridge the plate contact sites and cause the heating portion to "burn on." As a result, they are employed for operations such as pasteurizing milk, beer, and juices at high temperatures for short

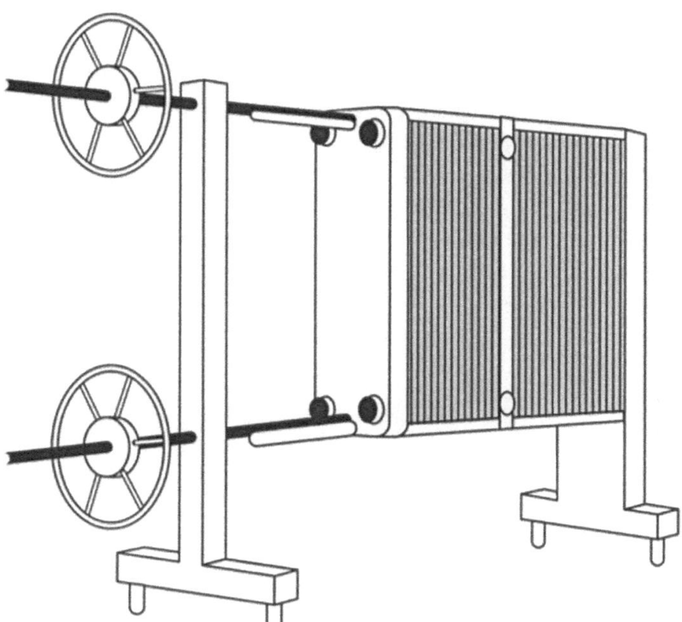

Fig. 11.1 A schematic representation of a plate heat exchanger.
Source: Singh, R. P., & Heldman, D. R. (2014). Heat transfer in food processing. *Introduction to food engineering, Vol. 5*, 265–417.

periods of time. PHEs are used in the manufacturing of stirred yoghurt to chill the yoghurt and stop lactic acid fermentation once the required acidity is reached (Fernandes et al., 2006).

11.2.2 Tubular and shell and tube heat exchangers

A double-pipe heat (tubular) exchanger is the simplest noncontact heat exchanger, consisting of a pipe concentrically inside the other pipe. In the annular gap and the inner pipe, the two fluid streams run in opposite directions. The streams may stream in the same way or in opposing directions. The product flows in the inner annular space of this heat exchanger, whereas the heating/cooling medium flows in the inner tube and outer annular space.

A shell-and-tube heat exchanger (Fig. 11.2) is another typical form of heat exchanger used in the food industry for purposes such as heating liquid food in evaporating processes. One fluid stream flows within the tube, whereas the other is pushed through the shell and over the tubes. We can attain better heat-transfer rates by directing the fluid stream in the shell side to flow over the tubes rather than parallel to them. The cross-flow pattern is enabled by baffles on the shell side. Depending on the design, one or more tube passes can be made (Singh & Heldman, 2014). This design of a heat exchanger is made up of a series of tubes that hold the fluid that needs to be heated or cooled, with a second fluid flowing over the tubes which need to be heated or cooled. Because of its toughness, shell and tube heat exchangers are often utilized for high-pressure applications (pressures more than 30 bar and temperatures greater than >260°C) (Hanson, 1989). It is best to employ a shell and tube heat exchanger with helically corrugated walls. Helically curved heat exchangers are chosen over smooth walls because they improve heat transmission by convection, boosting the efficiency. Tubes come in a range of shapes and sizes to accommodate various product properties. Single tubes with an outer jacket, double or many concentric tubes or shells, and tube kinds are examples of these. A multitube design is used in most ultra-high-temperature (UHT) facilities. Compared to PHEs, they can endure

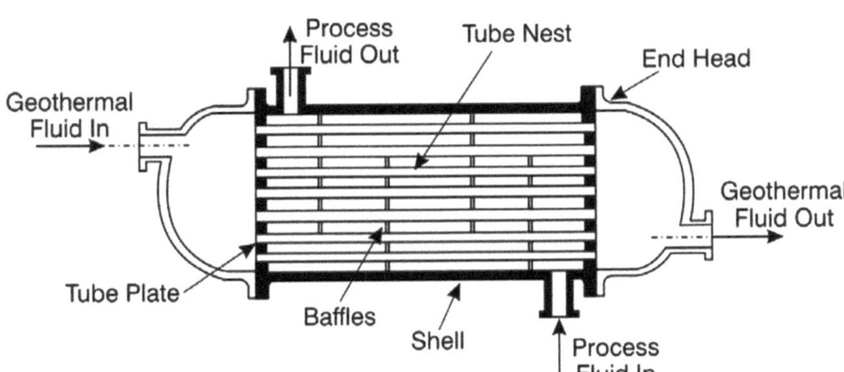

Fig. 11.2 A typical shell and tube heat exchanger (Agarwal et al., 2014).

higher pressures. High pumping pressures can be employed to overcome flow constraints, despite the fact that they are still subject to fouling. With viscous fluids, tubular heat exchangers provide longer processing durations (Söylemez, 2008).

11.2.3 Scraped surface heat exchangers

In conventional types of tubular heat exchangers, heat transfer to a fluid stream is affected by hydraulic drag and heat resistance because of film buildup or fouling on the tube wall. This heat resistance can be minimized if the inside surface of the tube wall is scraped continuously by some mechanical means (Fig. 11.3). The scraping action allows rapid heat transfer to a relatively small product volume. A scraped surface heat exchanger (SSHE) is being used in food processing (Awais & Bhuiyan, 2019). The food contact areas of a scraped-surface cylinder are fabricated from stainless steel (type 316), pure nickel, hard chromium-plated nickel, or other corrosion-resistant material. The inside rotor contains blades that are covered with a plastic laminate or molded plastic. The rotor speed varies between 150 and 500 rpm. Although a higher rotation speed allows better heat transfer, it may affect the quality of the processed product by possible maceration (Rao & Hartel, 2006). Thus, we must carefully select the rotor speed and the annular space between the rotor and the cylinder for the product being processed. The cylinder containing the product and the rotor is enclosed in an outside jacket (Abichandani, Sarma, & Heldman, 1986). The heating/cooling medium is supplied to this outside jacket. Commonly used media include steam, hot water, brine, and refrigerants.

Typical temperatures used for processing products in SSHEs range from 190 to 235°C. The constant blending action accomplished in SSHE is often desirable to enhance the uniformity of product flavor, color, aroma, and textural characteristics (Dehkordi, Fazilati, & Hajatzadeh, 2015). In the food processing industry, the applications of SSHEs include heating, pasteurizing, sterilizing, whipping, gelling, emulsifying, plasticizing, and crystallizing. Liquids with a wide range of viscosities that can be pumped are processed in these heat exchangers; examples include fruit juices, soups, citrus concentrate, peanut butter, baked beans, tomato paste, and pie fillings.

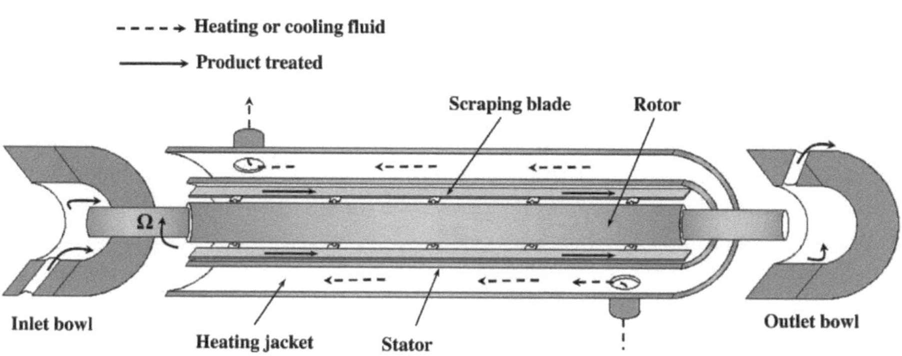

Fig. 11.3 Schematic diagram of the scraped surface heat exchanger (Blel, Legentilhomme, Bénézech, & Fayolle, 2013).

11.2.3.1 Dynamic scraped surface heat exchangers

Some processes in the food and beverage industry utilize multiphase flows in dynamic scraped surface heat exchangers (DSSHEs) to mix colloidal solutions that are delivered into containers with specific preserving conditions for transportation or to extend the shelf life (Blel et al., 2013). DSSHEs are widely used to homogenize solutions containing multiple substances at a controlled mixing velocity in a pressurized and temperature monitored environment. A DSSHE is a type of heat exchanger that uses an array of scraping blades attached to a rotor to create uniform mixing of viscous fluids within a chamber at specific settings to achieve desired consistencies, ultimately based on heat-transfer mechanisms within the region of interest (Chen, Wu, Ji, Lu, & Wang, 2017). Current blade designs and geometrical configurations are typically generated based on manufacturing feasibility and material cost because of the absence of a general theory to predict flow behaviors correlated with the geometry and speed of the rotor-blade assembly (Yataghene, Francine, & Jack, 2011). This leaves room to improve the system by better understanding the fluid and heat-transfer characteristics inside the heat exchanger. Although uniform mixing is desired in practical applications, it is difficult to achieve in both radial and axial directions in closed systems because there are no significant pressure gradients given said pressurized conditions and the rotating nature of the Taylor-Couette flow inside the chamber (Akonur & Lueptow, 2003).

The DSSHE enhances the heat transfer across a rigid wall between two fluids at different energetic states to perform engineering processes and achieve desired products by means of scraping blades. Controlling schemes in typical DSSHEs are rudimentary and inefficient because monitoring points are obtained indirectly because of the complicated nature of the system with rotating parts rendering in situ measurements difficult to retrieve (Yataghene & Legrand, 2013). A DSSHE is essentially a cylindrical steel chamber used for stirring substances by means of a rotor and mixing blades. The outer wall of the chamber is heated or cooled externally by a working fluid to initiate the multiphase crystallization process throughout the space enclosed by the inner wall. An array of mixing blades are used to continuously rotate around the periphery of the chamber scraping the inner wall to remove ice layers forming because of the liquid freezing developing a multiphase Taylor-Couette flow (Härröd, 1990). However, the scraping blades create vortices which directly contribute to a turbulent flow transition, a situation rather difficult to measure experimentally. Convection and conduction mechanisms used to transfer heat in DSSHE applications are nontrivial to analyze because they involve complicated flow dynamics of temperature-dependent viscous multiphase fluids. Analytical techniques to study fluid and heat-transfer mechanisms in a DSSHE are limited and offer approximated solutions based on restrictive assumptions. The latter are planes of symmetry and homogeneous conditions with primitive geometries needed to solve the Navier-Stokes equations for turbulent multiphase flows in three dimensions.

From a theoretical point of view, the mass flow across all components in a refrigeration system is assumed to be the same and a large portion of the heat loss is associated with the inefficiencies in the evaporator (Xu et al., 2020). Therefore, efforts to evaluate the efficiency of a DSSHE typically yield to the study of the macroscopic heat-transfer coefficients of the system by assuming uniform conduction across the rigid wall and constant convection profiles. A simplified approach ignores fluid velocity details, scraping effects, crystallization rates, and the complex rheological nature of multiphase fluids by assuming steady heat conduction through the wall is the main contributor. Unfortunately, experimental methods to measure in situ variables such as microscopic flow characteristics and material phase transition are often challenging to implement given the restrictions imposed by the rotor-blade assembly in the system (Fayolle, Mabit, & Legrand, 2005). Furthermore, common techniques such as electromagnetic tomography or particle tracking velocimetry offer limited information in nonisothermal investigations because of the obstruction imposed by a cooling jacket installed on the periphery of the mixing chamber, which allows the refrigerant to boil and transfer heat. Previous studies have researched the steady-state heat-transfer characteristics in a two-dimensional geometry with a laminar revolving flow and constant material properties (Rainieri, Bozzoli, Mordacci, & Pagliarini, 2012), successfully reporting the relationship between the heat-transfer coefficient and the rotor speed. More intensive computational efforts have studied the steady-state isothermal turbulence behavior of various viscous liquids in a three-dimensional model showing the complicated nature of vortices generated in such systems.

11.3 Applications of heat exchangers in the dairy industry

Heat exchangers can be used in the dairy industry as a process of heating up or cooling down various products in the industry. A large number of products such as milk, ice cream, and other dairy beverages are required to be cooled down or heated up in order to be processed further. For these processes, heat exchangers can be used.

11.3.1 Liquid milk (pasteurized, extended shelf life (ESL), sterilized, UHT)

Thermal processing procedures in milk are characterized as low temperature long time (LTLT) pasteurization, high temperature short time (HTST) pasteurization, and UHT pasteurization based on treatment duration and heat load applied. Although pasteurization remains the most effective method for increasing the shelf life and maintaining microbiological safety of milk till now, it frequently results in negative quality changes. As a result, the dairy industry has expressed interest in embracing alternative milk processing methods in order to enhance quality attributes (Ziuzina & Misra, 2016).

Milk and dairy products are now regarded fundamental aspects in everyone's life and a basic requirement for all humans. Throughout history, the human society has always looked for ways to store food for more difficult times. Food preservation technology has been adapted for many years since then. These innovations not only extended the shelf life of the product but also gave it a new set of physiochemical and sensory properties. Liquid/beverage milk, cheese, milk powders, concentrated milks, fermented milk products, butter, ice cream, newborn formula, creams, and protein-rich goods are all examples of milk products. One of the most important processing methods in the dairy business is heat treatment. In the dairy sector, heat treatments are widely used. Pathogenic and spoilage microorganisms as well as several enzymes that might produce taste and texture flaws in heated items are inactivated by them. Some heat treatments are intended to induce specific effects that improve the processed product's quality or shelf life (Deeth, 2022).

Pasteurized buffalo milk (at 72°C for 15 s), when kept at 101°C, has a storage life of up to 10 days (Nasr & Elshaghabee, 2019). Pasteurized buffalo skim milk has a shelf life of up to 12 days at 4–8°C after being pasteurized at 65°C for 30 min. When kept at 4–8°C, sterilization (110°C for 10 min) considerably improved the appearance/color and the aroma of buffalo skim milk and improved its shelf-life up to 28 days (ul Haq et al., 2014). Likewise, Hussain (2011) observed that after 90 days of storage at 10 and 25°C, commercially processed UHT buffalo skimmed milk had acceptable sensory attributes (Hussain, 2011). Innovative steam injection (ISI) heating, developed by Huijs et al., can heat milk at 150–180°C for 0.1 s. The technique proved the ability to produce economically sterile milk with minimum chemical alteration, causing just 20%–25% denaturation of lactoglobulin (Huijs, Van Asselt, Verdurmen, & De Jong, 2004). Heat treatments during cheese production cause denaturation of particular proteins, which improves the texture of the cheeses and the technological output. Thermal treatment of milk and dairy products can also improve their color, taste, and flavors (e.g., cheese browning, cheese melting). The degree of whey protein denaturation is normally controlled by the preheat treatment performed to liquid skim milk before spray drying (Guinee, 2017). Thermal treatment of milk prior to fermentation enhances denaturation of whey proteins (primarily lactoglobulin and lactalbumin) and their interaction with α-casein on the casein micelle surface through disulfide bonds. Thermally processed milk yoghurts have a thin micelle matrix, which reduces hardness and promotes whey retention. Stabilizers and thickeners are added to the milk to maximize the rheological and whey retention qualities, despite the fact that heat treatment enhances hardness and minimizes syneresis (Harte, Luedecke, Swanson, & Barbosa-Cánovas, 2003).

11.3.2 Cream

The industry trends to process whipping cream by UHT sterilization, which improves its shelf-life (Smith, Goff, & Kakuda, 1999). Cream cheese is a good model for studying the effect of the process. The intensity of heat treatment (72°C/20 s or 94°C/40 s), the final pH of acidification (pH 5.2 or 4.9), and the homogenization pressure (0, 5, 20, or 60 MPa) were studied. This result was modulated by heat treatment temperature and the subsequent whey protein denaturation. The cream cheese final pH between 5.2 and 4.9 had a low impact on rheological properties but was the most discriminating factor for sensory perception by strongly affecting product appearance, especially its brightness and its shade (Coutouly, Riaublanc, Axelos, & Gaucher, 2014). In heat-treated cream (90°C for 15 min) stored at 8°C, growth was dominated by naturally occurring *Bacillus* strains such as *Bacillus pumilus* and *B. licheniformis*. The high number of

Bacillus strains surviving the heat treatment represent a risk for heat-treated food that contains cream (Nissen, Holo, Axelsson, & Blom, 2001). Whipping cream was heat-treated in a PHE at about 120°C. The main fouling components of whipping cream appeared to be protein and ash (Hiddink, Maas, Lalande, Streuper, & Van Hooydonk, 1986). The shelf life of cream products depends, in addition to the factors mentioned for liquid milk, on the product application, for example, whipping cream, coffee cream, or cooking cream (Dhungana & Bhandari, 2020).

11.3.3 Yogurt

In the production of stirred yoghurt, PHEs are used for the cooling of yoghurt to stop lactic acid fermentation upon reaching a desired acidity (Gut & Pinto, 2003). The industrial manufacture of yoghurt is a sequential process. It involves collection of raw material, pasteurization, homogenization, fermentation, set or stirred yoghurt, cooling, storage, packaging, and distribution. The cooling process of stirred yoghurt, a non-Newtonian fluid, is carried out in a PHE. This cooling process in PHE is greatly dependent on geometrical properties of chevron plates, area enlargement factors, and channel aspect ratio as well as the thermal behavior of yoghurt in PHE (Agarwal et al., 2014; Hernández, Raventós, Auleda, & Ibarz, 2009). Fernandes et al. (2006) represented the thermal behavior of stirred yogurt during cooling in PHEs. The yogurt viscosity was being described by the Herschel-Bulkley model. Higher Nusselt numbers than the typical for Newtonian fluids were obtained, which was explained by the shear thinning behavior of the studied fluid (Fernandes et al., 2006). The use of a PHE promoted a decrease in syneresis, whereas a tubular heat exchanger had a greater effect in terms of increasing firmness and consistency of yogurt (Guénard-Lampron, St-Gelais, Villeneuve, & Turgeon, 2019).

11.3.4 Ice cream

Ice cream viscosity is one of the properties that most changes during crystallization in SSHEs. The result was a coupled model based on both the temperature and stirring power of SSHE, which allowed for online viscosity estimation with errors below 10% for crystallized systems with a 30% ice fraction at the exit of SSHE. The model obtained is a first step in the search for control strategies for crystallization in SSHE (De la Cruz Martínez et al., 2020). SSHEs (freezers) are used for whipping and freezing of an ice cream premix. With an outlet temperature of −5 to −8°C, these devices operate at relatively low ice cream viscosities. Here, the ice cream is processed in a low-temperature extruder (LTE) after a classic freezer, which cools it further to below −12°C. This way, shear forces exceeding those in the conventional system by 2–3 orders of magnitude can be applied. These structural changes have a significant impact on the foam-related quality characteristics of ice cream as is proven by measurements of the rheological behavior during thawing and melt down tests (Eisner, Wildmoser, & Windhab, 2005).

11.4 Conclusions

Heat exchangers have become a critical component of dairy processing, and without them, many milk products would be deemed unsafe for consumption because of the growth of harmful microorganisms. Heat exchangers are used for multiple treatment processes, including pasteurization, sterilization, and other food processing needs. Most dairy producing companies use a PHE to perform these tasks because of their versatility in design and size. Different dairy products require different treatment temperatures and times, so it is extremely important for a producer to ensure that they are selecting the heat exchangers best suited for the task to effectively treat the milk product before it is consumed.

References

Abichandani, H., Sarma, S., & Heldman, D. R. (1986). Hydrodynamics and heat transfer in liquid full scraped surface heat exchangers—A review. *Journal of Food Process Engineering*, *9*(2), 121–141.

Abu-Khader, M. M. (2012). Plate heat exchangers: Recent advances. *Renewable and Sustainable Energy Reviews*, *16*(4), 1883–1891.

Agarwal, P., Sikand, A., & Shanthi, V. (2014). Application of heat exchangers in bioprocess industry: A review. *International Journal of Pharmacy and Pharmaceutical Sciences*, *6*(1), 24–28.

Agyenim, F., Hewitt, N., Eames, P., & Smyth, M. (2010). A review of materials, heat transfer and phase change problem formulation for latent heat thermal energy storage systems (LHTESS). *Renewable and Sustainable Energy Reviews*, *14*(2), 615–628.

Ahmedsham, M., Amza, N., & Tamiru, M. (2018). Review on milk and milk product safety, quality assurance and control. *International Journal of Livestock Production*, *9*(4), 67–78.

Akonur, A., & Lueptow, R. M. (2003). Three-dimensional velocity field for wavy Taylor–Couette flow. *Physics of Fluids*, *15*(4), 947–960.

Awais, M., & Bhuiyan, A. A. (2019). Recent advancements in impedance of fouling resistance and particulate depositions in heat exchangers. *International Journal of Heat and Mass Transfer*, *141*, 580–603.

Bansal, B., & Chen, X. D. (2006). A critical review of milk fouling in heat exchangers. *Comprehensive Reviews in Food Science and Food Safety*, *5*(2), 27–33.

Blel, W., Legentilhomme, P., Bénézech, T., & Fayolle, F. (2013). Cleanabilty study of a scraped surface heat exchanger. *Food and Bioproducts Processing*, *91*(2), 95–102.

Chen, J., Wu, J., Ji, X., Lu, X., & Wang, C. (2017). Mechanism of waste-heat recovery from slurry by scraped-surface heat exchanger. *Applied Energy*, *207*, 146–155.

Coutouly, A., Riaublanc, A., Axelos, M., & Gaucher, I. (2014). Effect of heat treatment, final pH of acidification, and homogenization pressure on the texture properties of cream cheese. *Dairy Science & Technology*, *94*(2), 125–144. https://doi.org/10.1007/s13594-013-0148-z.

De la Cruz Martínez, A., Delgado Portales, R. E., Pérez Martínez, J. D., González Ramírez, J. E., Villalobos Lara, A. D., Borras Enríquez, A. J., et al. (2020). Estimation of ice cream mixture viscosity during batch crystallization in a scraped surface heat exchanger. *Processes*, *8*(2), 167.

Deeth, H. C. (2022). Heat treatment of milk: Principles and purpose. In P. L. H. McSweeney, & J. P. McNamara (Eds.), *Encyclopedia of dairy sciences* (3rd ed., pp. 655–658). Oxford: Academic Press.

Dehkordi, K. S., Fazilati, M. A., & Hajatzadeh, A. (2015). Surface scraped heat exchanger for cooling Newtonian fluids and enhancing its heat transfer characteristics, a review and a numerical approach. *Applied Thermal Engineering*, *87*, 56–65.

Dhungana, P., & Bhandari, B. (2020). Dairy creams and related products. In *Dairy fat products and functionality* (pp. 431–452). Springer.

Eisner, M. D., Wildmoser, H., & Windhab, E. J. (2005). Air cell microstructuring in a high viscous ice cream matrix. *Colloids and Surfaces A: Physicochemical and Engineering Aspects*, *263*(1), 390–399.

Fayolle, F., Mabit, J., & Legrand, J. (2005). Determination of heterogeneities in a scraped surface heat exchanger using electrochemical sensors. *Journal of Applied Electrochemistry*, *35*(5), 487–498.

Fernandes, C. S., Dias, R. P., Nóbrega, J. M., Afonso, I. M., Melo, L. F., & Maia, J. M. (2006). Thermal behaviour of stirred yoghurt during cooling in plate heat exchangers. *Journal of Food Engineering*, *76*(3), 433–439.

Guénard-Lampron, V., St-Gelais, D., Villeneuve, S., & Turgeon, S. L. (2019). Individual and sequential effects of stirring, smoothing, and cooling on the rheological properties of nonfat yogurts stirred with a technical scale unit. *Journal of Dairy Science*, *102*(1), 190–201.

Guinee, T. P. (2017). Pasteurized processed and imitation cheese products. In *Cheese* (pp. 1133–1184). Elsevier.

Gut, J. A., & Pinto, J. M. (2003). Modeling of plate heat exchangers with generalized configurations. *International Journal of Heat and Mass Transfer*, *46*(14), 2571–2585.

Hanson, F. V. (1989). *Heat exchangers selection, design and construction: By EAD Saunders, Longman scientific and technical, Essex, England, 1988*. ISBN 0-47020870-8, pp. 568, price: US $99.95 (co-published in the United States with John Wiley & Sons, Inc., New York) Elsevier.

Härröd, M. (1990). Methods to distinguish between laminar and vortical flow in scraped surface heat exchangers. *Journal of Food Process Engineering*, *13*(1), 39–57.

Harte, F., Luedecke, L., Swanson, B., & Barbosa-Cánovas, G. V. (2003). Low-fat set yogurt made from milk subjected to combinations of high hydrostatic pressure and thermal processing. *Journal of Dairy Science*, *86*(4), 1074–1082.

Hernández, E., Raventós, M., Auleda, J., & Ibarz, A. (2009). Concentration of apple and pear juices in a multi-plate freeze concentrator. *Innovative Food Science & Emerging Technologies*, *10*(3), 348–355.

Hiddink, J., Maas, R., Lalande, M., Streuper, A., & Van Hooydonk, A. (1986). Heat treatment of whipping cream: Fouling of the sterilization equipment. In *MILK the vital force* (p. 29). Springer.

Huijs, G., Van Asselt, A., Verdurmen, R., & De Jong, P. (2004). High speed milk. *Dairy Industries International*, *69*(11), 30–32.

Hussain, I. (2011). Effect of UHT processing and storage conditions on physico-chemical characteristics of buffalo skim milk. *Journal of the Chemical Society of Pakistan*, *33*(6), 783.

Karoui, R., & De Baerdemaeker, J. (2007). A review of the analytical methods coupled with chemometric tools for the determination of the quality and identity of dairy products. *Food Chemistry*, *102*(3), 621–640.

Lewis, M., & Jun, S. (2011). Thermal processing. In *Food processing handbook, 2 volume set* Wiley.

Nasr, N., & Elshaghabee, F. (2019). Enhancement of shelf life of pasteurized milk using NisplinŪ essential oils emulsion. *International Journal of Current Microbiology and Applied Sciences*, *8*(9), 257–266.

Nissen, H., Holo, H., Axelsson, L., & Blom, H. (2001). Characterization and growth of Bacillus spp. in heat-treated cream with and without nisin. *Journal of Applied Microbiology*, *90*(4), 530–534.

Palaniandy, Y., Adam, N. M., Hung, Y. P., & Naning, F. H. (2020). Potential of steam recovery from excess steam in sterilizer at palm oil mill. *Journal of Advanced Research in Fluid Mechanics and Thermal Sciences*, *79*(1), 17–26.

Rainieri, S., Bozzoli, F., Mordacci, M., & Pagliarini, G. (2012). Numerical 2-D modeling of a coaxial scraped surface heat exchanger versus experimental results under the laminar flow regime. *Heat Transfer Engineering*, *33*(13), 1120–1129.

Rao, C. S., & Hartel, R. W. (2006). Scraped surface heat exchangers. *Critical Reviews in Food Science and Nutrition*, *46*(3), 207–219.

Singh, R. P., & Heldman, D. R. (2014). *Heat transfer in food processing. Introduction to food engineering. Vol. 5* (pp. 265–417). Scribd.

Smith, A., Goff, H., & Kakuda, Y. (1999). Whipped cream structure measured by quantitative stereology. *Journal of Dairy Science*, *82*(8), 1635–1642.

Söylemez, M. (2008). Optimum length of finned pipe for waste heat recovery. *Energy Conversion and Management*, *49*(1), 96–100.

ul Haq, I., Khaskheli, M., Salman, M., Marri, M. Y., Marri, M. A., Lochi, G. M., et al. (2014). Effect of heat treatments on sensory characteristics and shelf-life of skimmed milk. *African Journal of Food Science*, *8*(2), 75–79.

Xu, Q., Liu, L., Feng, J., Qiao, L., Yu, C., Shi, W., et al. (2020). A comparative investigation on the effect of different nanofluids on the thermal performance of two-phase closed thermosyphon. *International Journal of Heat and Mass Transfer*, *149*, 119189.

Yataghene, M., Francine, F., & Jack, L. (2011). Flow patterns analysis using experimental PIV technique inside scraped surface heat exchanger in continuous flow condition. *Applied Thermal Engineering*, *31*(14–15), 2855–2868.

Yataghene, M., & Legrand, J. (2013). A 3D-CFD model thermal analysis within a scraped surface heat exchanger. *Computers & Fluids*, *71*, 380–399.

Ziuzina, D., & Misra, N. (2016). Cold plasma for food safety. In *Cold plasma in food and agriculture* (pp. 223–252). Elsevier.

Heat exchangers in the beverages and sugar industry

12

Felipe Richter Reis
Instituto Federal de Educação, Ciência e Tecnologia do Paraná, Campus Colombo, Colombo, PR, Brazil

12.1 General aspects of the beverages industry

12.1.1 Soft drink industry

The soft drink processing involves the following steps: water treatment, syrup preparation, carbonation, filling, and packaging. The water treatment comprises the following processes: removal of color, suspended particles, microorganisms, and organic compounds that may cause or transmit off-tastes and odors, reduction of alkalinity, and warranty of a consistent product (Mitchell, 1990). Additionally, removal of air from water is performed for avoiding oxidation, which would destroy the aromatic substances of soft drink, and for avoiding frothing over of the soft drink when the package is opened (TetraPak International S.A., 2020). Syrup preparation consists in mixing all the required ingredients in an appropriate vessel. Carbonation comprises injecting carbon dioxide into a product or water stream or adding it to a liquid that is static in a pressurized compartment. Filling is commonly carried out by counter-pressure, but there is an alternative system, which is based on liquid level in the container; both are made at a cool temperature to minimize CO_2 loss. Packaging is traditionally performed in glass bottles, although cans are also extensively used (Ashurst, 2016).

12.1.2 Beer industry

The brewing process, i.e., the process of making beer, consists of the following steps: malting, milling, mashing, lautering, boiling, fermenting, conditioning, filtering, pasteurization, and packaging. First, raw barley, wheat, oats, or rye is germinated in a malt house for up to 6 days to release enzymes that convert starch into sugars. Then, the grain is dried in a kiln, after which it is called malt. The malt is milled to expose the carbohydrates and sugars. Mashing consists in mixing milled malt with hot water for converting starch into fermentable sugars (saccharification), while lautering comprises the separation of the liquid present after mashing (wort) from the grains by draining. Boiling is performed after addition of hops to the wort in a large copper kettle, aiming at sterilizing the beer and providing it with bitterness. In fermentation,

Thermal Processing of Food Products by Steam and Hot Water. https://doi.org/10.1016/B978-0-12-818616-9.00010-9

yeast is added, and the conversion of sugars in alcohol and carbon dioxide begins. Conditioning is performed by yeasts and aims at the absorbance of off-flavors that were created during fermentation, while filtering aims at stabilizing the flavor and providing beer with a clean shine. Pasteurization kills any remaining yeast. Packaging is usually made in bottles, cans, or kegs (McHugh, 2017).

12.2 General aspects of the sugar industry

The sugar processing companies process sugar cane and sugar beets for manufacturing edible sugar. Sugar cane is the source of more than 60% of the world's sugar production and the remaining amount comes from sugar beets. Sugar cane contains around 70% moisture, 14% fiber, 13.3% sucrose, and 2.7% impurities (Cheremisinoff, Rosenfeld, & Davletshin, 2008). Fig. 12.1 shows a flowchart representing the production of raw sugar.

Juice extraction is usually achieved by means of rotating knives, shredders, or tandem mills associated with water addition. The aim at this step is to break the cells to release the juice. Shredding can lead to a 94% cell breakage, while milling can lead to a 96%–97% cell breakage (Delgado & Casanova, 2001). An alternative technique for juice extraction is diffusion, which consists of removal of sugar from the plant cells by a concentration gradient between the intracellular environment and the extracellular environment (Pamu, 2014). Purification involves the addition of lime to remove mud (calcium phosphate precipitate) and heating to precipitate albumin from the juice. Evaporation, carried out by means of steam, is aimed at concentrating the juice. Crystallization is designed to produce sucrose crystals and is based on the increase of the supersaturation of the syrup by boiling in vacuum pans followed by cooling. Molasses known as "final molasses" are separated from crystalline sugar by a process of centrifugation. Finally, storage of sugar is accomplished in bags or in bulk (Delgado & Casanova, 2001).

12.3 Use of heat exchangers in the beverages industry

12.3.1 Use of heat exchangers in the soft drink industry

In the soft drink industry, plate heat exchangers (PHEs) are used for dissolving sugar and pasteurizing the sugar solution. The minimum holding time commonly used is 40 s. Additionally, a heat exchanger is used for flash pasteurization of the soft drink, which can be a PHE or a tubular heat exchanger when dealing with large fruit cells. It is more usual to carbonate the soft drink after flash pasteurization. It is also possible to only flash pasteurize the sensitive components of a beverage and then mix them with the nonsensitive ones, usually water, which if treated by UV will become microbiologically safe, resulting is a cheaper process from both capital and operating viewpoints. The usual conditions of flash pasteurization are 85–93°C for 30–60 s. If a PHE is used for flash pasteurization, the plates carry product and heating or cooling medium, the latter being commonly propylene glycol. The plates are held together in a

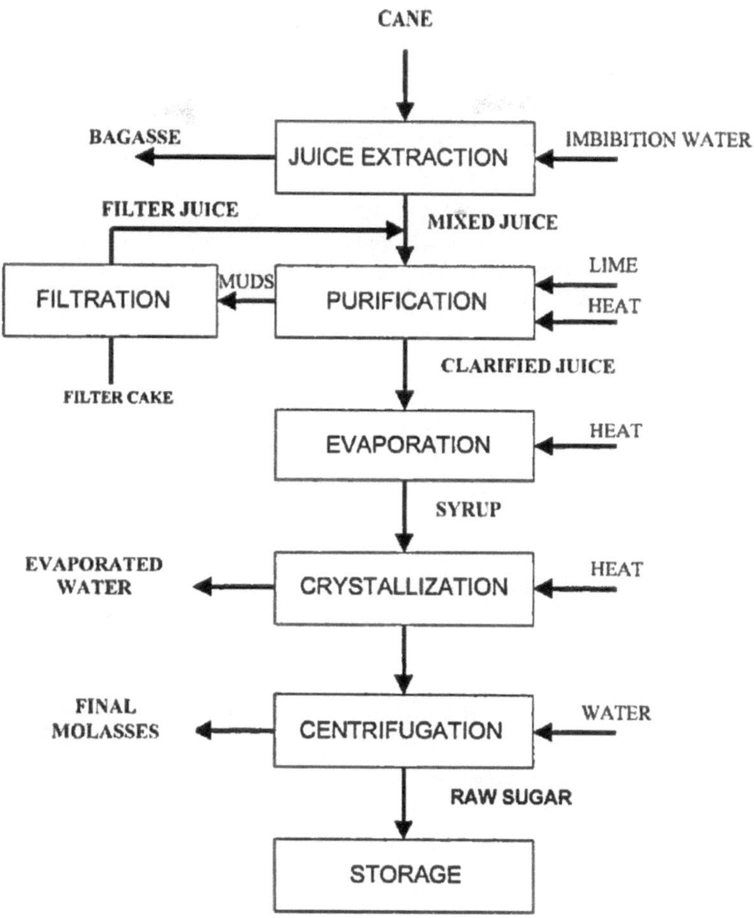

Fig. 12.1 The production of raw sugar.
Credit: Delgado, A.V., Casanova, C.A. (2001). *Sugar processing and by-products of the sugar industry.* Rome: FAO.

frame by welding or gasketing, and the gaps between the frames can reach 12 mm, allowing for fruit cell products to be treated. Other application of heat exchanger in the soft drink industry is during the clean-in-place process, to heat the lye to 80°C (Steen & Ashurst, 2006).

12.3.2 Use of heat exchangers in the beer industry

In the beer industry, heat exchangers are used to condense the vapors for recovering the energy of the remaining vapors from wort boiling and make hot water, which can be stored in a vertical tank to be used in an in-line heat exchanger to heat the wort between mash separation and the wort kettle. A PHE is used for wort cooling, which can be made with one or two stages. The temperature that must be achieved in the product ranges between 8°C and 12°C for lager beers and between 16°C and 20°C for ale beers, while the cooling water, which is brewing water at ambient temperature, achieves 80°C after heat transfer (Bamforth, 2006). Additionally, beer packaged in kegs or other containers that cannot be tunnel pasteurized passes by flash pasteurization in bulk in PHEs, which work at temperatures up to 78°C with beers that contain a ratio of 2.2/1 volumes of CO_2 per volume of beer, thus assuring that CO_2 does not escape from the solution, which would decrease critical residence times in the circuit (Fricker, 1984; Vaughan, O'Sullivan, & van Sinderen, 2005). According to Riverol and Napolitano (2002), flash pasteurization is a continuous method used for pasteurizing beer prior to filling into containers, which is normally carried out between 71.5°C and 74°C for 15–30 s. The same authors state that flash pasteurization of beer is normally conducted with a three-stage heat exchanger and uses hot water as the hot fluid.

12.4 Use of heat exchangers in the sugar industry

According to Barriquand Technologies Thermiques (2020), PHEs are used for heating juices before clarification (purification), for heating juices before evaporation, for heating particle-free syrups from crystallization, and for cooling molasses issued from the final stage of crystallization for storage. Alfa Laval (2015) states that raw juice and limed juice, which contain fibers, are heated by PHEs, which use pan vapor or vapor from the last evaporation effect. Additionally, the clarified juice is heated by PHEs, exploiting vapor bleed-offs and condensate from the evaporators as heat sources. Finally, syrups and molasses, which are high-viscosity liquids, are heated and cooled by standard PHEs (Alfa Laval, 2015). Solex (2020) describes a heat exchanger used for cooling powdered sugar crystals before storage and packaging: the indirect heat exchanger consists of a bank of vertical, closely spaced, hollow, stainless steel plates; the sugar flows slowly by gravity between the plates in mass flow; cooling water flows counter-current through the plates and heat is transferred through the sugar particles by conduction, cooling the sugar indirectly; there is not in contact between sugar and air; at the bottom of the heat exchanger unit, a vibrating discharge feeder is used for creating mass flow and for regulating the sugar throughput.

12.5 Reports on the application of heat exchangers in the beverages industry

Riverol and Napolitano (2002) successfully predicted the values of overall heat transfer coefficient during flash pasteurization of beer using neural networks. A tubular heat exchanger was used in the study and pasteurization conditions were a temperature of 72°C for approximately 30 s. Gut and Pinto (2005) optimized the parameters of a PHE used for beer processing, finding that a branching multiobjective optimization procedure, where the optimization variables are arranged on a tree structure, used for the configuration design of a PHE for continuous pasteurization of 3200 L/h of beer, was able to yield solutions for simultaneous low cost and low lethality, as shown in a graph of parcel of the pasteurization cost dependent on the PHE configuration (PC) versus lethality (F^{PHE}), where the multiple optima is highlighted (Fig. 12.2).

Mamvura, Iyuke, Cluett, and Paterson (2011) wrote a review on the deleterious effects of soil films in the beverage industry, affirming that they deteriorate water, corrode water distribution systems, contaminate food, reduce beverage industry product quality, and decrease heat transfer in heat exchangers. According to the authors, soil films are capable of fouling heat exchangers, which results in poor hygiene and in impaired product quality or even in food contamination. Gawande, Wankhede, Yerrawar, Sonawane, and Ubarhande (2012) designed a shell-and-tube heat exchanger for the beverages industry. According to the authors, the designing procedure comprised the thermal design, which included consideration of fluids in shell and tube side; selection of required temperatures; limitation of the pressure drop in both shell and tube sides; setting limits of velocity in both shell and tube sides; and calculating the area of heat transfer and the fouling factor. In addition, the mechanical design was also performed, which included selecting TEMA (Tubular Exchangers Manufacturers Association) layout—based on thermal design; selecting tube material, layout, thickness, pitch, size; limitation of the upper and lower design on tube length; selecting shell side clearances, baffle spacing and material; tube material thermal conductivity; setting upper and lower limits of design on baffle spacing and shell diameter.

Hucker, Wakeling, and Vriesekoop (2014) studied the impact of mashing, lautering, clarification and boiling in the thiamine and riboflavin vitamer content. After boiling, the wort was rapidly cooled to 20°C by means of a PHE, which according to the authors stabilized the vitamer content, reducing the deleterious effect of boiling on vitamer concentration. Yin et al. (2017) compared the impact of flash

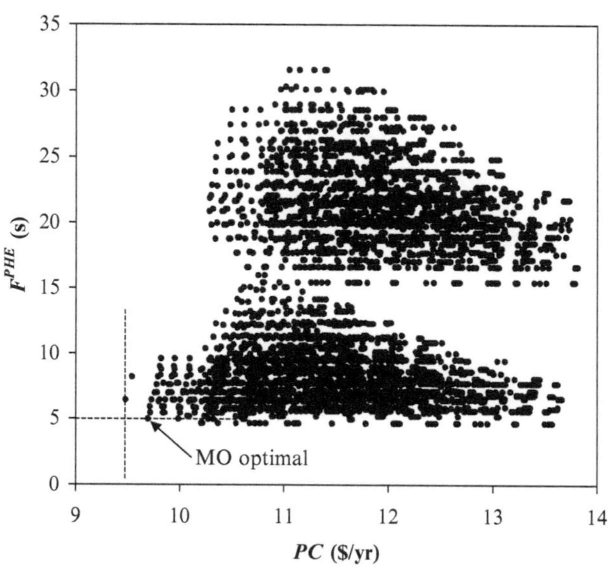

Fig. 12.2 Pareto chart for multiobjective optimization.
Credit: Gut, J.A.W., Pinto, J.M., (2005). *Optimal design of continuous sterilization processes with plate heat exchangers, European symposium on computer aided process engineering-15*. Amsterdam: Elsevier.

pasteurization on a tubular heat exchanger at two conditions with the impact of tunnel pasteurization of wheat beers on quality parameters, finding that flash pasteurization at 73°C/30 s or 76°C/30 s (37 or 100 pasteurization units—PU, respectively) is superior to tunnel pasteurization at 60°C/15 min (15 PU) with regard to color changes, oxidative stability, sensory quality and haze intensity.

12.6 Reports on the application of heat exchangers in the sugar industry

Smaïli et al. (1999) optimized the cleaning schedule of heat exchangers subjected to fouling in a sugar industry. According to the authors, heavy fouling occurred in several of the 11 heat exchangers used. As a result, vapor stream of higher temperature had to be used in some of the exchangers to achieve the target temperature in raw juice. They were successful in using heat transfer data to establish the fouling behavior of the heat exchangers and generate optimal cleaning schedules. Urbaniec, Zalewski, and Zhu (2000) proposed a retrofit in a beet sugar processing unit. They suggested maintaining the PHEs in use and substituting the shell-and-tube heat exchangers by PHEs, except from one unit in which the value of the temperature approach is less important. Among the PHEs, one of them should have its heat transfer surface reduced, another should be repiped and other three should have their heat transfer surfaces extended. The remaining shell-and-tube heat exchanger should have its duty (160 m^2) changed. They estimated the reduction in specific fuel consumption at 29%.

Gumbo, Mlilo, Broome, and Lumbroso (2003) performed a retrofit in a sugar refining industry by introducing a heat exchanger for reducing water intake. The heat exchanger received cooling water from the cooling tower and hot condensate (sweet water) from the condenser and sent hot cooling water to the cooling tower and cooling water (29°C) to the condenser. Before its installation in the processing line (Fig. 12.3), it was necessary to feed the system with 58 m^3 of borehole water per 100 t of raw sugar, and after its installation (Fig. 12.4), this value decreased to 26 m^3 of borehole water per 100 t of raw sugar. Additionally, the amount of effluent decreased from 35 m^3 of blow down per 100 t of raw sugar to 3.5 m^3 of blow down per 100 t of raw sugar. Finally, in the new configuration, 56 m^3 of buildup (sweet water) per 100 t of raw sugar were forwarded to other processes such as melting muds and de-sweetening.

Pacheco and Frioni (2004) evaluated the evaporation of sugar cane juice in a climbing/falling film evaporator with an intermating corrugated plate-type heat exchanger, obtaining data that could be used in future studies aimed at establishing a more general correlation for heat-transfer coefficients of climbing/falling film plate evaporators. La Madrid, Marcelo, Orbegoso, and Saavedra (2016) applied computational fluid dynamics (CFD) tools to simulate the flow and heat transfer characteristics of the jaggery furnace, composed by open heat exchangers, finding that CFD was a suitable tool for modeling the selected parameters of the jaggery furnace. Three studies were published on the mathematical modeling of the fouling formation in PHEs in the sugar industry (Kapustenko, Arsenyevaa, Matsegora, Kusakov, & Tovazhnianskyi, 2017; Kapustenko, Demirskiy, et al., 2018; Kapustenko, Klemeš, Arsenyeva, Matsegora, &

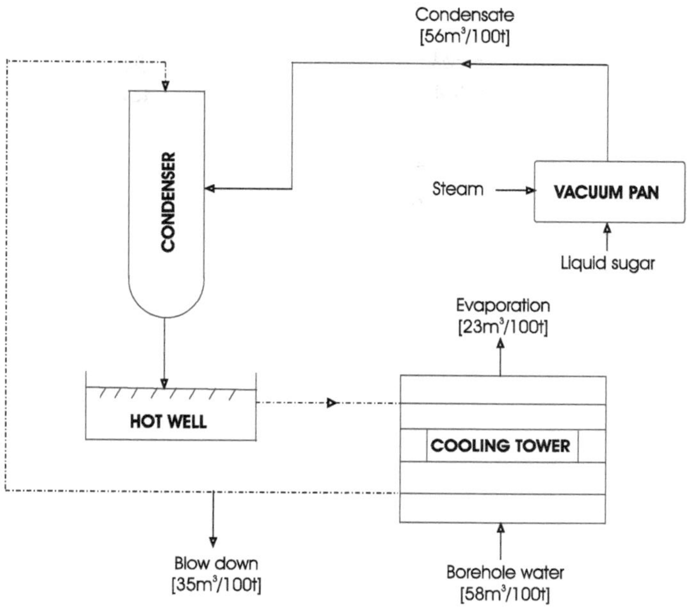

Fig. 12.3 The existing cooling water and condensate system.

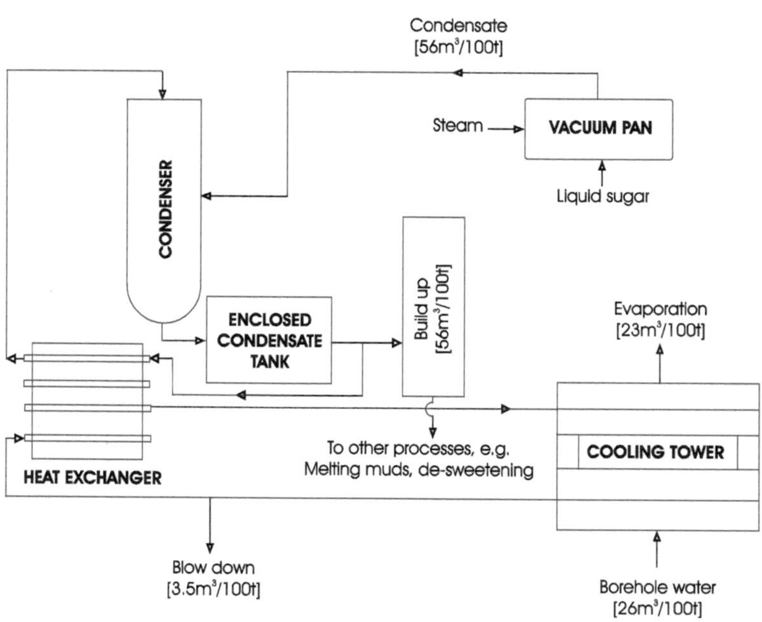

Fig. 12.4 The proposed cooling water and condensate system.
Credit: Gumbo, B., Mlilo, S., Broome, J., Lumbroso, D. (2003). Industrial water demand management and cleaner production potential: A case of three industries in Bulawayo, Zimbabwe. *Physics and Chemistry of the Earth*, 28(20–27), 797–804.

Vasilenko, 2018). According to the authors, the fouling formation in heat transfer surfaces is mainly influenced by temperature, but also by shear stress, in a way that the higher the temperature and the shear stress, the lower the probability of fouling formation.

In an attempt to obtain jaggery with a golden yellow color, Rane and Uphade (2018) preconcentrated sugar cane juice by using four tube-and-tube heat exchangers, namely, a three stream liquid–liquid heat exchanger, employed to precool raw sugar cane juice, a latent heat exchanger which worked as an evaporator, employed to freeze water from juice, a latent heat exchanger which worked as a condenser, employed to melt ice, and a heat exchanger employed for heating up the concentrated juice using the excess heat of condenser (latent heat exchanger). All of them were equipped with SS304 circular tubes of 9.52 mm of outside diameter and 1 mm thickness. The authors concluded that the thermal cycling loss in the latent heat exchangers was reduced with an increase in tube diameter and that the process used helped to save bagasse.

Farahbod (2019) investigated the effect of process parameters on the quality and quantity of outlet sugar from a forced circulation crystallizer coupled to a vertical tubular exchanger in the output line, which was used for heating the effluent liquor from the crystallizer. He found that the feed temperature increased from 27.3°C to 68.9°C after it passed by the heat exchanger, thus decreasing its density from 1.155 to 1.130 kg/m^3. The minimum energy consumed by the heat exchanger was about 5 kWh at a pressure of 421.9 mmHg. Tiwari, Gandharv, Dharamvir, Kumar, and Verma (2019) affirmed that organic and inorganic impurities contained in cane juice are responsible for scale formation in heat exchangers, including those that are part of evaporators, ultimately causing clogging of pipes, loss of heat transfer efficiency and process, and damage to the equipment. They found that bagasse fly ash effectively adsorbed sugar cane impurities. In addition, the clarification process was dose dependent and contact time dependent.

12.7 Conclusion and future outlook

The importance of heat exchangers for the beverages and sugar industry is undoubtable. From production processes to microbiological safety assurance processes, heat exchangers are present, playing a crucial role for the manufacturing soft drinks, beers, and sugar. In beer processing companies, even though the pasteurization of beer can be made in the bottle by tunnel pasteurization, it seems that flash pasteurization by means of heat exchangers presents advantages regarding product quality. In sugar processing companies, retrofit has been extensively put in place, and the heat exchanger network is always altered somehow ultimately improving the energetic efficiency of these industries. Fouling is also a concern that has been dealt with in several studies, especially in the sugar industry, due to the impurities contained in sugar juice, and it seems that clarification processes that target these impurities are a promising solution. Finally, the use of PHEs is certainly a trend in the beverages and sugar industry, given the efficient heat transfer obtained through this equipment.

References

Ashurst, P. R. (2016). *Carbonated beverages.* Amsterdam: Elsevier.

Bamforth, C. (2006). *Brewing: New technologies.* Cambridge: Woodhead Publishing.

Barriquand Technologies Termiques. (2020). *Heat exchangers for sugar factories and distilleries. Available from.* https://www.barriquand.com/en/heat-exchangers-sugar-ethanol#:~:text=Heat%20exchangers%20dimensioned%20to%20reduceof%20our%20clients's%20major%20objectives. [15 August 2020.

Cheremisinoff, N. P., Rosenfeld, P., & Davletshin, A. R. (2008). *Responsible care: A new strategy for pollution prevention and waste reduction through environment management.* Amsterdam: Elsevier.

Delgado, A. V., & Casanova, C. A. (2001). *Sugar processing and by-products of the sugar industry.* Rome: FAO.

Farahbod, F. (2019). Experimental evaluation of forced circulation crystallizer performance in production of sugar crystals. *Journal of Food Process Engineering, 42*(3), e13017.

Fricker, R. (1984). The flash pasteurization of beer. *Journal of the Institute of Brewing, 90*(3), 146–152.

Gawande, S. H., Wankhede, S. D., Yerrawar, R. N., Sonawane, V. J., & Ubarhande, U. B. (2012). Design and development of shell & tube heat exchanger for beverage. *Modern Mechanical Engineering, 2*(4), 121–125.

Gumbo, B., Mlilo, S., Broome, J., & Lumbroso, D. (2003). Industrial water demand management and cleaner production potential: A case of three industries in Bulawayo, Zimbabwe. *Physics and Chemistry of the Earth, 28*(20–27), 797–804.

Gut, J. A. W., & Pinto, J. M. (2005). *Optimal design of continuous sterilization processes with plate heat exchangers, European symposium on computer aided process engineering-15.* Amsterdam: Elsevier.

Hucker, B., Wakeling, L., & Vriesekoop, F. (2014). Vitamins in brewing: The impact of wort production on the thiamine and riboflavin vitamer content of boiled sweet wort. *Journal of the Institute of Brewing, 120*(3), 164–173.

Kapustenko, P. O., Arsenyevaa, O. P., Matsegora, O. I., Kusakov, S. K., & Tovazhnianskyi, V. I. (2017). The mathematical modelling of fouling formation along PHE heat transfer surface. *Chemical Engineering Transactions, 61*, 247–252.

Kapustenko, P. O., Demirskiy, O. V., Tovazhnyanskyi, V. I., Klemeš, J. J., Matsegora, O. I., Arsenyev, P. Y., et al. (2018). Mathematical modelling of the thermal and hydraulic behaviour of plate heat exchanger in the fouling conditions. *Chemical Engineering Transactions, 70*, 109–114.

Kapustenko, P., Klemeš, J. J., Arsenyeva, O., Matsegora, O., & Vasilenko, O. (2018). Accounting for local features of fouling formation on PHE heat transfer surface. *Frontiers of Chemical Science and Engineering, 12*(4), 619–629.

La Madrid, R., Marcelo, D., Orbegoso, E. M., & Saavedra, R. (2016). Heat transfer study on open heat exchangers used in jaggery production modules—Computational fluid dynamics simulation and field data assessment. *Energy Conversion and Management, 125*, 107–120.

Alfa Laval. (2015). *Cane processing.* Available from: https://www.alfalaval.com/industries/food-dairy-beverage/agriculture-processing/sugar-production/cane-processing/ (15 August 2020).

Mamvura, T. A., Iyuke, S. E., Cluett, J. D., & Paterson, A. E. (2011). Soil films in the beverage industry: A review. *Journal of the Institute of Brewing, 117*(4), 608–616.

McHugh, T. (2017). *How beer is processed. Available from.* https://www.ift.org/news-and-publications/food-technology-magazine/issues/2017/november/columns/processing-beer (20 August 2020).

Mitchell, A. J. (1990). *Formulation and production carbonated soft drinks*. Glasgow: Blackie and Son Ltd.

Pacheco, C. R. F., & Frioni, L. S. M. (2004). Experimental results for evaporation of sucrose solution using a climbing/falling film plate evaporator. *Journal of Food Engineering*, *64*(4), 471–480.

Pamu, V. S. (2014). Effect of different diffusion strategies on extraction of sugar from sweet sorghum, M.Sc thesis, Oklahoma State University, Rane, MV and Uphade DB 2018, Heat pump for energy-efficient sugarcane juice freeze pre-concentration. *Sugar Tech*, *20*(5), 591–598.

Rane, M. V., & Uphade, D. B. (2018). Heat pump for energy-efficient sugarcane juice freeze pre-concentration. *Sugar Tech*, *20*, 591–598.

Riverol, C., & Napolitano, V. (2002). Estimation of the overall heat transfer coefficient in a tubular heat exchanger under fouling using neural networks: Application in a flash pasteurizer. *International Communications in Heat and Mass Transfer*, *29*(4), 453–457.

Smaïli, F., Angadi, D. K., Hatch, C. M., Herbert, O., Vassiliadis, V. S., & Wilson, D. I. (1999). Optimization of scheduling of cleaning in heat exchanger networks subject to fouling: Sugar industry case study. *Transactions of the Institute of Chemical Engineers*, *77*. p. C, pp. 159–164.

Solex. (2020). *Improved operation with solex technology in sugar cooling*. Available from: https://www.solexthermal.com/resources/case-studies/sugar-cooling-poland/ (15 August 2020).

Steen, D., & Ashurst, P. R. (2006). *Carbonated soft drinks: Formulation and manufacture*. Oxford: Blackwell Publishing.

TetraPak International S.A. (2020). *Carbonated soft drinks—Get the blend just right*. Available from: https://www.tetrapak.com/findbyfood/juice-and-drinks/carbonated-soft-drinks#:~:text=Production%20and%20processingcarbonation%20and%2C%20finally%2C%20packaging (16 August 2020).

Tiwari, R. N., Gandharv, C. A., Dharamvir, K., Kumar, S., & Verma, G. (2019). Scale minimization in sugar industry evaporators using nanoporous industrial bio-solid waste bagasse fly ash. *Sugar Tech*, *21*(2), 301–311.

Urbaniec, K., Zalewski, P., & Zhu, X. X. (2000). A decomposition approach for retrofit design of energy systems in the sugar industry. *Applied Thermal Engineering*, *20*(15–16), 1431–1442.

Vaughan, A., O'Sullivan, T., & van Sinderen, D. (2005). Enhancing the microbiological stability of malt and beer—A review. *Journal of the Institute of Brewing*, *111*(4), 355–371.

Yin, H., Deng, Y., He, Y., Dong, J., Lu, J., & Chang, Z. (2017). A preliminary study of the quality attributes of a cloudy wheat beer treated by flash pasteurization. *Journal of the Institute of Brewing*, *123*(3), 366–372.

Retorts in the food industry

13

Hamed Vatankhah, Lovepreet Singh Sidhu, and Hosahalli S. Ramaswamy
Department of Food Science and Agricultural Chemistry, Macdonald Campus of McGill University, Ste Anne de Bellevue, QC, Canada

13.1 Introduction

Canning or thermal processing is not only an age-old technology but also very mature and has undergone significant developments since it was first introduced by Nicholas Appert in the early 1800. This process is generally carried out in pressure vessels commonly recognized as "retorts or autoclaves" and commonly uses steam or steam-heated water as the heating medium to heat or process the food. Since the beginning of the first commercial retorts that were introduced in the early 1900, many different types of retorts have been designed and introduced for commercial applications. Over the years, the improvement in the retort design has been focused on many aspects: efficient operation, efficient use and distribution of the heating media, efficient loading and treatment of containers, improving the energy efficiency, computerization, product agitation, providing batch versus continuous operations, and also the use of nonconventional media such as flame, microwaves, and so on. The first generation of retorts was the conventional retorts with a cylindrical shell. Most older versions were in the vertical format, and the more recent ones are horizontal. In most large establishments, there is usually a battery of such retorts for synchronized batch operations, providing some continuity for the handling of cans.

A sequence of filling, closing, retorting, and emptying is performed to provide a sufficient supply of cans to maintain continuous production, packaging, processing, labeling, and marketing. Canned products are arranged into crates/baskets for vertical retorts and rectangular cars for placement into horizontal retorts. The retort shell, which is a pressure vessel, has specified openings for the inlet of water, air, and steam and outlets for the venting air during retort come up and for retort draining. Retorts are equipped with a housing to hold temperature sensors, pressure probes, and pressure gauges and always have suitable controller/recorder devices. To ensure sufficient steam movement, the housing is fitted with a continuous vapor bleed around the temperature sensors. The lid is hinged at the top of vertical retorts and is secured to the shell by several strong bolts during processing. In horizontal retorts, the door is at front the end of the shell, which remains locked during the operation. The equipment is always equipped with safety valves or pressure relief valves for protection and safety of operation (Ramesh, 2007).

A majority of commercial canning applications do not involve or gain from product agitation, and thus these are static retorts. Traditional static heating has some

Thermal Processing of Food Products by Steam and Hot Water. https://doi.org/10.1016/B978-0-12-818616-9.00011-0

limitations because of slow heat penetration in foods especially while using conduction heating foods in larger sizes. This often results in significant overprocessing when processed long enough to ensure that a safe product is obtained (Ates, Skipnes, Rode, & Lekang, 2016). Overall, static retorts can be used for all sterilizing or pasteurizing needs in the food industry, especially for liquid or particulate liquid foods from a limited and medium efficiency to the high production volumes.

A retort is a pressure vessel which is used for a thermal processing or commercial sterilization of foods placed into hermetically sealed containers. These days, retorts can be classified into four categories. They can be batch or continuous type and static or agitating type. Static batch retorts have been used since the mid-19th century and can sterilize food containers at a selected temperature above 100°C. These early models used the steam as the heat-transfer medium as most of the retorts are using it today. Nowadays, water is introduced to the system through different ways such as showering, cascading, raining, or spraying. Steam/air combination is also used for the processing of containers which requires counter pressure to retain their integrity and shape (Ramaswamy & Marcotte, 2006). Agitating batch retorts are the later generation retorts among the four categories (Walden, 2008). Technically, static batch retorts are used for highly viscous food products because any kind of movement does not aid in the improvement of heat transfer during processing of these products. Typically, agitating style retorts are used for the liquid, semisolid, or flowable products where product movements such as stirring and mixing can take place effectively. Agitation of retorts depends on the orientation of product containers. If containers are placed vertically, then they are agitated in the rotary style generally in an end-over-end format. However, horizontally placed containers generally use axial agitation in continuous turbo cookers or reciprocation methods such as Shaka and Gentle Motion (Singh, Yen, Ramaswamy, & Singh, 2018).

13.2 Batch retorts

Batch retort systems (Fig. 13.1) provide the most versatility when it comes to meeting expectations to process a vast range of foods and package styles and sizes, using a variety of process distribution methodologies, providing the accurate temperature and pressure control, and adapting rapidly to new business opportunities. These are easy to operate and maintain as compared to the continuous retorts. In batch retorts, loading and unloading the baskets or crates or cars is performed manually, making it a labor-intensive and time-consuming process (Simpson, 2007).

The retort is generally tested to be safe and equipped with several safety features. Most retorts operate with steam or steam-heated water at elevated temperatures, and therefore, temperature and pressure need to be monitored and controlled. There needs to be at least one indicating thermometer on the retort. The mercury in glass (MIG) thermometers are the industry standard in most retorts, but other temperature monitoring devices are acceptable. In most retorts, the electronic temperature, pressure, and other process parameters are continuously monitored and controlled. Other important features include steam/water/air inlet and control valves for venting and draining,

Fig. 13.1 Standard batch retort.
Photo: Courtesy of Dixie Canning Company, USA.

steam spreaders and bleeders for directional distribution of steam, and pumps for distributing water. There must be a safety valve to protect the equipment as well.

13.3 Agitating retorts

The use of rotation to increase the heat-transfer rate was first suggested by Clifcorn, Peterson, Boyd, and ONEIL (1950). There are several different batch agitating retorts used for the agitation by rotating the containers. As shown in Fig. 13.2, there are two general modes of agitating the retorts: end-over-end (EOE) and axial rotation. In the EOE mode, cans are rotated in a vertical plane around a circle. In the axial mode, cans pass through a revolving helical coil, continuously traveling from one end of the retort to the other.

Fig. 13.2 A pilot agitating retort technology suitable for static, axial, and end-over-end processing.
Photo: Courtesy of DFT Technology GmbH, Germany.

Clifcorn et al. (1950) introduced the EOE technology to improve the thermal processing efficiency in cans. First, a full water immersion agitating retort with an overpressure was developed by the Stock company in Germany. After that, semi-immersion sterilization technology was introduced by improvements made by including water circulation arrangements. The end-over-end rotation causes a high rate of heat transfer to cans, resulting in a shorter processing time and higher product quality. In such agitation modes, the headspace helps in improving mixing and turbulence. The headspace's motion depends on different physical parameters such as system geometry, rotational speed, headspace volume, product viscosity, axis of rotation, presence of particles, and particle density (Knap & Durance, 1998).

In axial rotation, the basket containing sealed cans is rotated in a horizontal plane around the circle in a single direction. When cans rotate, the headspace moves along the length of the can, which causes the agitation of the food inside the can as well (Dwivedi & Ramaswamy, 2009). In the FMC Turbo cookers, biaxial rotation is also facilitated by reverse rotation caused by rolling of cans in the lower third of the helix when cans contact the cooker wall.

A third mode of rotation is a newer technique that uses reciprocal agitation to decrease the processing time and increase the quality of the final product. It is a very powerful system that uses vibrations of 100–200 cycles per min frequency (Holdsworth & Simpson, 2015). This technique is recognized in Shaka retorts. The first pilot Shaka retort was manufactured in 2006. It was developed in the United Kingdom by Richard Walden. This is a major improvement in retorts and is mainly different from others that use reciprocal agitation and gravity (Angalet, 2011).

Studies have shown that the Shaka process (Angalet, 2011) cuts the processing time more than 20-fold and more than 10-fold relative to a still cycle and a rotary process, respectively (Holland, 2008). The latest model of these retorts will be wide enough to accommodate at least 50, 73, or 110mm cans. They are also equipped with a storage mechanism that separates each container from its neighbor. The storage mechanism is flexible to accommodate a range of container sizes and forms. The drive for stroke capacity is from 25 to 300mm, and the reciprocation speed is up to around 250rpm. Such retorts vent and reach the temperature quickly and have cooling systems using spray or flood cooling. The reciprocal system is a crank and slider mechanism. It pushes the basket through a drive shaft that enters the retort through a boss with seals and bearings. The basket is carried on rails and is driven by flanged wheels (Walden, 2008). The heat treatment in the Shaka system is similar to that of the UHT system, so the product quality is high (Walden, 2008). More recently, Singh, Singh, and Ramaswamy (2015) characterized the efficiency of reciprocation retorts focusing on enhanced heat transfer effects of shaking on the temperature distribution within canned products. It was shown that a 52%–87% reduction in the equilibrium time was achieved up to 4Hz reciprocation frequency. A critical factor in the utilization of such systems is that the quality of canned products should be closely monitored, and proper optimization should be conducted because an improper shaking process would damage the product. A moderate agitation speed ($\approx$1Hz) is suggested for cut vegetables to minimize textural loss and increased brine turbidity (Singh et al., 2018; Singh, Singh, & Ramaswamy, 2015a, 2015b).

For effective agitation, cans must have sufficient headspace. When the can rotates, the headspace bubble travels along the length of the can and creates agitation inside the container (Holland, 2008). In rotary retorts, the product must be clamped within the crates, preventing any damage during the movement. The rotary process is often used in combination with operation at a temperature higher than that of the static process. The higher temperature is less harmful to the product because the movement, particularly at the container wall, prevents overheating and helps in homogenous energy dissipation (Richardson, 2001).

One of the agitating retorts is the orbital rotary pressure stabilizer (FMC Corp., Madera, CAI), also called Orbitort. In this type, loading and unloading happen simultaneously. The orbital sterilizers' innovative configuration enables the containers to be loaded and unloaded simultaneously by a spiral and reel assembly. The inner reel configuration helps in reduction of the time needed for venting the orbital sterilizer (Kimball & Heyliger, 1990).

Agitating retorts can increase the heat-transfer rate through forced convection within the liquid or semiliquid foods. Therefore, the temperature distribution within the container will be more uniform, resulting in a higher product quality during a shorter processing time. The disadvantage of rotational agitation is that the overall heat-transfer coefficient and the fluid-to-particle heat-transfer coefficient start to decrease after a certain rotation speed. Thus, further optimization studies are always needed considering the product nature and operation parameters. Rotational processes are not suitable for some products, for instance, where the product's texture is sensitive to physical impact or high shear forces, for example, cream or soft fruit. The benefits of rotating processes are marginal for those that heat up rapidly by natural convection, for example, thin soups or vegetables in brine (Ates et al., 2016) (Fig. 13.3).

Fig. 13.3 The industrial Multiprocess Rotomat MRX agitating retort.
Photo: Courtesy of DFT Technology GmbH, Germany.

13.4 Steam–air retorts

The concept of using steam/air mixture was questioned for decades. Thus, early research focused on the heat-transfer efficiency of steam/air used as a heating medium (Ramaswamy, Tung, & Stark, 1983; Tung, Britt, & Ramaswamy, 1990). In these retorts, an overpressure environment is provided by deliberate mixing of steam and air in the retort vessel. Such a condition is suitable for processing of pressure-sensitive containers such as jars, semirigid trays, and pouches (Weintraub, Ramaswamy, & Tung, 1989). The steam/air retort concept is different from the conventional saturated steam retort, where air removal is a necessary precursor to starting the holding cycle. In steam/air retorts, a proportion of air is retained or added in a retort to balance the internal pressure inside sterilized containers. Uniform mixing of steam/air is necessary because the presence of large air packets leads to understerilization. Moreover, occurrence of steam condensation results is insulating layers of air around containers. In large retorts, the breakdown of the residual air layer and the mixing of the air and steam is performed with large turbo fans, which pull the air–steam mixture through the load and recirculate it to the other end of the system (these retorts are mostly horizontal). It was proposed that mixing could also be accomplished by continuous ventilation, but this is not used in industrial systems because it is not energy-efficient. The temperature uniformity of these steam/air systems is maintained by the efficient operation of the fan and ventilation systems. Like saturated steam retorts, steam/air retorts are potentially vulnerable to steam collapse when cooling water is added; however, because they are prehooked with air supply systems, the collapsing steam is readily replaced with air when the same retort pressure levels are maintained (pressure cooling).

13.5 Crateless retorts

The vertical retort was improved by increasing their size, removing cabinet configurations (crateless), and more automation in the operation system. Crateless retorts are a scale-up version of traditional vertical retorts and have 4 to 5 times higher space capacity. Crateless retorts are equipped with an automated cooking operation consisting of a series of individual retorts, each running in a batch mode. The loading, unloading, and process operation are fully automated (Teixeira, 2019).

The operation starts with the retort being loaded with warm water, which eventually would act as a cushion for the cans falling from the 10–20 in. opening on top of the chamber. After loading, the door is closed, the spreader introduces steam to the chamber, and water is removed from the bottom draining valve. Once the venting is started, the bottom drain, bleeder, and well bleeder remain opened. Steam enters from the top of the retort through the regulator valve. Venting would take about 4 min, and the temperature at the end of this stage is about 230°F, and after that, the process is started while the drain valve is closed, but the bottom bleeder remains open, acting as a vent valve (Downing, 1996). At the end of the process, warm water enters the chamber and collapses the steam, and then cans are subsequently discharged to a cooling channel. The process of handling the cans is designed and conducted gently to avoid physical damages. There is a possibility of reaching continuous canning by linking multiple retorts with the same water feed and discharge conveyer mechanism (Teixeira, 2019). Because of being fully automatic with the lack of crates, this system works with a minimum number of labors. Also, it is more efficient than still vertical retorts when it comes to steam consumption. This system is flexible regarding different sizes of containers and can operate at temperatures of up to 274 °F (Downing, 1996).

13.6 Continuous cookers

In continuous cookers, containers are first filled and sealed automatically. Then, they are continuously moved to a pressurized steam environment where they are held or conveyed for a particular processing time. Finally, they are returned to the normal atmospheric conditions for further handling operations. Crateless retorts, continuous rotary cookers, and hydrostatic sterilizers are the best-known commercially designed configurations with these arrangements (Teixeira, 2019).

13.6.1 Continuous turbo cooker/coolers (FMC)

A rotary retort is a continuous retort. Also, some models provide agitation to containers. The continuous cycle requires both heating and refrigeration. In these retorts, the continuous cycle is achieved by an axial (or side-over-side) rotation made during the processing. Such a rotation causes movement of the head space bubble to allow a better heat dissipation and a decrease in the processing time. Moreover, the product quality and flavor would be maintained through such homogenous thermal processing (Angalet, 2011; Downing, 1996).

The process time is controlled based on the residence time of the products at a fixed reel spinning speed. Obviously, the reel speed would determine the whole production speed as well. Introduction of cans happens through an inlet conveyor which enters cans to a rotary transfer valve mechanism. This would control the feeding process and stops steam escape and an unwanted operating pressure drop. Such retorts are made of several shells that can be used for either additional cooking or the cooling stage. Such reel and shell systems provide a perfect flexibility in terms of coupling different operation methods and cooking/cooling strategies; however, in case of changing the size of the can, this system is not very flexible and would need further adjustments. This is one of the limitations with respect to such a system (Teixeira, 2014).

13.6.2 Hydrostatic sterilizers

Hydrostatic retorts offer an innovative design which works with the help of the pressure provided by water columns. The configuration usually has high water towers (known as water legs) which are placed outside but close to the production line. The system is a continuous (but nonagitating) type of processor which can adjust the processing time based on its conveying speed. Cans are transformed using a carrier chain to the top of the water legs, and then they move down to the main water steam interface which provides the high temperature and thigh pressure as the main processing step (Angalet, 2011; Holdsworth & Simpson, 2015).

The hydrostatic sterilizer consists of three main parts for preheating, processing, and cooling. However, various configurations exist based on different products and production yields and temperature and pressure requirements (Teixeira, 2014). The advantages of such retorts are their space efficiency and their high flexibility in using different products and containers. In these retorts, glass jars and containers can also be processed; however, the total cost of such systems is still high (Downing, 1996).

13.7 Operation of retorts

Operation of retorts starts with proper positioning of filled and sealed cans in crates or cars (Fig. 13.4). These might be equipped with dividing layers necessary for the prevention of abrasion damage to containers. However, dividers must be perforated to ease the steam circulation and reach a homogeneous heat transfer. After loading, the door can be closed, and the steam valve can be opened. Venting is necessary to remove the trapped air between containers. The presence of air would negatively affect the efficiency of the process because of its heat isolating effects. Thus, a thermal treatment is started only after a proper venting process. Various programmable devices are used to control temperature variation and ensure process efficiency; this remains a critical parameter and hence needs to be established by a process authority.

Rapid cooling of processed cans is necessary to prevent the growth of thermophilic spores that may have survived during the process. Usually, 40°C is the goal temperature during cooling. Because of high intensity of the thermal process, the temperature and pressure inside the can is high at the end of the heating stage. Thus, the cooling operation should be performed with extreme precautions. The rigidity of cans can tolerate

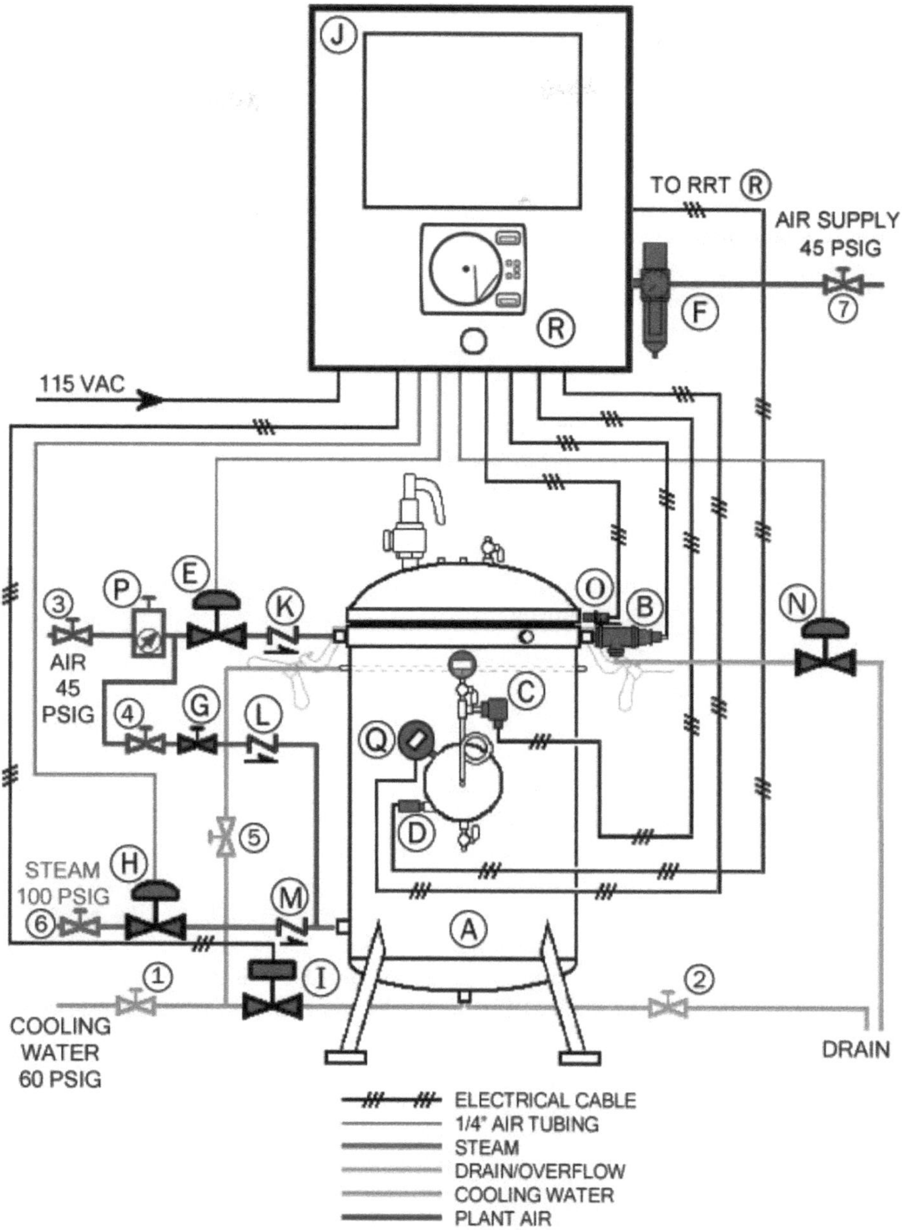

Fig. 13.4 Schematic of an automatic batch retort control system. Photo: Courtesy of Dixie Canner Company, USA.

high pressure to some extent. However, if the steam is switched of at the beginning of the cooling, the removal of external steam pressure (i.e., 10 kg/cm^2 at 121°C) will lead to permanent damage on the cans. The problem is solved by feeding air inside the retort to substitute the steam and build up external pressure. Another solution that is implemented in continuous retorts is that the cooling occurs in pressurized cooling water chambers. The cooling is followed by draining of cooling water and unloading (May, 2001).

The water used for the cooling process needs to have minimum microbial standards because it is in direct contact with cans and the temperature is optimum for microbial growth. In fact, the main concern is that the microbes would reach the seal or seam leaks and penetrate the food (Holdsworth, Simpson, & Barbosa-Cánovas, 2008; Podolak, Stone, & Black, 2010). Thus, the cooling water is treated with chlorine, usually in the form of hypochlorites or chlorine gas, 20 min before using it and should have a residual concentration of at least 2 ppm after the cooling (Suwanrangsi, Keerathiviriyaporn, Sophonphong, Briliantes, & Limpus, 1995). Complexes of iodine such as iodophors are also effective against vegetative bacteria and yeasts. However, they do not have strong effectiveness against spore-forming bacteria (Podolak et al., 2010). Nowadays, water disinfection can be implemented by the computer-aided programs and devices. They can monitor the disinfection using different oxidizers such as chlorine, iodine, and bromine based on water's oxidation–reduction potential (ORP) sensors (Podolak et al., 2010; Suslow, 2004).

13.8 Conclusions

The thermal processing industry seeks processing solutions that provide high flexibility concerning processing a wide range of canned foods. Also, these days, energy consumption and environmental challenges are essential concerns. Being able to use the steam source efficiently at a minimum level is getting critical. In this regard, using techniques such as agitation and would decrease the processing time. Besides, application of steam/air combination also brings the opportunity for less energy loss. These techniques not only optimize energy consumption but also will lead to achieving better food quality in terms of maintaining acceptable nutrient and sensorial properties.

References

Angalet, S. M. (2011). *Shaka: A new and novel processing technology to produce commercially sterile canned foods.*

Ates, M. B., Skipnes, D., Rode, T. M., & Lekang, O. I. (2016). Comparison of spore inactivation with novel agitating retort, static retort and combined high pressure-temperature treatments. *Food Control*, *60*, 484–492.

Clifcorn, L. E., Peterson, G. T., Boyd, J. M., & ONEIL, J. (1950). A new principle for agitating in processing of canned foods. *Food Technology*, *4*(11), 450–457.

Downing, D. L. (1996). *Complete course in canning and related processes.* Woodhead.

Dwivedi, M., & Ramaswamy, H. S. (2009). Comparison of heat transfer rates during thermal processing under end-over-end and axial modes of rotation. *Lwt-Food Science and Technology*, *43*(2), 350–360.

Holdsworth, S. D., & Simpson, R. (2015). Thermal food processing equipment. *Thermal Processing of Packaged Foods, 337–366.* https://doi.org/10.1007/978-3-319-24904-9_17.
Holdsworth, S. D., Simpson, R., & Barbosa-Cánovas, G. V. (2008). *Thermal processing of packaged foods. Vol. 284*. New York, NY: Springer.
Holland, C. (2008). *Advances in retort equipment and control systems. In in-pack processed foods* (pp. 71–85). Woodhead Publishing.
Kimball, R. N., & Heyliger, T. L. (1990). Verifying the operation of steam retorts. *Food technology (Chicago), 44*(12), 100–103.
Knap, R. P., & Durance, T. D. (1998). Thermal processing of suspended food particles in cans with end-over-end agitation. *Food Research International, 31*(9), 635–643.
May, N. S. (2001). *Retort technology. In thermal technologies in food processing* (pp. 7–28). Woodhead Publishing.
Podolak, R., Stone, W., & Black, D. G. (2010). Retort cooling water bacteriological load and possible mitigation strategies for microbial build-up in cooling water. *Food Protection Trends, 160.*
Ramaswamy, H. S., & Marcotte, M. (2006). *Food processing, principles and applications*. CRC Press. A Taylor and Francis Book.
Ramaswamy, H. S., Tung, M. A., & Stark, R. (1983). A method to measure surface heat transfer from steam/air mixtures in batch retorts. *Journal of Food Science, 48*(3), 900–904.
Ramesh, M. N. (2007). Canning and sterilization of foods. In *Handbook of food preservation* (pp. 603–642). CRC Press.
Richardson, P. (Ed.). (2001). *Thermal technologies in food processing* Taylor & Francis.
Simpson, R. (2007). *Thermal processing of packaged foods. Food engineering series*. Springer.
Singh, A., Singh, A. P., & Ramaswamy, H. S. (2015). A refined methodology for evaluation of heat transfer coefficients in canned particulate fluids under rapid heating conditions. *Food and Bioproduct Processing, 94*, 169–179.
Singh, A. P., Singh, A., & Ramaswamy, H. S. (2015a). Modification of a static steam retort for evaluating heat transfer under reciprocation agitation thermal processing. *Journal of Food Engineering, 153*, 63–72.
Singh, A., Singh, A. P., & Ramaswamy, H. S. (2015b). Effect of processing conditions on quality of green beans subjected to reciprocating agitation thermal processing. *Food Research International., 78*, 424–432.
Singh, A. P., Yen, P. P. L., Ramaswamy, H. S., & Singh, A. (2018). Recent advances in agitation thermal processing. *Current Opinion in Food Science, 23*, 90–96.
Suslow, T. V. (2004). Oxidation-reduction potential (ORP) for water disinfection monitoring. *Control, and Documentation*. https://doi.org/10.3733/ucanr.8149.
Suwanrangsi, S., Keerathiviriyaporn, S., Sophonphong, K., Briliantes, S., & Limpus, L. G. (1995). *Canned tuna quality management manual.*
Teixeira, A. A. (2014). Thermal food preservation techniques (pasteurization, sterilization, canning and blanching). *Conventional and Advanced Food Processing Technologies*, 115–128.
Teixeira, A. A. (2019). Thermal processing for food sterilization and preservation. In *Handbook of farm, dairy and food machinery engineering* (pp. 499–523). Academic Press.
Tung, M. A., Britt, I. J., & Ramaswamy, H. S. (1990). Food sterilization in steam/air retorts. *Food technology (Chicago), 44*(12), 105–109.
Walden, R. (2008). The Zinetec Shaka™ retort and product quality. In *In-pack processed foods* (pp. 86–101). Woodhead Publishing.
Weintraub, S. E., Ramaswamy, H. S., & Tung, M. A. (1989). Heating rates in flexible packages containing entrapped air during overpressure processing. *Journal of Food Science, 54*(6), 1417–1421.

Index

Note: Page numbers followed by "*f*" indicate figures, "*t*" indicate tables, and "*b*" indicate boxes.

FSC
www.fsc.org
MIX
Papier | Fördert
gute Waldnutzung
FSC® C083411

Kolibri 360 GmbH
Ferdinand-Jühlke-Straße 7
99095 Erfurt, Deutschland
produktsicherheit@kolibri360.de